T0331992

ARCHITECTS OF STRUCTURAL BIOLOGY

Architects of Structural Biology

Bragg, Perutz, Kendrew, Hodgkin

John Meurig Thomas

Foreword
by
Venki Ramakrishnan

UNIVERSITY PRESS

OXFORD
UNIVERSITY PRESS

Great Clarendon Street, Oxford, OX2 6DP,
United Kingdom

Oxford University Press is a department of the University of Oxford.
It furthers the University's objective of excellence in research, scholarship,
and education by publishing worldwide. Oxford is a registered trade mark of
Oxford University Press in the UK and in certain other countries

First Edition published in 2020

Impression: 1

Published in the United States of America by Oxford University Press
198 Madison Avenue, New York, NY 10016, United States of America

British Library Cataloguing in Publication Data
Data available

Library of Congress Control Number: 2019945688

ISBN 978–0–19–885450–0

DOI: 10.1093/oso/9780198854500.001.0001

Printed and bound by
CPI Group (UK) Ltd, Croydon, CR0 4YY

To My Dearest Wife Jehane,
With Love

Foreword

In the very first of his famous '*Lectures on Physics*', Richard Feynman said, 'If, in some cataclysm, all of scientific knowledge were to be destroyed, and only one sentence passed on to the next generation of creatures, what statement would contain the most information in the fewest words? I believe it is the atomic hypothesis that *all things are made of atoms—little particles that move around in perpetual motion, attracting each other when they are a little distance apart, but repelling upon being squeezed into one another.* In that one sentence, you will see, there is an enormous amount of information about the world, if just a little imagination and thinking are applied.'

Atoms do not generally exist by themselves. They bond with each other to form defined groupings called molecules, and it is molecules that give all matter its properties and infinite variety. One of the most amazing feats of the human intellect was not only to deduce the existence of atoms and molecules, but also to infer the structure of simple molecules. All of this was done by the end of the nineteenth century, before anyone had even seen a molecule directly. Moreover, once the distance between atoms in a molecule were known to be several thousand times smaller than the wavelength of light, the theorem of Ernst Abbe meant that it would be impossible to see molecules using ordinary light, no matter how powerful a microscope was used.

The direct visualization of molecules, from common salt, which contains only two types of atoms, to biological molecules that contain half a million atoms or more, is another great triumph in the study of matter. It has led to revolutions in both materials science and biology, but it did not just happen. It was work that spanned a century and owes its development to a few bold and brilliant individuals. Many of them, such as the William and Lawrence Bragg, Linus Pauling, Max Perutz, Dorothy Hodgkin, and John Kendrew, have been the subject of extensive biographies before.

How is this book different? The author John Meurig Thomas (affectionately known as JMT) is himself a distinguished chemist who personally knew many of the principals involved in this endeavour. Moreover, the pioneering advances were made mainly in a few places, and JMT spent much of his research life working in two of them. Finally, JMT is known as an erudite scholar with a love of the history of science. The result is a book which is authoritative, but, unlike most scholarly books of that sort which tend to be very dry and could be used as a treatment for insomnia, this book is engaging and exciting and makes the characters come alive. The extensive quotes from many of them allow their voice to be heard directly. We learn not only about what they did, but also about what they were like as human beings, and the social, intellectual, and physical environment which made their work possible. Importantly, they are not treated as isolated characters, as is often the case in dedicated biographies. Rather, we learn of their interactions, how they stimulated each other, and how ideas and expertise spread to lead to further advances. I particularly enjoyed the comparison between Lawrence Bragg and Linus Pauling. Both were brilliant and colourful rivals, but the two could not have been more different in their personalities. Yet each in his own way was responsible for huge advances in structural chemistry, all the way from simple inorganic compounds to complex biological molecules.

Science has often been told as a series of stories about great men. But real science does not work that way. It depends on lots of advances, both big and small, some of which are ideas, but some of which are quite practical and technical. Here too, JMT does justice to the actual developments that made various advances possible. He gives credit to the many (mostly) unsung heroes, without whose work major advances would have stalled or never happened. The book has a series of pen portraits of people who are largely unknown, even to people in the field. By resurrecting them, JMT has given us a much better understanding of how science happens.

Apart from people who contributed ideas and technical expertise, we read about people who had a vision for how to fund and nurture science and also how to build institutions. We learn what makes some people successful as leaders and what makes some institutions so productive. Sadly, we also learn in a cautionary tale how institutions can decline. Maintaining excellence is a hard and eternal task.

Finally, today the term 'molecular biology' has been coopted to mean the study of information and control in living systems, principally having to do with DNA and genes. This book is thus a much needed corrective to restore the term molecular biology to its original meaning of the study of biology at the molecular level or of molecules of biological significance. After all, the remarkable advances in our understanding of genetics and control would not have been possible without the structure of DNA, which, in turn, would not have been possible without decades of study of the structure of molecules using X-ray diffraction.

Venki Ramakrishnan
Cambridge, 2019

Preface

Photograph of the 22-year-old Lawrence Bragg (1890–1971) taken at the time when he wrote his seminal paper (provided by his son the late Stephen Bragg). Alongside are the drawings made by Sir Lawrence, in 1964, of Max Perutz (centre) and John Kendrew. (bottom) Dorothy Hodgkin (née Crowfoot), taken in 1986 when she collaborated with Guy and Eleanor Dodson in Oxford.

(Kind permission of Prof Eleanor Dodson)

The pioneering work of the four scientists that constitute the subject of this book transformed modern medicine and profoundly influenced the health and well-being of the human race. They were mainly responsible for the burgeoning of the new science of structural molecular biology where the properties and behaviour of molecules of the living world are interpreted at the atomic level.

Take, for example, haemoglobin, one of the most interesting chemical substances in the world. Each human being carries around a supply—in the red corpuscles of the blood—of approximately one per cent of the body weight. A single molecule of haemoglobin is composed of nearly 10,000 atoms of carbon, nitrogen, oxygen, sulfur, and hydrogen, but there are also four atoms of iron which, almost magically, carry oxygen from the lungs to the tissues and carbon dioxide and other products of the breakdown of food back to the lungs.

Arguably, the greatest ever authority on haemoglobin and its nature and function in health and disease was Max Perutz (1914–2002), one of the architects that constitute the subject of this book. He, along with his colleague John Kendrew (1917–1997), evolved a fundamentally new approach to the determination of the structures, and hence the mode of action, of two vitally important giant biological molecules—haemoglobin and myoglobin, the proteins that transport and store oxygen in all animals. Other proteins are *the* fundamental building blocks of living things—tendons, muscles, antibodies, and enzymes—and are, consequently, of central importance in the whole corpus of biology, as outlined below. Dorothy Hodgkin (1910–1994) solved the structures of cholesterol (which is both the steroid precursor of sex hormones, such as testosterone and oestradiol, and a structural component of cell membranes), the antibiotic penicillin, vitamin B_{12}, and the key hormone in diabetes, insulin.

Perutz and Kendrew were both protégés of Sir Lawrence Bragg, who, by exploiting a discovery made in Munich in 1912 by Max von Laue involving the scattering of X-rays by solids, a phenomenon known as X-ray diffraction, was responsible, over a hundred years ago, for the creation of a new branch of science. This was the determination of the atomic composition and detailed molecular structure of materials from the patterns displayed when X-rays are shone upon them. These three and Dorothy Hodgkin (the first woman, after Marie Curie and her daughter Irene, to win the Chemistry Nobel Prize) are the principal architects of structural molecular biology (three chemists, one physicist).

In writing this book, I have had several aims in mind. It is intended as a memoir of some of the most gifted and remarkable experimental scientists of the twentieth century. Although I am not a biologist, I knew three of the protagonists well and was acquainted with several other key figures (and their biological and scientific progeny) described in the text. I describe the importance of their work and also outline how they achieved their successes, but I also dwell on their personalities, idiosyncrasies, ambitions, and single-mindedness, as well as those of their contemporaries and collaborators such as J. D. Bernal (friend of Picasso, Paul Robeson, and Earl Mountbatten). Their interactions, controversies, and rivalries with the charismatic US polymath Linus Pauling of the California Institute of Technology are also chronicled.

Second, I have attempted to convey, in a manner intelligible to the general reader, and especially to pre-university students, an insight into the great current excitement associated with structural molecular biology. Unlike most of physics, for example, molecular biology directly informs our understanding of ourselves. In addition, it has helped guide human beings towards healthier, cleaner, and safer lives.

Third, I wanted to explore how Perutz, Kendrew, and Hodgkin revolutionized structural molecular biology, and how Perutz and Kendrew, assisted by Bragg, established in Cambridge in 1962 what is now universally regarded as one of the most successful ever scientific research centres—the Laboratory of Molecular Biology (LMB)—under the aegis of the UK Medical Research Council (MRC).

To date, twenty-three of the scientists who have worked there have been awarded Nobel Prizes, and new medicines developed there are among the most extensively used worldwide for the treatment and cure of breast cancer, leukaemia, arthritis, and life-threatening respiratory conditions.

John Kendrew and Max Perutz worked jointly with others to form the European Molecular Biology Organisation (EMBO), and Kendrew was the founding Director of the European Molecular Biology Laboratory (EMBL) at Heidelberg. He also founded, in 1959, the first, and still the premier, *Journal of Molecular Biology*.

In the late 1950s, Perutz and Kendrew, with early deployment of the most powerful computers then being developed, were highly successful in their development of new procedures—first hinted at by Bernal and Hodgkin in 1934—and not hitherto used by chemists, biochemists, and biologists in opening up a new era in biological science that led to the creation in the United States, and now worldwide, of the Protein Data Bank (PDB) which currently holds more than 150,000 structures and constitutes the universal repository of the intricate folded structures of proteins. This Data Bank is of crucial importance in modern medicine, as it guides scientists in understanding the causes of misfolding of certain proteins, a phenomenon that is highly correlated with the onset of Alzheimer's disease, Parkinson's disease, and related neurodegenerative diseases. The PDB is also useful for the study of the function and inhibition of proteins.

Fourth, I also wanted to explore why two among the most renowned research centres in the UK—the world-famous LMB and the Davy-Faraday Research Laboratory (DFRL) of the Royal Institution (RI)—achieved such iconic status. In my professional life, I have served as Head of three research centres: the DFRL at the RI; the Department of Physical Chemistry at the University of Cambridge; and the Chemical Laboratories of the University College of Wales in Aberystwyth. A question that I frequently ponder is: *How does university life add depth and quality and also opportunity to professional research?* In answering this question, I am fortunate to be able to draw on my experience as Master of an Oxbridge college—Peterhouse, which has an uncommonly good record for nurturing teachers and researchers of exceptional skill and ability, and which had unique links with the founding and function of the LMB. I have also had the privilege of being intimately involved as a researcher at the DFRL of the RI for some twenty years.

Max Perutz was my close friend and neighbour for a quarter of a century. We interacted almost daily and frequently went on long walks. He revealed to me his views on how he created and ran the LMB and commented on numerous contemporary scientific, political, and general topics. Kendrew was, likewise, a friend for over a decade. They each were Fellows of Peterhouse, to which they made many transformative contributions. Sir Lawrence Bragg was my predecessor-but-one as Director of the RI of Great Britain in London where, apart from pursuing research as Fullerian Professor of Chemistry (the chair first created for Michael Faraday), he continued the tradition of presenting science to young people and non-expert adults. Sir Lawrence's key collaborator David Phillips (a long-standing friend and co-author of mine), who was also a major contributor to structural biology, was the first, with his student Louise Johnson (later Professor of Molecular Biophysics at Oxford), to solve the structure of the enzyme lysozyme and to explain its mode of action.

Perutz and Kendrew, like their mentor Bragg, were also expert popularizers of science. At different times, all three gave riveting lecture demonstrations (many of them televised by the BBC) to young and general audiences at the RI where both Perutz and Kendrew were visiting scientists and from where Dorothy Hodgkin derived, indirectly, her early interest in science. The expository gifts of all four are illustrated in my narrative.

Dorothy Hodgkin never worked at either the LMB or the DFRL, except for one short week, in 1937, as a guest of W. H. Bragg (Sir Lawrence's father, while he was Director of the RI), but she is indisputably one of the leading architects of structural molecular biology. Her work had an enormous impact on medical science, biochemistry, chemistry, and crystallography. When Lawrence Bragg, in January 1960, wrote to the Nobel Committee from the DFRL (see Figure 8.12), he proposed that Perutz, Kendrew, and Hodgkin should share the Nobel Prize in Physics (at the same time, he indicated that Crick, Watson, and Wilkins be accorded either the Chemistry or Physiology Prize).

It was at the DFRL in W. H. Bragg's day as Director in the late 1920s and early 1930s that the earliest investigations of 'living' molecules (hair, wool, fingernails, skin, and horns) were first investigated by X-ray diffraction. Later work (on vitamins, steroids, and viruses) was pursued at the Department of Mineralogy in Cambridge (where Perutz and Dorothy Hodgkin did their PhD under J. D. Bernal) and at the Textile Physics Unit at the University of Leeds by W. T. Astbury, who, along with Warren Weaver of the Rockefeller Foundation, New York, was one of the first to use the term molecular biology.

It was also at the RI in 1954 that Lawrence Bragg encouraged Rosalind Franklin and Aaron Klug, who were based at Birkbeck College, to investigate the diffraction patterns of a range of viruses using the high-power X-ray sources available there. This subsequently led Klug and his colleagues to pioneer a fundamentally new quantitative approach, using electron microscopy, to determine the detailed structure of viruses and numerous macromolecular assemblies, including transfer RNA. Klug was one of the progenitors of the technique of electron microscopy, now sweeping through structural biology. His prolific contributions to structural molecular biology, for which he, like Hodgkin in 1964, was the sole recipient of a Chemistry Nobel Prize (1982), are outlined in this text, along with his interactions with his mentors Bragg, Perutz, and Kendrew.

As well as her major achievements as a scientist, Dorothy Hodgkin, sometimes referred to as the Queen of crystallography, wrote engagingly about her contemporaries and teachers, notably J. D. Bernal and Kathleen Lonsdale, the first scientist to establish by X-ray diffraction that molecules of benzene (C_6H_6) were flat and hexagonal—which she did at the DFRL where she worked for twenty years.

Selections of her writings are reproduced here. Dorothy Hodgkin also had a lifelong friendship and participated in deep exchanges of structural biological information with Max Perutz. They each were fond of quoting Albert Schweitzer's dictum: '*Example is not the main thing in influencing others, it is the only thing.*'

Perutz and Kendrew's colleagues Francis Crick and James Watson solved, in the early 1950s, the double-helical structure of DNA, widely acclaimed as one of the greatest breakthroughs in twentieth-century science. This led to major advances in genetics, pathology, and almost all other biological subjects, as well as in biotechnology, genetic engineering, and latterly gene editing. In due course, Crick's work at the LMB produced the so-called 'central dogma' of molecular biology that explains how the atomic information contained in a one-dimensional structure (DNA) gives rise to the three-dimensional structure of proteins that consists of a specific sequence (unique for every protein) of the twenty amino acids that occur in all living matter.

In this book, I do not delve into the vast repercussions of the understanding of the structure and role of DNA in life processes. These have been extensively described many times before. Instead, I focus on the structural biology of proteins, which are of central importance in life sciences. The human body has the capacity to generate tens of thousands of different proteins. All the molecules in our immune system are proteins; so is collagen, the structural material used by all animals, and all enzymes are proteins. Enzymes are involved in the conversion of one biochemical to another, for example in the conversion of sucrose (sugar) to alcohol. Some enzymes possess the ability to convert molecules at rates that are so rapid that they exceed by hundreds of millions the conversion rate in their absence.

An additional feature of this book is that it traces the origins of new ideas, like the emergence of structural molecular biology from its crystallographic–mineralogical background, and the way advances in science gain momentum. It also dwells on the human value of science, and the element of chance is also discussed. For example, had Bragg not dissuaded Rutherford from dismissing Bernal because of his personal lifestyle and political activities, Kendrew would not have met Perutz and Crick would not have met Watson.

The four scientists described here are amongst the greatest in modern times. Not only were their accomplishments of lasting value, but they also conducted themselves in an exemplary way. The following is an excerpt from Max Perutz's address at Dorothy Hodgkin's memorial service in Oxford, in 1994, with her former student Margaret Thatcher in the congregation:

> '*She radiated love for chemistry, her family, her friends, her students, her crystals and her college to which she generously gave part of her Nobel Prize, for the support of students and for the college staff. There was a magic about her person. She had no enemies, not even among those whose scientific theories she demolished or whose political views she opposed. Just as her X-ray cameras bared the intrinsic beauty beneath the rough surface of things, so the warmth and gentleness of her approach to people uncovered in everyone, even the most*

> *hardened scientific crook, some hidden kernel of goodness. It was marvellous to have her drop in on you in your lab, like Spring. Dorothy will be remembered as a great chemist, a saintly, tolerant and gentle lover of people and a devoted protagonist of peace.'*
>
> G. Dodson, *Biographical Memoirs of Fellows of the Royal Society, D. M. C. Hodgkin*, **2008**, *48*, 3

In view of the necessarily Anglo-centric flavour of this work, the Glossary of technical terms is supplemented by descriptions and explanations of institutions and universities and other British establishments, and tradition.

Sir John Meurig Thomas
April 2019

Acknowledgements

I am greatly indebted to several friends who have critically scrutinized what I have written and who have made valuable improvements to my original text. Foremost, among these are Dr Richard Henderson (LMB, Cambridge); Professor Lubert Stryer (Stanford University); Professor Steve Harrison (Harvard University); and Professor Venki Ramakrishnan, President of the Royal Society, London, all four of whom have read drafts of every chapter.

Several others have read substantial parts of the book: Professor Dudley Herschbach, Harvard University; Professor John H. C. Spence, Arizona State University and Berkeley; Professor Jack Dunitz, ETH Zurich, who worked with Pauling, Dorothy Hodgkin, and Lawrence Bragg and provided me with unique insights and guidance in the writing of several chapters, as did Professor Dame Jean Thomas, Cambridge; Dr Kiyoshi Nagai, Cambridge; Professor Sophie Jackson, Cambridge; Professor Archie Howie, Cavendish Laboratory; and Professors Krister Holmberg, Sven Lidin, and Bengt Norden of Sweden. To all these kind persons, as well as to the following, who also read and commented upon some of my original compositions, I express my profound thanks:

Ellen Ambelton, Royal Society
Mary Archer, Science Museum
Wolfgang Baumeister, Martinsried
Richard Beales, New York
Ian Blatchford, Science Museum
M. S. Bretscher, LMB, Cambridge
A. D. Buckingham, Cambridge
P. J. Butler, Cambridge
C. R. Calladine, Cambridge
D. Caspar, Florida
Soraya de Chadarevian, UCLA
M. Chalfie, Columbia University
Graham Christie, Biotechnology, Cambridge
R. A. Crowther, LMB, Cambridge
Nick Cumpsty, Imperial College
Gideon Davies, York
T. K. Dickens, Cambridge

R. E. Dickerson, UCLA
A. K. Dixon, Cambridge
Christopher Dobson, Cambridge
Eleanor Dodson, York
W. A. Eaton, NIH, Bethesda
Peter P. Edwards, Oxford
David Eisenberg, UCLA
Sharif Ellozy, New York
Gerhard Ertl, Berlin
Phil Evans, Cambridge
Annette Faux, LMB
Georgina Ferry, Oxford
Alan Fersht, Cambridge
F. Gadala-Maria, South Carolina
Robert M. Glaeser, Berkeley
A. M. Glazer, Oxford and Warwick
H. B. Gray, California Institute of Technology
John Gurdon, Cambridge
K. D. M. Harris, Cardiff
S. Hasnian, Liverpool
Jacques Heyman, Cambridge
Roald Hoffmann, Cornell
Judith Howard, Durham
Colin Humphreys, Cambridge
Herbert Huppert, Cambridge
Frank James, Royal Institution
D. N. Johnstone, Cambridge
Yvonne Jones, Oxford
M. Karplus, Harvard and Strasbourg
Olga Kennard, Cambridge
David Klug, Imperial College
Roger Kornberg, Stanford
A. M. L. Lever, Cambridge
Malcolm Longair, Cavendish Laboratory
Ben Luisi, Cambridge
Master and Fellows of Peterhouse

I. W. Mattaj, EMBL, Heidelberg
P. A. Midgley, Cambridge
Keith Moore, Royal Society
Charlotte New, Royal Institution
A. C. T. North, Leeds
Linda Pauling Kamb, Pasadena
Robin Perutz, York
Vivien Perutz, Cambridge
G. A. Petsko, Cornell and Harvard
Martin Pope, New York
Randy Read, Wellcome Trust, Cambridge
Dai Rees, Medical Research Council
Martin Rees, Cambridge
Ann Rhys, Cardiff
Michael Rossmann, Purdue University
Umar Salam, Cambridge
Ana Talaban-Bailey, Cambridge
Song Tan, Pennsylvania State University
Lisa Thomas, London
Janet Thornton, EBI Hinxton
P. M. Wassarman, Mount Sinai, New York
Richard Welberry, Canberra
Benjamin Widom, Cornell
Dillwyn Williams, Cambridge
Gregory Winter, Cambridge
P. Wright, *Journal of Molecular Biology*

In collecting the various illustrations and photographs, I have been greatly helped by Dr Duncan Johnstone and Professor Kenneth Harris. They have worked like Trojans to seek out many of the appropriate illustrations that I have used. Professor Harris also composed the Appendix to Chapter 2. In retrieving photographs from the archives of Peterhouse, Dr Philip Pattenden has given me excellent support. And in accessing two illustrations pertaining to Sir John Kendrew, Sir Ian Blatchford of the Science Museum, London, and Professor Richard Ovenden, Director of the Bodleian Libraries, Oxford, were very helpful. The photograph of Alexander Todd and Linus Pauling (Chapter 7) punting on the Cam river in 1948, as well as others involving Linus Pauling, were accessed through the Director of the Archives at Oregon State University, Corvallis, and at Churchill College, Cambridge.

Ms Annette Faux at the LMB has also been exceptionally helpful, as have Charlotte New and Frank James at the Royal Institution. Permission to reproduce illustrations is gratefully extended to the American Philosophical Society; the LMB; The Nobel Foundation; the Royal Institution; the Royal Society; the Royal Society of Chemistry; Elsevier; Wiley-VCH; Springer Nature; and the Master and Fellows of Peterhouse.

I am especially thankful to Robin and Vivien Perutz for the encouragement they gave me and for providing personal reminiscences of their father. Likewise, Patience Thomson (Sir Lawrence Bragg's daughter) reminded me of many key facts about her father, as did (the late) Dr Stephen Bragg, her brother. The book by Patience Thomson and M. Glazer '*Crystal Clear*', published by Oxford University Press, provided fascinating insights into the private life of W. L. Bragg. The late Hugh Huxley, Michael Rossmann, and the late Sir Aaron Klug gave me much important information pertaining to both John Kendrew and Max Perutz.

The idea of tracing the remarkable achievements of Bragg, Perutz, Kendrew, and Hodgkin was first suggested to me in 2016 by Professor Krister Holmberg, Chalmers University, Gothenburg, Sweden, where, for several years, I have given annual public lectures on the contributions to the growth of science by notable individuals: Rumford, Davy, Faraday, Rutherford, Bernal, and others. I gladly accepted his invitation, as I had six months earlier accepted an invitation from Sir Dillwyn Williams, former President of the British Medical Association, to give a similar talk to the *History of Medicine in Wales Society* at Cardiff in December 2015. But it was the lecture that Dr Richard Henderson, former Director of the LMB, Cambridge, and the 2017 Nobel Prize winner in Chemistry, invited me to give at the LMB in November 2017, and the encouraging response it elicited, that prompted me to expand it during the course of 2018.

This text would have taken much longer to write, were it not for the secretarial and library help at Peterhouse by Sarah Anderson, Helen Cross, Alison Pritchard-Jones, and Liz Wake and by the superb and expeditious work of my personal assistant Mrs Linda Webb, who accepted with equanimity and good cheer my numerous requests for syntactical and other changes. In addition, I would never have reached the end of the road, were it not for the thoughtful comments of my wife Jehane (an emeritus Professor of Chemistry at the American University of Cairo), who acted as a sounding board for the content of most of the book. To all these kind folk, as well as to my daughters Lisa and Naomi, I renew my deep indebtedness. The responsibility for any errors or faults in this book rests entirely with me.

Contents

1

Max Perutz, John Kendrew, Peterhouse, and the Davy-Faraday Research Laboratory

1.1 Introduction

Two Fellows of Peterhouse, the oldest and smallest conventional college of the University of Cambridge—Max Ferdinand Perutz and John Cowdery Kendrew—are acknowledged as the founders (in 1962) of one of the world's leading scientific establishments—the Laboratory of Molecular Biology (LMB), Cambridge, of the Medical Research Council (MRC) of the United Kingdom (UK). Perutz and Kendrew were also crucially involved in establishing the European Molecular Biology Organization (EMBO), and it was Kendrew who became the first Director of the European Molecular Biology Laboratory (EMBL) founded in Heidelberg in 1974. It, too, is now renowned as a world-leading centre in molecular biology.

Kendrew, as a teaching Fellow of Peterhouse, was Director of Studies for students reading the natural sciences and the history and philosophy of science. In 1959, from his rooms in college, he founded the first and, for a long time, leading journal in his field (*Journal of Molecular Biology*) (see Figure 1.1). Yet its inception was greeted with scepticism by many eminent contemporary biologists. Thus, the erudite Conrad Waddington, a British developmental biologist, geneticist, embryologist, and philosopher, wrote a strongly antipathetic letter to the journal *Nature* (8 April 1961), claiming that the name 'molecular biology' was unfortunate and only marginally appropriate to any aspect of the then existing area of biology. This journal now has a global readership of over 2.1 million (downloads) per month and it reaches some thirteen million scientists each month.

Starting their joint scientific work in 1945—Kendrew was Perutz's first PhD research student—these two pioneered the use of X-ray crystallography to determine the structures of proteins, in a manner that had not been undertaken by chemists and biochemists hitherto. It was in 1962 that they were jointly awarded the Nobel Prize in Chemistry for accomplishing their work on haemoglobin and myoglobin, proteins that carry and store oxygen.

Journal of
MOLECULAR
BIOLOGY

EDITOR-IN-CHIEF
J. C. KENDREW

EDITORIAL BOARD
P. Doty
A. F. Huxley
R. L. Sinsheimer
J. D. Watson
M. H. F. Wilkins

VOLUME I
1959

Figure 1.1 *The first issue of the* Journal of Molecular Biology *showing also (left) Old Court, Peterhouse, where John Kendrew's room was, and (right) the Royal Institution (RI) of Great Britain, in London. Old Court is very much as it was in 1440, and the RI as it was in 1840.*
(Copyrights: JET Photographic; Master and Fellows of Peterhouse; J. Mol. Biol.*, Elsevier; RI; Courtesy of RI)*

During the course of their studies, they continued to collaborate with their mentor Sir Lawrence Bragg, who, from 1953 to 1966, was Director of the Davy-Faraday Research Laboratory (DFRL) at the Royal Institution in London, where Perutz and Kendrew had appointments from 1953 to 1966 as Honorary Readers in Crystallography.

It was at the Cavendish Laboratory, Cambridge, that Sir Lawrence Bragg was able to establish for Perutz the MRC Unit for Biological Structures, which, after 1962, became the Laboratory of Molecular Biology, on a site close to Addenbrooke's Hospital in Cambridge. It was also in 1962 that the Nobel Prize for Medicine or Physiology was awarded, for the determination of the structure of DNA, to two members of the Perutz–Kendrew group, namely Francis Crick, who was Perutz's PhD student, and James D. Watson, an American geneticist, who had joined Kendrew as a postdoctoral research fellow.[1]

A young member of the MRC Unit for the study of the molecular structure of biological systems who came to work as a PhD student, in Peterhouse and the Cavendish Laboratory, with John Kendrew, was Peter Pauling, Linus Pauling's son, the foremost molecular biologist and chemist in the United States (US). Both Perutz and Kendrew were skilled in recruiting co-workers of exceptional intellectual ability. It was Perutz who persuaded Fred Sanger—after he had won the first of his two Nobel Prizes in Chemistry for his work on insulin—to join the LMB from the Department of Biochemistry, Cambridge. He also recruited the Argentinian Cesar Milstein, who later won the Nobel Prize for his work on monoclonal antibodies. Yet another recruit that Perutz and Kendrew attracted (from Oxford formerly and subsequently from South Africa) was the future Nobel Prize winner Sydney Brenner. John Kendrew's first research student, a physicist by training, was Hugh Huxley, who was later to make a major contribution to our understanding of the mechanism of muscle contraction. Many of these remarkable individuals were entrusted by the chairman of the LMB (Max Perutz) to guide scientific activities there (see Figure 1.2).

When John Kendrew retired as Director of Studies at Peterhouse in 1962, his replacement was Aaron Klug, who established his research base at the LMB. Klug had earlier been working alongside Rosalind Franklin on the structure of viruses at Birkbeck College, University of London, which was headed by the polymathic J. D. Bernal, whose influence on Perutz, Kendrew, and Dorothy Hodgkin was, as described later, profound. Klug became another Nobel Laureate in 1982, and another former graduate student at Peterhouse in his days at the LMB—Michael Levitt—was a joint winner of the Nobel Prize in Chemistry in 2013. Numerous other eminent molecular biologists, notably John Walker (Nobel Prize in 1997),

Figure 1.2 *The Governing Board, which consisted initially of all LMB Fellows of the Royal Society. From the left: H. E. Huxley, J. C. Kendrew, M. F. Perutz, F. H. C. Crick, F. Sanger, and S. Brenner (1967).*

(Courtesy MRC-LMB, Cambridge)

John Sulston (Nobel Prize in 2002), John Gurdon (Nobel Prize in 2012), David Blow, and Ulrich Arndt (formerly at the Davy-Faraday Research Laboratory), were all recruited by Perutz (mainly) and Kendrew, Brenner, or Sanger. The recent recipient of the Nobel Prize in Chemistry—Dr Richard Henderson (2017)—was interviewed and appointed by Perutz and Kendrew. And Sir Greg Winter, who won the 2018 Nobel Prize in Chemistry, did his PhD and a research fellowship (sponsored by Trinity College, of which he is now Master) at the LMB. (The recipient of the 2009 Nobel Prize in Chemistry—Dr Venki Ramakrishnan—was appointed when Henderson was Director of the LMB.) It was John Kendrew who recruited Jacques Dubochet, joint winner of the 2017 Nobel Prize in Chemistry, to pursue electron microscopic studies at EMBL.

1.2 Popularization of Science

In addition to their towering contributions to science, especially in founding the vitally important and far-reaching subject of protein crystallography, both Perutz and Kendrew communicated scientific knowledge—their own and others'—to members of the general public, as well as to the students of Peterhouse, in elegant and gripping ways. Perutz, especially, became adept in reaching out to literary circles. His contributions to such journals as *The New York Review of Books*, *The New Yorker*, *The London Review of Books*, *New Scientist*, and other outlets have been widely quoted and read. Kendrew, likewise, was spectacularly successful in his series of BBC broadcasts entitled '*The Thread of Life*', which was transmitted in 1964 (and published in book form in 1966). They each combined felicity of literary expression with brilliant original scientific research. Moreover, they were able, as also was Dorothy Hodgkin, to communicate with lay audiences at Friday Evening Discourses at the Royal Institution—in the unique manner laid down in 1826 by the founder of these Discourses Michael Faraday and sustained there ever since.

1.3 J. D. Bernal's Influence upon Perutz, Kendrew, and Hodgkin

Perutz graduated in chemistry at the University of Vienna in 1936. His mentor there Hermann Mark had earlier carried out research at the Kaiser Wilhelm Institute in Berlin where he investigated, using X-rays, the properties of 'living' materials such as silk. Mark drew Perutz's attention to the work of Gowland Hopkins (1861–1947) in Cambridge on vitamins, and also to the work of John Desmond Bernal (1901–1971). In a *Scientific American* article of 1978, Perutz wrote:

> '*When I was a student, I wanted to solve a great problem in biochemistry. One day I set out from Vienna, my home town, to find the Great Sage at Cambridge. He taught me that the*

> *riddle of life was hidden in the structure of proteins, and that X-ray crystallography was the only method of solving it. The sage was John Desmond Bernal, who had just discovered the rich X-ray diffraction patterns given by crystalline proteins. We really did call him Sage, because he knew everything, and I became his disciple!'*

So, later in 1936, Perutz registered as a research student in the Department of Mineralogy and Petrography in Cambridge, and Peterhouse as his college. A few years earlier, Bernal with his PhD student Dorothy Crowfoot (later Hodgkin) (see Figure 1.3) had discovered the rich X-ray diffraction patterns given by crystalline enzymes, patterns that implied to the physicist (see Chapter 2) that the atomic detail of the three-dimensional structure of enzymes and other proteins (all of which are giant molecular entities) could be determined directly by the new, purely physical method of X-ray diffraction. (Both Perutz and Dorothy Hodgkin, as well as countless others, persisted in referring to Bernal as 'Sage', because he seemed to know everything, not just his science.)

Figure 1.3 *Photograph of I. Fankuchen, Dorothy Crowfoot-Hodgkin, J. D. Bernal, and Dina Fankuchen.*

(Courtesy Prof Eleanor Dodson)

Dorothy Hodgkin, in her memoir of Bernal for the Royal Society, captures the impact that Bernal's vast range of knowledge had upon her and other collaborators: '*Every day, one of the group would go and buy bread from Fitzbithies,*[1] *fruit and cheese from the market, while another made coffee on the gas ring in the corner of the bench. One day there was talk about anaerobic bacteria at the bottom of a lake in Russia and origin of life, another about Romanesque architecture in French villages, or Leonardo da Vinci's engines of war, about poetry or painting. We never knew to what enchanted land we would next be taken...*'

In Sweden, late in 1964, where she was lecturing after the award of the Nobel Prize to her, a journalist asked Dorothy Hodgkin what the first thing would be she would do if someone handed her a piece of moon rock. Her immediate reply was '*Give it to Bernal!*'. According to Perutz, Bernal was a bohemian, a flamboyant Don Juan, and a restless genius, always searching for something more important to do. Perutz said that he could not have wished for a more inspiring supervisor. Yet Bernal was erratic; his desk was chaotic; he left no time to finish things: '*He was a Communist and a woman-chaser and let his scientific imagination run wild.*' It was largely because of Bernal's personal lifestyle that the conservative Head of the Cavendish Laboratory (a member of which Bernal had become in 1936) Lord Rutherford wanted to dismiss him. It was only because Sir Lawrence Bragg (henceforth WLB) dissuaded Rutherford from doing so, that Bernal retained his position.

Bernal was also instrumental in converting Kendrew from being a chemical reaction kineticist to a protein scientist. After graduating in physical chemistry at Cambridge, Kendrew was greatly attracted by the impressive mathematics skills and scholarship of E. A. Moelwyn-Hughes[2] (1905–1978) and he began his research on the muta-rotation of sugars before World War II (published in *Proceedings of the Royal Society* in 1940).[3] Wartime duties diverted Kendrew to radar, and later to operational, research and he ended, along with Bernal, as advisor to Lord Mountbatten in the South Eastern Command in Ceylon (now Sri Lanka).

Writing later in the *Dictionary of National Biography*, John Kendrew said of Bernal:

> '*He had an infectious delight in new ideas, whether his own or another's, the question of credit did not arise, for all that mattered was that the idea was exciting and that it had to be pursued. Other people's results gave him as much pleasure as his own. He had an immensely stimulating influence on scientists of his own and younger generations, which was far beyond, and possibly more important than, his own personal contributions!*'

A similarly handsome tribute was paid to Bernal by Earl Mountbatten:

> '*Desmond Bernal was one of the most engaging personalities I have ever known. I became really fond of him, and enjoyed my discussions and arguments immensely. He had a very clear analytical brain; he was tireless and outspoken. Perhaps his most pleasant quality was*

[1] A local shop close to the centre of Cambridge.

his generosity. He never minded slaving away at other people's ideas, helping to decide what could or could not be done, without himself being the originator of many of the major ideas on which he actually worked. This may be the reason why his great contribution to the war effort has not been properly appreciated, but those of us who really knew what he did, have an unbounded admiration for his contributions to our winning the war.'

While they were together in the jungles of Sri Lanka, Bernal convinced John Kendrew that his PhD should be concerned with the three-dimensional structure of proteins. He advised Kendrew to call on his way home to the UK, to talk with Linus Pauling at the California Institute of Technology in Pasadena. This visit by Kendrew[4] further convinced him of the merit of attacking the structure of proteins. Both Bernal and Pauling advised him, on his return to Cambridge, to seek out Max Perutz.

1.4 Perutz First Meets Kendrew

Perutz has given a charming account of his first encounter with Kendrew:[5]

'In October 1945, a young man in a Wing Commander's uniform walked into my room at the Cavendish Laboratory in Cambridge and said he wanted to become my research student…I was flattered because I had never had a research student, let alone one hardly my junior who had distinguished himself in the war, but I also felt embarrassed.'

Perutz himself had completed his PhD in 1940, working with Sir Lawrence Bragg, who had succeeded Rutherford as the Cavendish Professor and had obtained a small grant for him from The Rockefeller Foundation to enable his research to continue. However, the project was interrupted by Perutz's internment in 1940, along with several hundred German and Austrian refugee scholars, mostly Jewish and anti-Nazi. They were rounded up and sent to an internment camp in Quebec, Canada—see Perutz's fascinating essay '*Enemy Alien*'.[6] Fortunately, Perutz's academic friends secured his release.

He returned to his studies and to work of national importance, which involved him in an unsuccessful scheme to make ships of sawdust and ice for refuelling aircraft in the North Atlantic—see Perutz's account.[7]

Even before the war, after publishing an important paper with Bernal and Fankuchen[8] in 1938 on X-ray studies of the digestive enzyme chymotrypsin and the blood protein haemoglobin, but largely as a result of conversations with a relative in Prague (Dr Felix Haurowitz), Perutz had already decided to work on the massive globular protein haemoglobin, which was easy to purify and crystallize. Moreover, it was an extremely important biomolecule that had already elicited great interest among the giants of science, notably the Master of Pembroke College, Cambridge G. G. Stokes (1819–1903), the President of Harvard J. B. Conant (1893–1978); and, from the mid 1930s, Linus Pauling at Caltech.

Shortly after he had accepted Kendrew as his first PhD student, Perutz, while walking out of his office in the Cavendish Laboratory, had a fortunate encounter with Sir Joseph Barcroft (1872–1947), the great respiratory physiologist, who was Head of the Cambridge Physiological Laboratory at the time. Barcroft suggested that Kendrew might make a comparative study of adult and fetal sheep haemoglobin, for which Barcroft would supply the blood. Perutz was much relieved by this suggestion, which Kendrew gladly accepted. Indeed, within a few years, Kendrew and an eminent Cambridge scientist F. J. W. Roughton (Head of the Department of Colloid Science) were the principal organizers of a symposium in 1948 entitled '*Haemoglobin*' (see later), in memory of Sir Joseph Barcroft. In the meanwhile, Lawrence Bragg, whose enthusiasm for Perutz's work on the X-ray study of haemoglobin was, and remained thereafter, very high, obtained a three-year ICI Fellowship for Perutz. This supported Perutz until 1947, when Kendrew's research grant also ran out.

Sir Edward Mellanby, Secretary of the MRC, and WLB formulated a scheme for the maintenance of the work by Perutz and Kendrew. After discussions at the Athenaeum Club (London), Mellanby, prompted by WLB, put forth a proposition for the establishment of a MRC Research Unit at the Cavendish Laboratory on the molecular structure of biological systems. (Such was the mundane set of events that led to the birth of the MRC's most famous research unit—the LMB in Cambridge!)

1.5 The Trajectory of the Perutz–Kendrew Collaboration

The first paper, written jointly by Perutz and Kendrew, appeared in the *Proceedings of the Royal Society* in 1948,[9] entitled '*A comparative X-ray study of foetal and adult sheep haemoglobin*'. We note, in passing, that the haemoglobins of mother and fetus are not the same. Contrary to most people's belief, the mother's blood does not flow through the fetus. The fetus has its own circulatory system, which is connected by the umbilical cord to the placenta where it terminates in a multitude of fine capillaries. There, the blood of the fetus comes into close contact with the mother's blood, which is pumped through the placenta in a similar, but entirely separate, system of capillaries, accompanied by an exchange of oxygen and carbon dioxide.

The paper by Kendrew and Perutz, submitted in January 1948, was, by the standards that these two authors achieved subsequently, a rather mundane contribution to the science of proteins. Unsurprisingly, this paper merely reported facts pertaining to the optical properties, unit-cell dimensions—the smallest repeat volume in a crystal structure—and crystallographic space groups, which is a classification system, based on symmetry, for various types of arrangements of structural units in a crystal (see Chapter 2) of wet and dried crystals, and they again established the non-identity of fetal and adult haemoglobin.

By June 1948, however, Perutz and Kendrew had already made significant progress in their long and labyrinthine path to elucidate the internal structure of the proteins haemoglobin and myoglobin—the precise nature, function, and mode of action of these two vitally important proteins will be described in Chapter 5. Suffice it to say here that the haemoglobin molecule contains some 5000 non-hydrogen atoms, and myoglobin some 1260. As will emerge later, haemoglobin, which reacts reversibly with oxygen, carries that gas from the lungs to the various components of the body (and also transports carbon dioxide produced by metabolic processes in the body back to the lungs). Myoglobin, which is a fourth of the size of haemoglobin (molecular weight 16,700), contains one so-called haem group (to be described later), which is responsible for its main function of carrying oxygen molecules to muscle tissues. The haem group is what confers the red colour to both myoglobin and haemoglobin. In his Nobel Lecture,[10] Kendrew described why he chose myoglobin, and, in particular, the animal species whose myoglobin formed crystals suitable, both morphologically and structurally, to suit his goals. Eventually, he was led to the sperm whale *Physeter Catodon*—the material that he used came either from Peru or from the Antarctic.

In June 1948, the important symposium on '*Haemoglobin*', in memory of Sir Joseph Barcroft, organized by John Kendrew and F. J. W. Roughton (the Professor of Colloid Chemistry), was convened in Cambridge. At this meeting, three existing Nobel Prize winners spoke: E. D. Adrian, H. Dale, and A. V. Hill. It is remarkable that six of the other speakers later became Nobel Prize winners: Linus Pauling, R. R. Porter, F. Sanger, C. de Duve, M. F. Perutz, and J. C. Kendrew (see Figure 1.4).

In the Proceedings of this symposium, there were three important papers, one each by Perutz and Kendrew, and a joint one, the latter being entitled '*The Application of X-ray Crystallography to the Study of Biological Macromolecules*'. (Perutz's paper was on recent developments in the X-ray study of haemoglobin, and Kendrew's on the crystal structure of horse myoglobin.)

The joint paper by Kendrew and Perutz,[11] dealing with the X-ray study of biological macromolecules, is of great significance. In their own words, it was:

> '*...'a brief conspectus of the applications of X-ray crystallographic techniques, particularly to the study of large molecules (but excluding fibrous structures), for the benefit of workers in other fields unfamiliar with these methods.*'

So important is this early joint article by Kendrew and Perutz, that it is no exaggeration, in hindsight, to regard it as their manifesto. It rehearses all the key points needed to meet the challenge of determining the atomic structure of the giant molecules that are so much a feature of what is now the corpus of structural molecular biology.

It is prudent first to distinguish, as did Kendrew and Perutz, between biochemistry and molecular biology, a topic which is also discussed in Chapter 3. The biochemists started off using the traditional techniques of chemistry. These techniques can readily handle small molecules, containing, say, thirty or forty atoms—these days, organic chemists and biochemists (equipped with sophisticated new

Figure 1.4 *F. Sanger (right) with R. R. Porter, on the steps of the Department of Biochemistry, Cambridge, 1947.*

(Courtesy Wellcome Foundation)

techniques such as magnetic resonance spectroscopy and mass spectroscopy) can handle much larger molecules of several hundreds or thousands of atoms. But when Kendrew and Perutz wrote their classic conspectus,[11] they were aware that molecular biology entailed handling protein and other molecules consisting of up to several tens of thousands (non-hydrogen) atoms. Moreover, it was not just the constitution (the primary structure) of these biological macromolecules that was the objective; it was the three-dimensional structure that needed to be determined, as this is what governs the mechanism of enzymatic and similar processes that take place in the living cell. It was here that molecular biology, with its *new physical method* (based on X-ray diffraction, which we discuss in Chapter 2) for studying very large molecules, began to make a significant impact on biological science. And, as we shall see later, it was the critical experiment by Bernal and Crowfoot (Hodgkin) in 1934, when they reported the rich X-ray diffraction patterns produced by the enzyme pepsin (in its mother liquor), that convinced physicists that X-ray analysis held the key to progress in the determination of macromolecular structure.

What Kendrew and Perutz knew, as largely self-taught X-ray crystallographers, was that, in principle, X-ray diffraction patterns can determine the precise electron

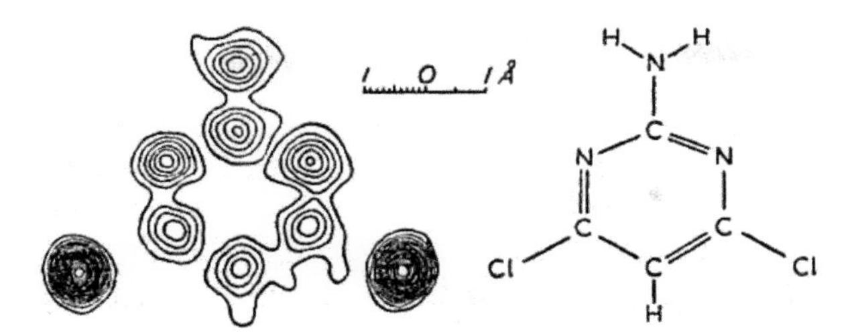

Figure 1.5 *Electron density projection of a dichloro-amino derivative of pyrimidine, published by Clews and Cochran in 1948. (An angstrom unit Å is 10^{-8} cm—one hundred millionth of a centrimetre.)*

(Reproduced with permission of the International Union of Crystallography)

density distribution in any given molecule, large or small, from its characteristic X-ray diffraction pattern.

Apart from introducing Bragg's Law ($n\lambda = 2d \sin \theta$, where θ is the angle of incidence and reflection, d the atomic spacing, λ the X-ray wavelength, and n an integer), the notion of space groups and the unit cell, and the manner in which X-ray diffraction patterns may be interpreted (see also Chapter 2), they specifically dwell on the way in which electron densities in crystals may be retrieved by X-ray crystallography (see, for example, Figure 1.5).

Figure 1.5 is a planar electron density picture of a material that their colleagues (Clews and Cochran)[12] had only recently derived of a dichloro amino derivative of a pyrimidine compound (see Glossary).

This paper, by Kendrew and Perutz,[11] also constitutes one of the best non-mathematical accounts of the so-called Fourier analysis (see Chapter 2) for the evaluation of *electron densities* of molecules from X-ray diffraction patterns of a crystal of that molecule. Another pedagogically oriented account was given later by Crick and Kendrew.[13] It spells out precisely what was to occupy the thoughts and actions of the Perutz–Kendrew teams for the next ten years, until they finally succeeded to map out the internal three-dimensional structure of both myoglobin and haemoglobin.

Perutz and Kendrew would have been made aware by Bernal and others that Kathleen Lonsdale had, in the late 1920s at the DFRL, shown that the benzene molecule was planar from her Fourier method of analysing—the first ever recorded—of the X-ray diffraction patterns of both hexachloro- and hexamethyl-benzene. The key point to note here is that the determination of structure by X-ray diffraction gives directly the electron density distribution within the molecule under study. From the resulting electron density, the nature of the bonding and of the constituent atoms can be confidently deduced. This fact became obvious from the work done in W. H. Bragg's group at the DFRL (see Chapter 3) in his time there as Director. Thus, his son W. L. Bragg used a series of examples from that work in his presentation to the Royal Society of Edinburgh in 1935 (see Figure 1.6). One of these is naphthalene,[15] the other oxalic acid.[16] The electron densities in this

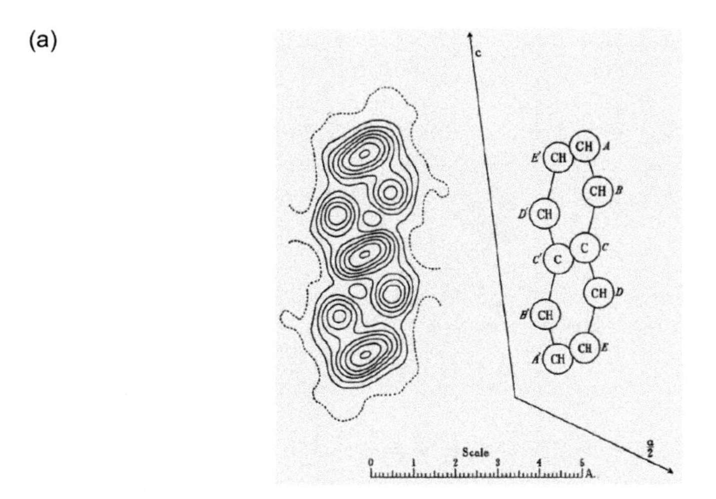

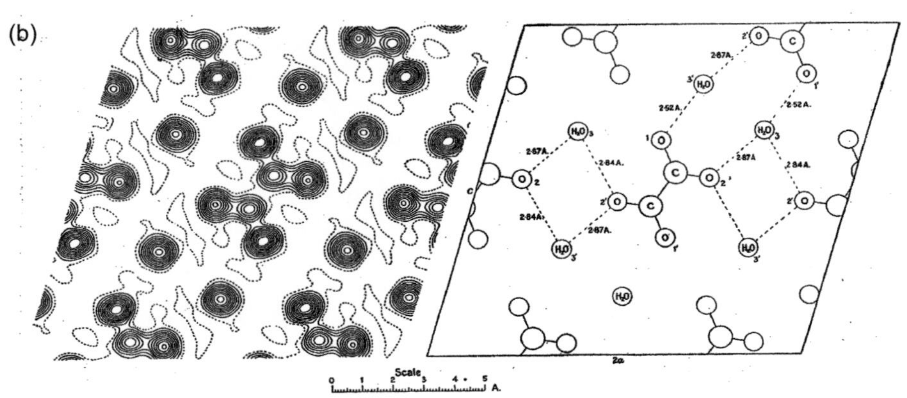

Figure 1.6 *Illustrations of the power of the electron density map shown by W. L. Bragg in Edinburgh, 1935.*[14] *(a) Naphthalene.*[15] *(b) Oxalic acid.*[16]

(Reproduced with permission of the Royal Society of Chemistry)

early result show the location of the carbon atoms clearly. In the later work on oxalic acid, the electron densities of the carbon and oxygen atoms are readily distinguishable.

To quote W. L. Bragg,[14] '*Investigations of this kind do not merely confirm and give precision to formulae which have already been determined; they lead on to the determination of structural formulae for far more complex molecules, whose constitution has not yet been fully established.*' X-ray analysis goes much further than the degradative methods of the organic chemist and biochemist.

When a heavy atom (rich in electrons) is present in a molecule, its presence in the electron density distribution that can be derived from the X-ray diffraction pattern stands out clearly. A good example of how the structure of the vitamin calciferol—the hormonally active metabolite of vitamin D, which stimulates the release of calcium from bones and promotes absorption of dietary calcium from the gastrointestinal tract—was determined[17] from its electron density distribution is shown in Figure 1.7. Further reference is made to calciferol in Chapter 3. Iodine, which has fifty-three electrons, was used as a tag in the molecule to boost the electron density in that region of the structure. This is a tactic that both Perutz and

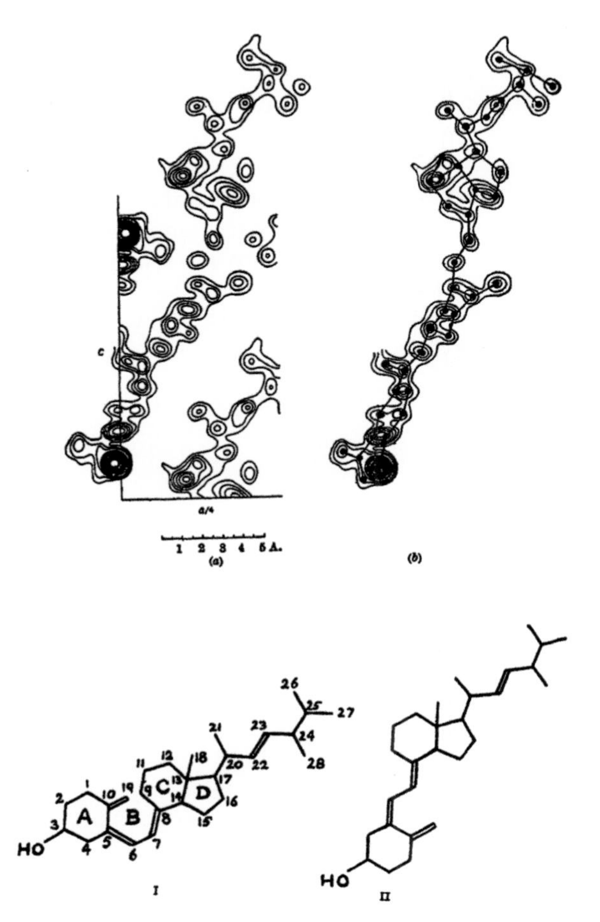

Figure 1.7 *(a) Electron density projection of an iodo-substituted derivative of the vitamin calciferol determined by Crowfoot and Dunitz*[17] *in 1948. The actual compound is calciferol-4-iodo-5-nitrobenzene. (b) The same, superimposed, with the positions derived for the atoms in the molecules. The ring of the sterol skeleton is broken, and structure I, deduced on chemical evidence, is correct. In the crystal, the molecule does not maintain the curled-up configuration suggested by I but exists in an extended form, approximately represented by II.*

(Reproduced from Nature*)*

Kendrew were later to use in their pioneering studies of haemoglobin and myoglobin, as described in later chapters.

1.6 Envoi

Before the achievements of Perutz, Kendrew, and their associates in elucidating the nature and function of haemoglobin and myoglobin and of a host of numerous other substances that belong to the living world can be satisfactorily outlined, it is first necessary to discuss the birth of X-ray diffraction and to enquire into its quintessential nature and power. It is also necessary to outline the nature of some of the 'living molecules' that were first explored during the genesis of structural molecular biology. Most of the early exploration of these materials was done at the DFRL and later in Leeds, Cambridge, Oxford, and Pasadena. These materials included amino acids, fatty acids, proteins, wool, silk, muscles, hairs, horn, sex hormones, algae, flagella, vitamins, and enzymes.

Chapter 2 focuses on the birth and power of X-ray crystallography, and it is followed by an account of W. H. Bragg's achievements with his team of workers at the DFRL.

APPENDIX 1

Sponsorship of Scientific Research, Especially by the Medical Research Council

In Section 1.4 above, reference was made to the support of Bragg's research at the Cavendish Laboratory by The Rockefeller Foundation, by Imperial Chemical Industries (ICI), and by the Medical Research Council (MRC). Both The Rockefeller Foundation[18] and, to a slightly lesser degree, ICI were magnanimous in the ways that they supported fundamental research in British universities in the inter-war years and for several decades thereafter. They tended, however, to favour mission-oriented research, whereas the MRC, on the other hand, by granting long-term funding to exceptional research leaders, gave greater intellectual freedom to permit their recipients to explore a wider range of important scientific options.

Max Perutz frequently told me that, thanks to the policy of the MRC, he felt he could assemble outstanding individuals at the LMB, and then grant them the freedom to pursue the long-term research which they felt their abilities and interests could achieve.

It is noteworthy to outline the genesis of the MRC and the enlightened policy towards research that it implemented from the time of its inception. The origins of the MRC go back to 1915 when the Medical Research Committee (its forerunner) was created by David Lloyd George under the National Insurance Act that he introduced, as Chancellor of the Exchequer, in 1911. The first Chief Executive Officer (CEO) was a former Cambridge University physiologist Sir Walter Morley Fletcher (who was succeeded by Sir Edward Mellanby, with whom W. L. Bragg built up a warm personal relationship) (see Section 8.3.1).

At the cessation of World War I, Lloyd George, who was then Prime Minister, formed the MRC, with Fletcher as its CEO. The MRC was, from its outset, free from any control by the Ministry of Health and placed under the Privy Council, with a charter of its own and direct financial support from the Treasury. Over the years, the MRC has focused on high-impact research and has provided financial support and scientific expertise behind a number of medical breakthroughs, including the development of penicillin, the discovery of the structure of DNA, monoclonal antibodies, and the genetic code, as well as the foundation of molecular biology with its profound impact on medicine. It was work supported by the MRC that established that smoking causes cancer. Research funded by the MRC has produced thirty-two Nobel Prize winners to date.

The MRC is responsible for co-ordinating and funding medical research in the UK. It is now part of the UK Research and Innovation (UKRI), which came into operation on 1 April 2018 and brings together the UK's seven research councils, Innovate UK, and Research England.

Writing in '*A Half Century of Medical Research and Policy*', its author Sir A. Landsborough-Thomas refers to the First National Insurance Act of 1911 as epoch-making. He goes on to quote Sir Walter Fletcher who said that the three British Statesmen who had notably furthered the cause of medical science were: (i) King Henry VIII, who founded the Regius Chairs of Physic (medicine) at Oxford and Cambridge; (ii) King Charles II, who gave the Royal Society its charter; and (iii) Mr Lloyd George with his creation of the National Insurance Act.

APPENDIX 2

The Origin of the Phrase Molecular Biology

Notwithstanding the claims of W. T. Astbury and the debate between Kendrew and Waddington mentioned in the Introduction, many biologically oriented scientists ascribe the first public use of the term 'molecular biology' to Dr Warren Weaver. In 1938, Weaver wrote the following letter, which was re-published in *Science*, **1970**, *170*, 582: '*Very shortly after I was appointed, in 1932, the director for the natural sciences in the Rockefeller Foundation, I urged the Trustees, with the full backing of the then president of the Rockefeller Foundation, Max Mason, that the science program of the foundation be shifted from its previous preoccupation with the physical sciences, to an interest in stimulating and aiding the application, to basic biological problems of the techniques, experimental procedures, and methods of analysis so effectively developed in the physical sciences.*

This proposal was accepted and approved by the Rockefeller Foundation Trustees, and progress in the program was sufficiently prompt and promising so that when I drafted the "natural sciences" section of the Annual Report of the Rockefeller Foundation for 1938 this section began with a sixteen-page portion, pages 203–219, which was headed, in large type, MOLECULAR BIOLOGY, the first sentence being 'Among the studies to which the Foundation is giving support is a series in a relatively new field, which may be called molecular biology, in which delicate modern techniques are now being used to investigate ever more minute details of certain life processes.'

REFERENCES

1. Crick, born in 1916, was 12 years older than Watson and much more experienced as a physicist, but not in molecular biology or genetics. It is interesting to note that Crick's work that earned him his Nobel Prize was completed before his PhD was awarded (in 1954).
2. For a summary of E. A. Moelwyn-Hughes' work, see his obituary in *Nature*, **1979**, *277*, 334 (see also Section 8.3.2).
3. J. C. Kendrew and E. A. Moelwyn-Hughes, *Proc. Roy. Soc. A*, **1940**, *176*, 352.
4. Pauling's daughter Linda told the author in February 2016 that she vividly recalled John Kendrew's visit to their home (dressed in his Wing-Commander uniform).
5. M. F. Perutz, *Science*, **1978**, *201*, 1187–91.
6. M. F. Perutz, '*Enemy Alien*' in '*Is Science Necessary: Essays on Science and Scientists*' (by M. Perutz), Barrie and Jenkins, London, **1989**.
7. See ref [6], pp. 111–24.
8. J. D. Bernal, I. Fankuchen, and M. F. Perutz, *Nature*, **1938**, *141*, 523.
9. J. C. Kendrew and M. F. Perutz, *Proc. Roy. Soc. A*, **1948**, *194*, 375.
10. J. C. Kendrew, Nobel Lecture, published in *Science*, **1962**, p. 676.
11. M. F. Perutz and J. C. Kendrew in '*Haemoglobin*' (eds. F. J. W. Roughton and J. C. Kendrew), Butterworths, London, **1949**, p. 161.
12. C. J. B. Clews and W. Cockran, *Acta Cryst.*, **1948**, *1*, 4.
13. F. H. C. Crick and J. C. Kendrew, *Adv. Protein Chem.*, **1957**, *12*, 133.
14. W. L. Bragg, *Proc. Roy. Soc. Edinburgh*, **1934–5**, *Vol LV*, 64.
15. J. M. Robertson, *Proc. Roy. Soc. A*, **1933**, *142*, 674.
16. J. M. Robertson, *Rep. Prog. Phys.*, **1937**, *4*, 332.
17. D. Crowfoot and J. D. Dunitz, *Nature*, **1948**, *162*, 608.
18. D Fisher, '*The Rockefeller Foundation and the Development of Scientific Medicine in Britain*', *Minerva*, **1987**, *16*, 20.

2

The Birth and Initial Exploitation of X-ray Diffraction

Author's Note

For the non-scientist reader, parts of this chapter may be omitted. It is not essential, for example, to scrutinize Figure 2.8, and all that is necessary to note concerning Figures 2.3 and 2.4 is that, with the aid of Bragg's Law (shown in Figure 2.6), the interatomic distances in solids, such as graphite, diamond, and rock salts, may be readily determined from the position of the spots (the diffraction patterns) generated by X-rays that strike these solids. What is important to note in Figure 2.12 is that an electron density contour map of the internal structure of a solid can be deduced from an X-ray diffraction pattern.

At the outset of their monumental work on haemoglobin and myoglobin, Perutz and Kendrew realized (as described in Section 1.5) that they could, in principle, determine the electron density distributions, and hence the fine details, of their folded proteins, to such a degree that they could identify individual atoms and clusters of atoms of the chains that are present in giant macromolecular proteins. What was necessary for them to determine were the following three measurements: the precise location of each spot in the diffraction patterns; the absolute intensity of each spot; and its phase, as illustrated in Figures 2.16 and 2.17. Perutz and Kendrew realized at the outset of their investigations that electron densities of folded proteins could be retrieved from X-ray diffraction patterns, just as they had been determined for smaller molecules, like that shown in Figure 1.5.

Figures 2.12 to 2.17 are visual aids for those unfamiliar with the way X-ray crystallography constitutes a new branch of unconventional microscopy.

2.1 Introduction

In 1968, in the autumn of his career, Lawrence Bragg began a popular article[1] on X-ray crystallography with the words:

'Fifty six years ago a new branch of science was born with the discovery by Max von Laue of Germany that a beam of X-rays could be diffracted, or scattered, in an orderly way by the orderly array of atoms in a crystal. At first the main interest in von Laue's discovery was focussed on its bearing on the controversy about the nature of X-rays; it proved that they were waves not particles. It soon became clear to some of us, however, that this effect opened up a new way of studying matter, that in fact man had been presented with a new form of microscope, several thousand times more powerful than any light microscope, that could, in principle, resolve the structure of matter right down to the atomic scale. The development of X-ray crystallography since 1912 has more than fulfilled our early expectations. It not only has revealed the way atoms are arranged in many diverse forms of matter, but also has cast a flood of light on the nature of the forces between the atoms and on the larger scale properties of matter. In many cases this new knowledge has led to a fundamental revision of ideas in other branches of science. A culmination of sorts has been reached in the past few years with the successful structure analysis of several of the basic molecules of living matter – the proteins – each of which consists of thousands of atoms held together by an incredibly intricate network of chemical bonds.'

2.2 Sequence of Events Leading to the Birth of X-ray Diffraction

Although, in many respects, 1913 marks the year when the application of the power of X-ray diffraction was first demonstrated by the determination of the structure of sodium chloride (NaCl), potassium chloride (KCl), potassium bromide (KBr), and potassium iodide (KI) by W. L. Bragg[2] and that of diamond by him and his father W. H. Bragg[3] (see Figures 2.1 and 2.2), it is three critical events in 1912 that really marked the birth of X-ray diffraction.[4–6]

The first took place in May of that year when von Laue, Friedrich, and Knipping[7] submitted a paper to the *Bayerische Akademie der Wissenschaften* in Munich that

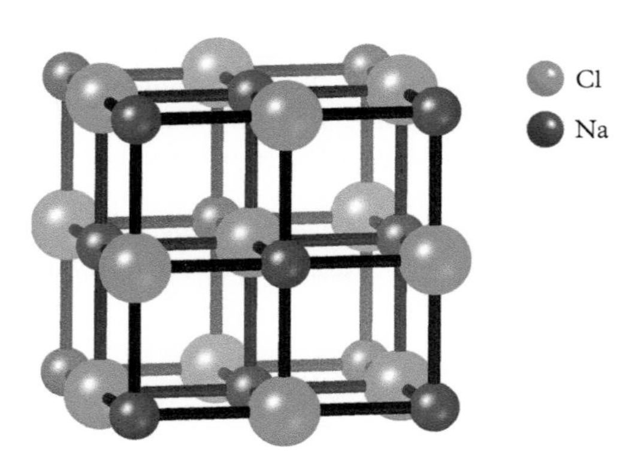

Figure 2.1 *The structure of sodium chloride, determined by W. L. Bragg.*[2]
(Courtesy Dr D. N. Johnstone)

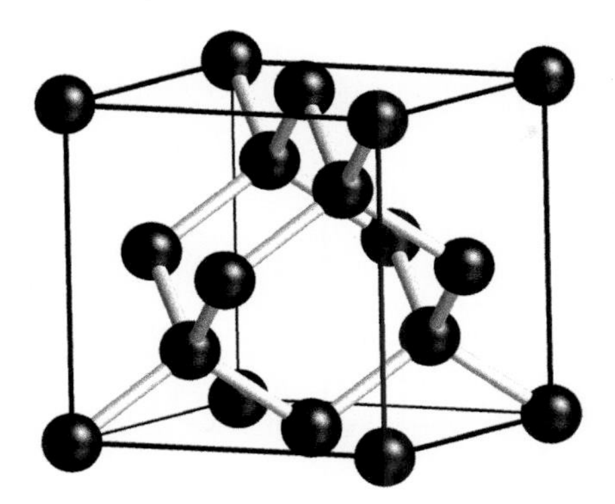

Figure 2.2 *The structure of diamond, determined by W. L. and W. H. Bragg.*[3]
(Courtesy Dr D. N. Johnstone)

contained experimental proof that X-rays could behave as waves, as well as corpuscles, thereby settling an enigma that had lasted seventeen years since their discovery by Roentgen.

The second occurred in summer of 1912 when the 22-year-old Lawrence Bragg, a graduate student supervised by J. J. Thomson at the Cavendish Laboratory, University of Cambridge, went on his holidays to Cloughton, a town on the Yorkshire coast, to join his parents (his father W. H. Bragg was, at that time, the Professor of Physics at the University of Leeds). Lawrence Bragg, in his own words, said that, on arrival in his parents' home, he was deeply unhappy. This was because the Cavendish Laboratory at the time '*was a sad place. There were too many young researchers ... too few ideas for them to work on, too little money, and too little apparatus.*'

When, however, Lawrence Bragg was shown a letter that his father had just received from an ex-student Lars Vegard (1880–1963), a Norwegian gentleman then working in Germany with the eminent physicist Wien in Wurtzburg, he became especially interested. Vegard described in detail[8] the contents of a seminar that von Laue had delivered in Wurtzburg.[9] He illustrated what happened when a crystal of the mineral zinc blende (ZnS) was exposed to a beam of X-rays. A well-defined, symmetrical spot pattern surrounding the central spot of the undeviated X-rays on a photographic plate placed beyond the crystal of the mineral (see Figures 2.3 and 2.4) was observed.[10]

The third event occurred in autumn of 1912 when Lawrence Bragg had an idea, while walking along the Backs of Cambridge, that[11] '*led immediately*', in the words of W. L. Bragg's Royal Society biographer Sir David (later Lord Phillips of Ellesmere), '*to a dramatic advance in physics and has since transformed chemistry, mineralogy, metallurgy, and, most recently, biology. He realised that the observation of X-ray diffraction by a crystal, which had been reported by von Laue and his associates earlier in the year*[7] *can be interpreted very simply as arising from reflections of the X-rays by planes of atoms in the crystal and hence that the X-ray observations provide evidence from which the arrangement of atoms in the crystal may be determined. A few*

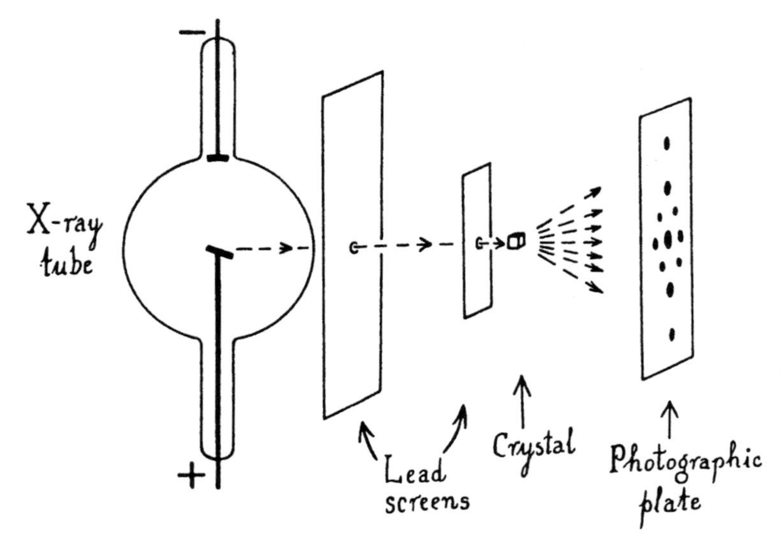

Figure 2.3 *Schematic diagram of the kind of set-up used by von Laue* et al. *in their discovery of X-ray diffraction.*

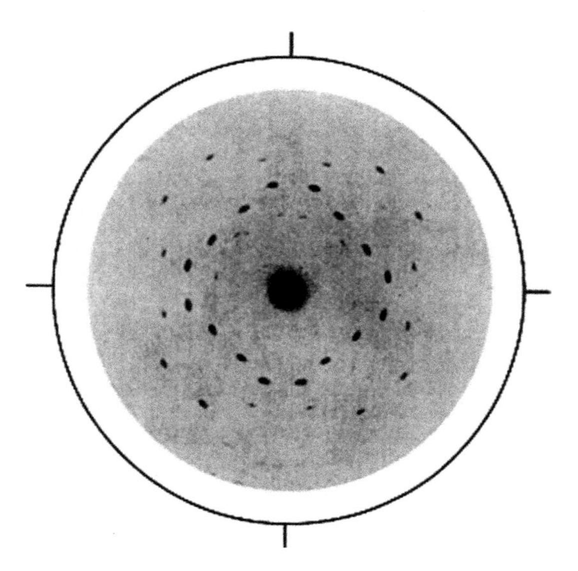

Figure 2.4 *A typical X-ray diffraction pattern of the kind discovered by von Laue* et al. *of the mineral zinc blende ZnS.*

weeks of intensive work on simple compounds were enough to demonstrate the correctness of these ideas but the development of the method, at first in association with his father (W. H. Bragg), and later as the leader or guiding influence of a host of workers (at Cambridge, Manchester, and the Davy-Faraday Laboratory of the Royal Institution, London) was the "labour of a lifetime".'

The paper by von Laue and his colleagues in 1912[7] was a major landmark in X-ray physics and in the subsequent development of crystallography and molecular biology. (In her review given at the New York Academy of Sciences in 1997, Dorothy Hodgkin[12] regarded this work as the vital starting point of the study of proteins.) While von Laue *et al.*'s paper revealed ways of recovering quantitative information about the internal structure of crystalline solids, it yielded a relatively complicated way of evaluating the separation distances between atoms. von Laue and his co-workers envisaged crystals in terms of a three-dimensional network of rows of atoms and based their analysis on the notion that a crystal behaves, in effect, as a three-dimensional grating. Figure 2.5 is the kind of pattern described by W. H. Bragg in his early publication[13] and illustrates the complexity of the picture portrayed by Laue's photographs such as those used to record the pattern for ZnS (see Figure 2.4).

von Laue and co-workers found it necessary to assume that there were five distinct X-ray wavelengths in their source used to record the pattern shown in Figure 2.4 above.

W. L. Bragg's picture of what gave rise to the diffraction pattern produced by X-rays is based on the notion that a crystal is composed of layers of sheets of atoms[1] (see Figure 2.6).

W. L. Bragg's picture (see Figure 2.6), in which a crystal is regarded as composed of layers of atoms that behave, in effect, as reflecting planes, for which the angle of incidence equals the angle of reflection. The beams of X-rays that are strongly 'reflected' (i.e. diffracted) are generated when the path difference between reflections from successive planes in a family is equal to an integral number of wavelengths. This idea gave rise to the extremely simple, well-known Bragg equation[15] [also known as Bragg's Law, equation (1)]:

$$n\lambda = 2d \sin\theta \tag{1}$$

where λ is the wavelength of the incident X-rays, n is the order of the reflection, d is the interplanar spacing, and θ is the angle of incidence to the planes.

Upon realizing this simple, fruitful fact, W. L. Bragg worked with feverish intensity, both with his father in the Department of Physics, Leeds University, and at the Cavendish Laboratory where, with his father's help, he was able to install a so-called X-ray spectrometer, identical to that which his father had constructed (with the aim of measuring wavelengths of X-rays). He submitted his work, via his supervisor (J. J. Thomson)—as he himself was not allowed by the rules to present his paper (as a graduate student)—to the Cambridge Philosophical Society in November 1912.[14]

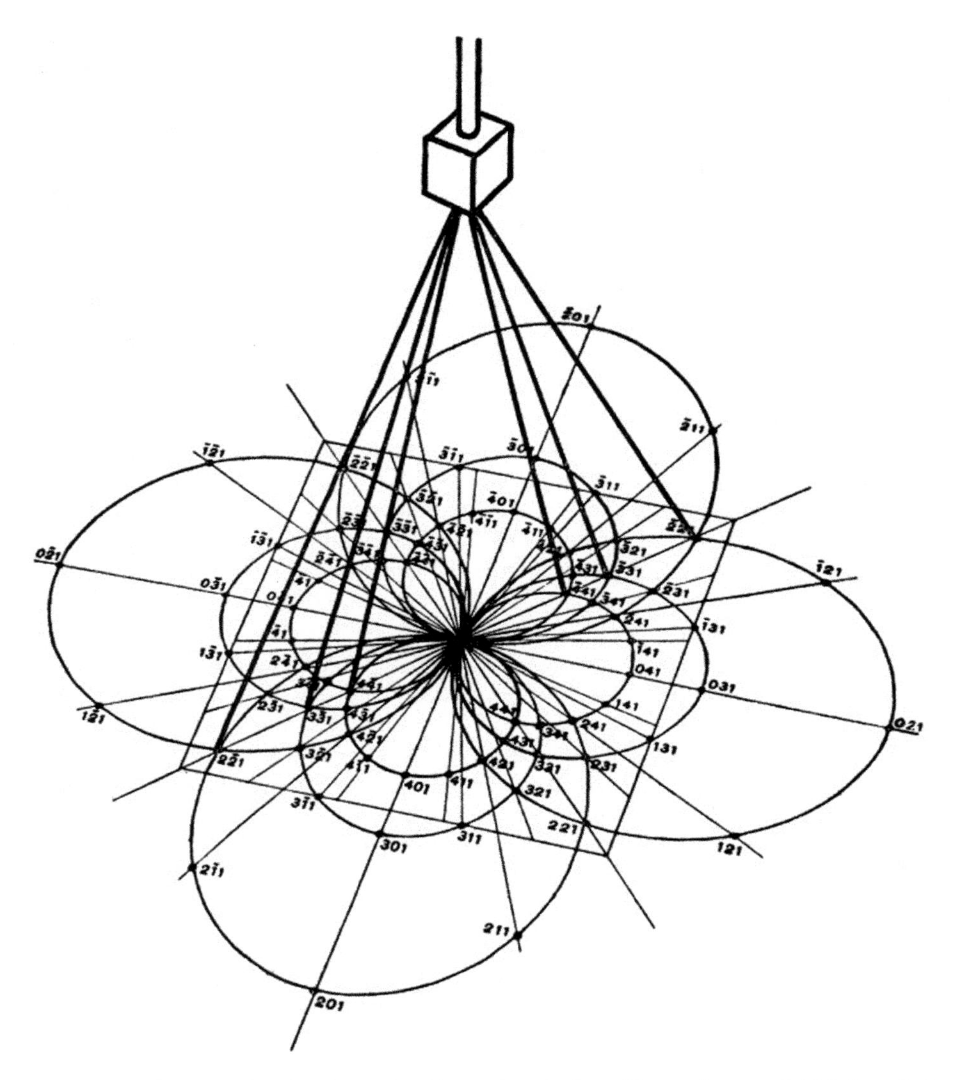

Figure 2.5 *Drawing showing the mode of formation of a Laue X-ray photograph. This is typical of the illustrations used by W. H. Bragg.*

('An Introduction to Crystal Analysis'*, G. Bell and Sons, London, 1928)*[13]

Reflecting in his article in the text '*Fifty Years of X-ray Diffraction*',[15] the distinguished German crystallographer P. P. Ewald stated that W. L. Bragg's paper to the Cambridge Philosophical Society could '*hardly be overestimated*' in its importance (see also Section 9.4 and Figure 9.3). It contained three major points:

1. The idea of explaining the von Laue spots as reflections of the incident ray on the internal planes;

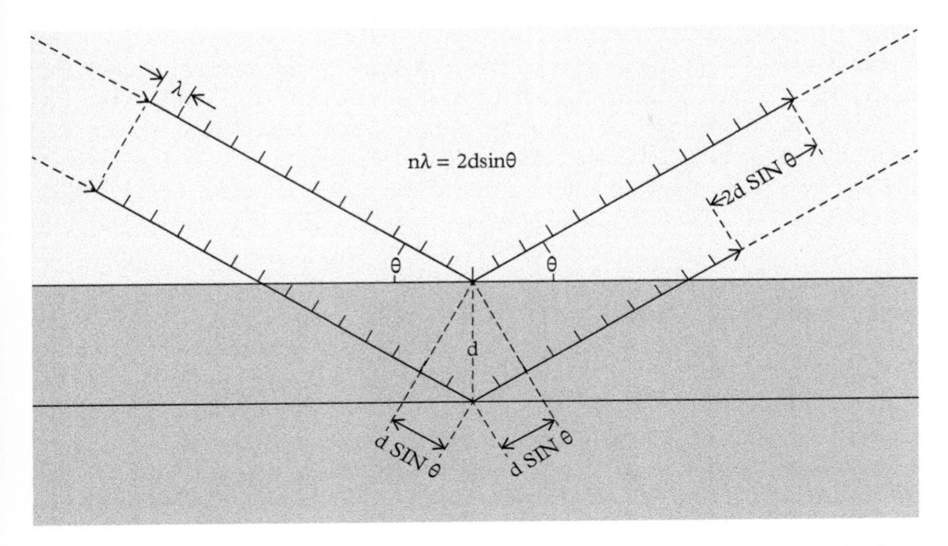

Figure 2.6 *W. L. Bragg's interpretation of how diffraction arises from reflection at atomic planes, and a statement of Bragg's Law.*

(Copyright Joan Starwood)

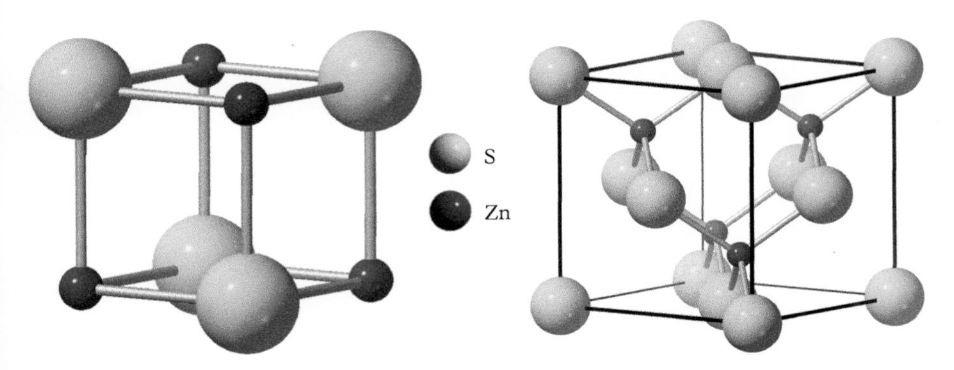

Figure 2.7 *Competing zinc blende (ZnS) crystal structures. (Left) Depiction of the simple cubic structure of ZnS proposed by von Laue* et al.[7] *(Right) The face-centred cubic structure of ZnS proposed by W. L. Bragg.*[2,14]

(Courtesy of Prof K. D. M. Harris)

2. The assumption of a continuous spectrum (i.e. of 'white' X-rays) of the incident ray and the *selective action* of the sets of reflecting planes in reinforcing only those wavelengths which fit into their distance of repeat;[16] and
3. The proof that the structure of ZnS is not a simple one (envisaged by von Laue), but a face-centred cubic one, as depicted in Figure 2.7.

In the audience in Cambridge in November 1912 was the physicist C. T. R. Wilson,[17] whose work using cloud chambers to track cosmic rays earned him the Nobel Prize in 1927. Wilson suggested that X-rays should also reflect from the external faces of crystals, provided the surfaces were sufficiently smooth. So Lawrence Bragg tested whether X-rays that reflected from the cleavage face of mica—known for its supposed flatness at the atomic scale—could be photographed. In December 1912, *Nature* published his paper '*The Specular Reflection of X-rays*'.[18]

W. H. Bragg quickly demonstrated that his X-ray spectrometer (see Figure 2.8) could detect diffracted monochromatic X-rays—not on photographic plate, but with a gas ionization detector.

The power of Bragg's Law, together with W. H. Bragg's spectrometer for recording the intensities of reflected X-rays of fixed wavelength, was spectacularly demonstrated in the two 1913 papers W. L. Bragg published on the halides of the alkali metals[2] (see Figure 2.9) and, with his father, on diamond.[3]

The approach of the Braggs in providing a reliable way to determine the internal architecture of all crystalline solids, and thus to explain many of their properties, was of prime importance. Once the structure of diamond was elucidated (see Figure 2.2)—with its infinite array of carbon atoms bonded strongly to others in three directions—its hardness could be understood (likewise, when X-ray crystallography, in the hands of Bernal at the Davy-Faraday Research Laboratory (DFRL), revealed the structure of graphite—see Figure 2.10—in 1924,[19,20] its softness made

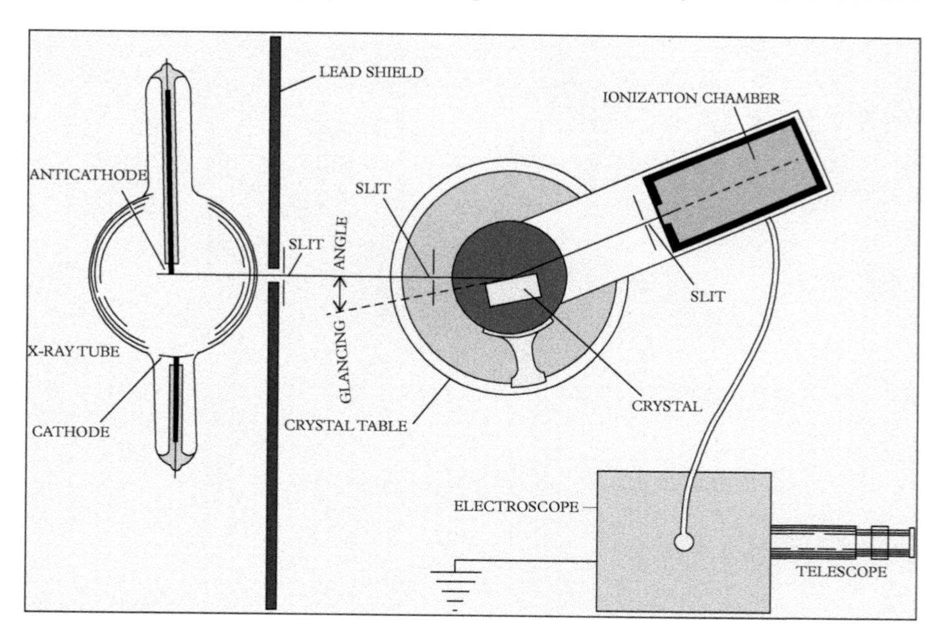

Figure 2.8 *Depiction by W. L. Bragg of the X-ray spectrometer designed by his father W. H. Bragg. (See J. M. Thomas*[5] *for further details of this set-up.)*

The Analysis of Crystals by the X-ray Spectrometer.

By W. LAWRENCE BRAGG, B.A.

(Communicated by Prof. W. H. Bragg, F.R.S. Received November 13,—Read November 27, 1913.)

In a former communication to the Royal Society,* an attempt was made to determine for certain crystals the exact nature of the diffracting system which produces the Laue X-ray diffraction photographs. The crystals chosen for particular investigation were the isomorphous alkaline halides NaCl, KCl, KBr, and KI. As in the original experiments of Laue and his collaborators, a thin section of crystal was placed in the path of a narrow beam of X-rays, and the radiation diffracted by the crystal made its impression on a photographic plate. By noticing what differences were caused in the photograph by the substitution of heavier for lighter atoms in the crystal, a definite arrangement was decided on as that of the diffracting points of the crystalline grating.

Figure 2.9 *Title and abstract of W. L. Bragg's landmark paper*[2] *in the* Proceedings of the Royal Society *in 1913.*

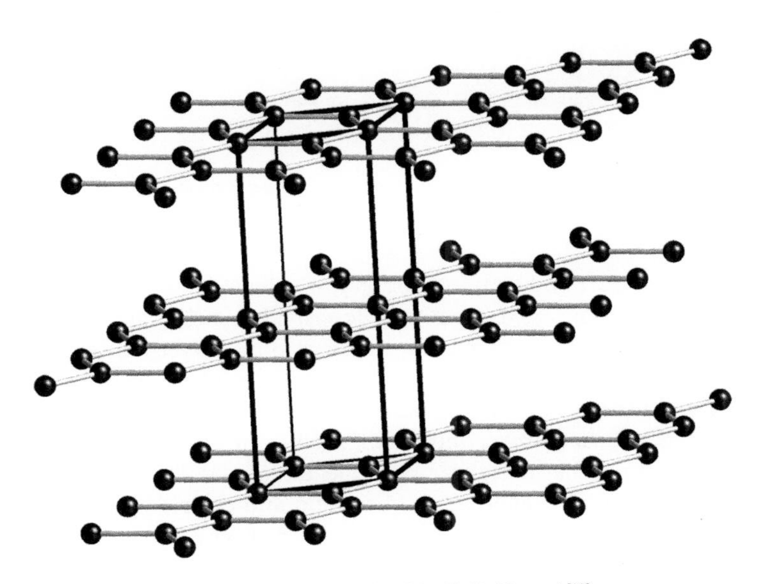

Figure 2.10 *The structure of graphite, determined by J. D. Bernal.*[19]

(Courtesy, Dr D. N. Johnstone)

Figure 2.11 *The protagonists involved in the discovery and development of X-ray diffraction.* *(Courtesy Royal Society; S. Bragg, Royal Institution, Wiley VCH)*

sense) (see Figure 2.11). Diamond and graphite have the same composition, but their structures are different—they are described as being allotropes—and this makes them mechanically, chemically, and electronically very different.

Not until after the First World War did shock and exhilaration greet the publication of these papers, when their content filtered through to the textbooks. Shock, because Bragg had incontrovertibly established, contrary to what many chemists thought at the time, that there was no molecule of sodium chloride inside rock salt—simply an extended alternation of sodium and chloride ions. A particularly intemperate attack was mounted by Henry Armstrong, former president of the Chemical Society of London. Writing in *Nature* in 1927, he described the '*chess-board pattern*' of atoms in sodium chloride as '*repugnant to common sense*' and '*absurd to the nth degree*'. Others were exhilarated because the structure of diamond confirmed the tetrahedral co-ordination of carbon, as envisaged by J. H. van't Hoff and others forty years earlier.

2.3 Determination of Molecular Weights from X-ray Diffraction Data

From Figure 2.7 above, it is straightforward to evaluate the molecular weight—nowadays described as molecular mass or molar mass—of the substance that has had its unit cell determined by X-ray studies. (The unit cell is the smallest unit of structure, which, on repetition in three dimensions, yields the total structure of the solid.) In the Appendix to this chapter, a simple calculation shows how the unit cell volume V is related to the density ρ of the solid via Avogadro's number (which is 6.022×10^{23} mol^{-1}). This equation holds for the structure in Figure 2.7 above.

$$\rho = (4 \times M_{\mathrm{m}}) / (N_{\mathrm{A}} \times V) \tag{2}$$

Here M_{m} = molar mass of the formula unit (ZnS), N_{A} is Avogadro's number, and V = volume of the unit cell.

The number of formula units in the unit cell in Figure 2.7 is 4. [In the Appendix to this chapter, a full calculation is shown where, in the case of ZnS, the measured unit cell repeat a is 5.406 Å (or 5.406×10^{-10} m) (1 Å is 10^{-8} cm). This yields a density of 4.097 g cm^{-3}.]

We may generalize this equation. For any molecule of molecular mass M_{m}, provided the density is accurately known and the unit cell volume V is measured from the X-ray diffraction data distances, it is possible accurately to derive the molecular mass—even for an insoluble material.

In earlier pre-X-ray diffraction days, molecular masses were determined from so-called colligative properties, such as the elevation of boiling point or depression of freezing point, or from osmotic pressure measurements. It was also possible to derive molecular masses of giant molecules using the Swedish worker Svedberg's ultra-centrifuge method.[21]

When Bernal and his co-workers W. T. Astbury, Kathleen Lonsdale, and J. M. Robertson, working under W. H. Bragg at the DFRL in the late 1920s and with Dorothy Crowfoot in Cambridge in the 1930s, were confronted with evaluating the molecular masses of their new 'living molecules' (such as sterols, proteins, and insulin), it was possible to derive values of molecular masses for these materials with unsurpassed precision from measurements of their corresponding densities. We shall return to the determination of molecular weights again in Chapter 3 (see Section 3.5.1) where we describe work (by Bernal and Crowfoot especially) on the structure and chemistry of sterols.

2.4 Atomically Resolved Structures of Inorganic Materials

In the immediate wake of the major breakthrough that von Laue and the Braggs initiated by introducing X-ray diffraction, the first materials to be elucidated structurally were minerals. Thus, Lawrence Bragg quickly determined the structures of: fluorspar, CaF_2; cuprite, Cu_2O; iron pyrites, FeS_2; sodium nitrate (Chile saltpeter), $NaNO_3$; and the calcite group of minerals, $CaCO_3$ in the form of calcite itself, and its polymorphs, aragonite, and vaterite. These studies he accomplished while still a member of the Cavendish Laboratory. And when his father W. H. Bragg moved from Leeds first to University College, London, and then to the Directorship of the DFRL of the Royal Institution (RI), he elucidated the structures of the mineral magnetite Fe_3O_4 and of the spinel $MgAl_2O_4$, a gemstone. Soon, in 1915, father and son published their first monograph '*X-rays and Crystals*' (1915), an influential text that went to five editions and was later translated into Russian and French.

2.4.1 W. L. Bragg's description of W. H. Bragg's suggestion concerning Fourier analysis

In 1915, during the course of his highly significant Bakerian Lecture to the Royal Society, W. H. Bragg[22] proposed a far-reaching idea, which, it later transpired, vitally influenced essentially all subsequent efforts to retrieve the structures of, and electron distributions within, crystals by X-ray diffraction. What W. H. Bragg said—and W. L. Bragg often repeated it verbatim—was that the periodic repeat of atomic patterns in crystals could be represented by Fourier series: '*If we know the nature of the periodic variation of the density of the medium we can analyse it by Fourier's method into series of harmonic terms. We may even conceive the possibility of discovering from the (relative X-ray) intensities the actual distribution of the scattering centres, electrons and molecules of the atom.*'

In explaining W. H. Bragg's insight concerning Fourier Series, Crick and Kendrew[23] recalled that it is convenient to think of a crystal as a three-dimensional pattern of *electron density* which reaches high values near the centres of atoms and low or zero values in the spaces between them. It can be shown that the X-ray picture represents a 'wave analysis' (sometimes called Fourier analysis) of this electron density. When we make a wave analysis of a crystal, we think of it as made up of a very large number of waves of electron density, running in many different directions through it. If we have carried out the analysis correctly, we shall find that when we add together all those waves—each of the correct size (amplitude) and to the right extent in or out of step with its neighbour (phase)—we get back to the actual electron density of the crystal. This is the three-dimensional wave analysis, often known as a Fourier analysis. The significance of a particular X-ray spot is that it corresponds to one of these (imaginary) sinusoidal waves of electron density. The position of the spot on the X-ray picture shows us both the direction of the wave and its wavelength. If it is far from the centre, it corresponds to a short distance in real space. Thus, the outer parts of X-ray diffraction patterns are concerned with fine details of structure.

2.4.2 A pictorial guide to the derivation of electron density from X-ray diffraction patterns

It is instructive to refer again to Bragg's Law: $n\lambda = 2d \sin \theta$. We see from this equation that the larger the value of $\sin \theta$ (and hence the larger the value of θ), the smaller the value of d when we have a fixed value of the wavelength. There is clearly a reciprocal relationship between real-space distances and those in what we term Fourier space (or simply reciprocal space). It therefore follows, as alluded to above, that the greater the number of diffraction spots that are at large values of reciprocal space, the greater the resolution one obtains in the electron density maps constituted (by Fourier analysis) from the original diffraction pattern. The veracity of this statement is illustrated in Figure 2.12 taken from the work of Dickerson.[24]

In the language of X-ray crystallography, it is stated that the Fourier transform of the diffraction pattern of a particular structure constitutes its actual structure. So, for heuristic purposes, it helps to acquaint ourselves with the Fourier transforms of a few scattering objects (see Figure 2.13). (It was P. P. Ewald, in his early work, carried out shortly after the discovery of diffraction of X-rays by von Laue

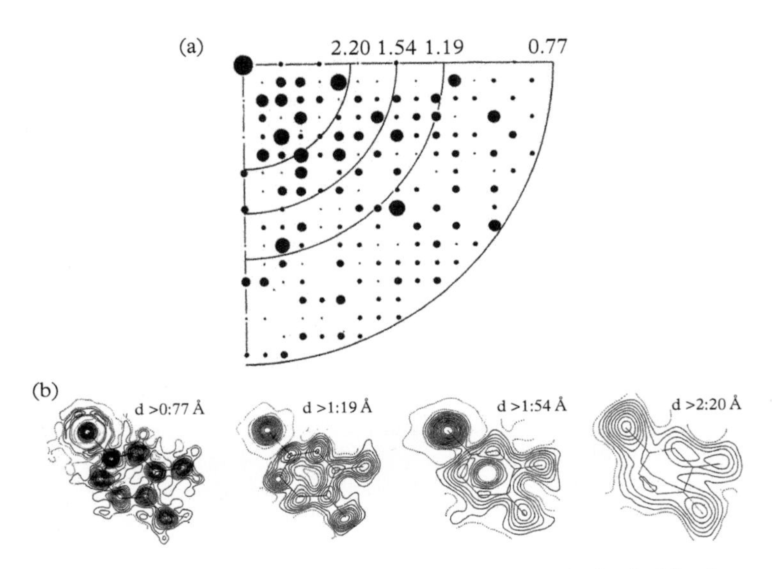

Figure 2.12 *Illustration (after R. E. Dickerson) of the improvement in detail of the electron density of 4,5 diamino-2-chloropyrimidine with the increased area of the diffraction pattern.*

(From 'Present at the Flood: How Structural Molecular Biology Came About', *Sinauer Associates, Sunderland, MA, USA, 2005)*

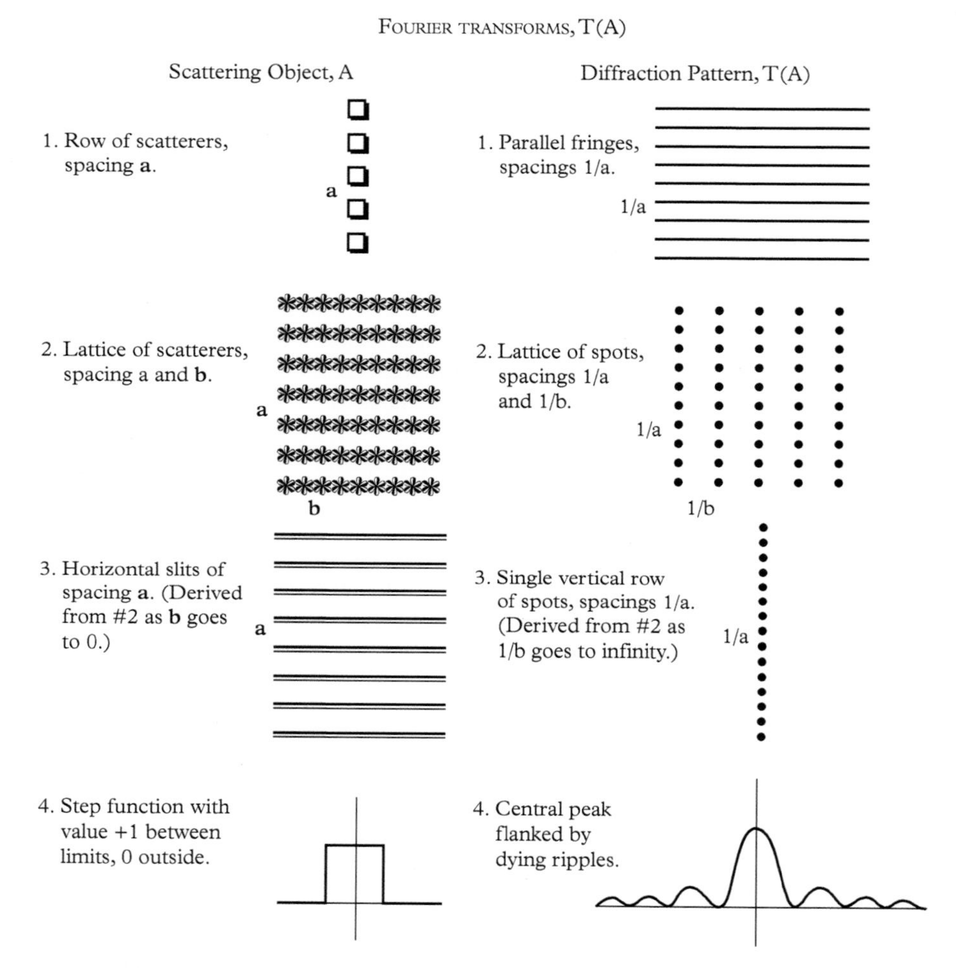

Figure 2.13 *Depiction of four scattering objects (column A, on left) and their diffraction patterns (Fourier transforms) (on right).*

(By kind permission of Dr R. E. Dickerson)

et al., who was the first to talk about reciprocal space—later, Bernal began to use the term, independently of the arguments of Ewald. It was Ewald who first pointed out that the Fourier transform of the diffraction pattern generated by the scattering of X-rays by the structure is the actual structure in real space.)

It is also instructive to utilize optical analogues (pursuing a suggestion made originally by W. L. Bragg) to understand better the relationship between an image and its diffraction pattern. C. A. Taylor and his colleagues[25] followed up Bragg's early work on this topic using 'optical' examples. In Figure 2.14, the top three objects give rise to their corresponding Fourier transforms (so-called optical transforms), shown immediately beneath them.

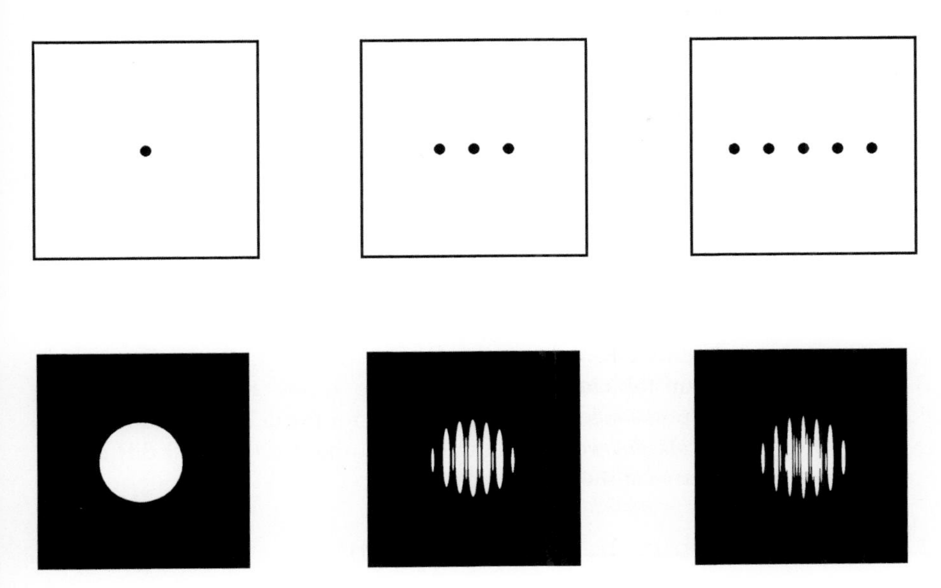

Figure 2.14 *(Top) Three scattering objects. (Bottom) Their corresponding diffraction patterns (from Harburn, Taylor, and Welberry).*[25]

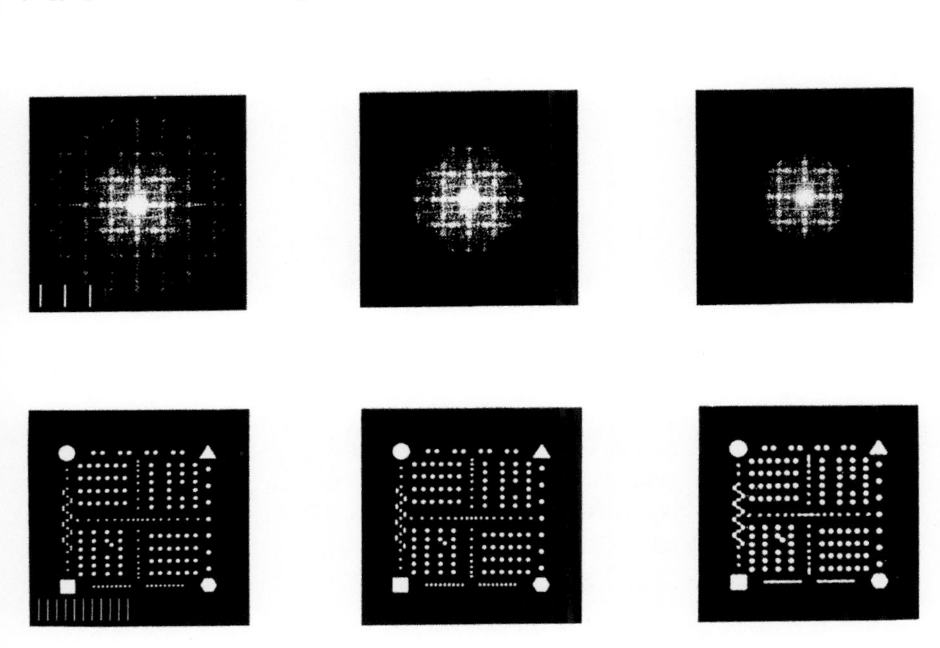

Figure 2.15 *(Top) Three diffraction patterns (optical transforms), with their corresponding objects (shown below).*[25]

More complicated diffraction patterns and the corresponding real-space objects that give rise to them are shown in Figure 2.15.

Note in particular how a diffraction pattern that does not extend fully in reciprocal (Fourier) space gives a much lower resolution picture of the object that gave rise to the diffraction pattern.

It has to be remembered that, were it possible to focus X-rays, then, as in an optical or electron microscope, the X-ray diffraction pattern could be straightforwardly converted to an enlarged image of the object by the Fourier transform effected by the putative X-ray lens (see Figure 9.6 of Chapter 9).

In the mid to late 1920s, and for a good while thereafter, the American worker R. J. Havinghurst[26] and UK workers G. C. Darwin[27] and W. L. Bragg and B. E. Warren[28] took up the challenge of measuring accurately the intensities of the X-ray diffraction spots—needed, as we see later, for the determination of electron densities in crystals. As well as the intensities, the phases of all diffraction spots are needed to arrive at the electron density.[27]

2.4.3 A word about the phase problem

A qualitative outline of how the phase of a diffraction spot may be determined has been given by Perutz in a popular article,[29] which is briefly summarized here. In order to obtain the right image of the crystal under investigation, one needs to place each set of X-ray reflections correctly with respect to some arbitrary chosen common origin. At this origin, the amplitude of any particular set of reflections may show a crest or a trough or some intermediate value. The distance of the crest of the wave from the origin is called the phase (see Figure 2.16).

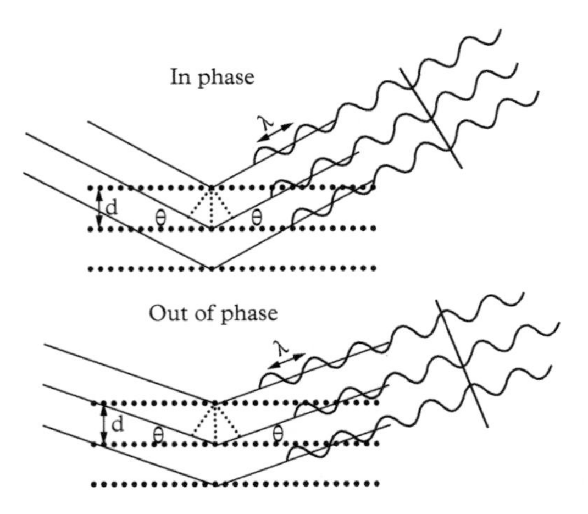

Figure 2.16 *Graphical illustration of the importance of phase in X-ray diffraction studies (compare with Figure 2.6 above).*

(Courtesy of Prof A. M. Glazer)

Figure 2.17 *Illustration of the importance of phase in deduction of images from diffraction patterns. (Courtesy Prof Randy Read)*

By itself, the X-ray pattern tells us only the amplitude (related to the intensity) and nothing about the phase. To appreciate the importance of the phase problem, we show below an illustration concerning the reproduction of the faces of Max Perutz and John Kendrew (see Figure 2.17).

When, in 1952, three eminent X-ray crystallographers Bijvoet, Bernal, and Patterson[30] wrote a celebratory article entitled '*Forty years of X-ray diffraction*', they concluded that the phase problem '*constituted the central one of X-ray analysis today*'. (We shall see in Chapter 5 that, in 1953, Max Perutz found a way of solving the phase problem, which had an enormous influence on subsequent structural studies by X-ray diffraction.)

2.4.4 Author's note regarding the phase problem

For readers interested in a rigorous discussion of how the phase problem is solved in the context of structural molecular biology, a lucid summary, using the necessary aspects of mathematical crystallography, has been given by one of Dorothy Hodgkin's close collaborators Eleanor Dodson.[31,32] Fuller descriptions are given by Blundell and Johnson,[33] by Phillips and Hodgson, and by Hendrickson.[35] An instructive qualitative account of how the phase problem may be solved is also given by Ramakrishnan.[36] In Dodson's approach, the so-called single-wavelength anomalous dispersion (SAD) method is advocated. It sometimes proves simpler than phasing from multi-wavelength (MAD) measurements.

2.4.5 The pioneering studies of W. L. Bragg and Linus Pauling

Both during his time at the University of Manchester (as successor to Rutherford), and after his return to the Cavendish Laboratory (again as successor to Rutherford), W. L. Bragg and his co-workers wrote (or edited) a number of books on crystal structures determined by X-ray diffraction, as well as numerous original articles. These were published at intervals between 1930 and 1965.[37–40] He concentrated heavily on the determination of the structure of silicate and alumino-silicate minerals; one especially beautiful study that was rapidly completed was the structure of the mineral cordierite,[41] the diffraction pattern, with its pseudo-hexagonal symmetry, being particularly helpful in aiding the study (Figure 2.18).

X-ray crystallographic studies of silicates and alumino-silicates revealed that their structures could be envisaged as made up of various combinations of SiO_4 tetrahedra, AlO_4 tetrahedra, and $(Mg,Fe,Al)O_6$ octahedra in which there are present corner-sharing tetrahedra (of SiO_4 and AlO_4), and of edge-sharing tetrahedra (of SiO_4) with edges of $(Mg,Fe)O_6$ octahedra or $(Mg,Al)O_6$ octahedra. Figure 2.18 reflects these various combinations. For the sake of clarity, the polyhedra (i.e. tetra- and octahedra) are used mainly in the structural representation, but isolated oxygen atoms are shown (in red) at the vertices of both the tetrahedra and octahedra.

In the early 1920s, Linus Pauling, at the California Institute of Technology, Pasadena, determined a large number of crystal structures of minerals and of inorganic solids of particular scientific interest. These included: molybdenite (MoS_2); haematite (Fe_2O_3); corundum (Al_2O_3); topaz ($Al_2(F,OH)_2SiO_4$); barite ($BaSO_4$); and NaN_3 and KN_3.

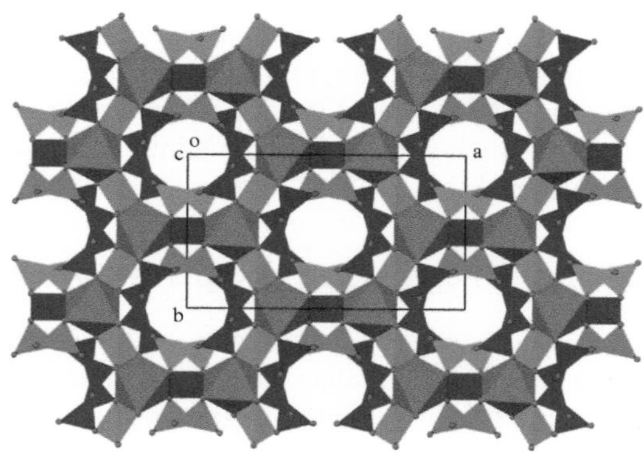

Red = O Green = Al Brown = Fe/Mg Blue = Si

$a = 17.23$ Å, $b = 9.83$ Å, $c = 9.29$ Å

Figure 2.18 *Depiction of the atomic structure of the mineral cordierite ($Al_3(Mg,Fe^{2+})(Si_5Al)O_{18}$), which is related to the gemstone beryl ($Be_3Al_2Si_6O_{18}$). Quasi-hexagonal pores run through the crystal.*[42]

(Courtesy Professor K. D. M. Harris)

Figure 2.19 *Linus Pauling, taken after graduation in 1922 at Oregon State University, Corvallis.*

(Courtesy of the Ava and Linus Pauling Archives)

Pauling also wrote a definitive article in 1927 where he tabulated the sizes of ions and the structure of ionic crystals[42,43] (see Figure 2.19). He also enunciated his influential *Pauling Rules*,[44] which are discussed in Section 7.2, where the rivalry between Bragg and Pauling is described.

Pauling used both the W. L. Bragg and von Laue methods for determining his structures. He also, from his observations, wrote other even more significant papers,[42,43] in which he was able to classify all the various kinds of silicate and alumino-silicate minerals, embracing, thereby, chain silicates (like pyroxenes and amphiboles), layered silicates (smectites like montmorillomite, mica, and talc), framework silicates (e.g. feldspars), and some very complicated zeolites. This paper greatly rationalized the inter-relationships between various rock minerals and also made logical sense of patterns of isomorphous replacements of ions by others, e.g. Al^{3+} could readily substitute for Si^{4+} in four-co-ordinated states, because of their similar size—likewise, Mg^{2+} for Al^{3+} in octahedral co-ordination. This work by Pauling led to his monumental monograph '*The Nature of the Chemical Bond*'.[45]

2.5 Initial Crystallographic Studies of Materials of Biological Interest

In parallel with W. L. Bragg's and Linus Pauling's extensive work (as well as that of others) on inorganic solids carried out from the 1920s onwards, a group of researchers was assembled by W. H. Bragg at the DFRL from 1923 onwards, and they began to investigate some 'living molecules', following on from studies that had been commenced at the Kaiser-Wilhelm Institute in Dahlem, Germany, by Hermann Mark and Michael Polanyi. These workers had reported results of their investigations on silk, cotton, cellulose, and fibrous proteins (see Figure 2.20).

And when Bernal returned from the DFRL to Cambridge in 1927 and W. T. Astbury migrated from there to Textile Physics in the University of Leeds, each of these workers (including collaborators such as Dorothy Crowfoot—later Hodgkin, Max Perutz, and I. Fankuchen) began to explore the vast field of materials of great biological significance such as amino acids, vitamins, sterols (sex hormones), and the ubiquitous single-cell algae *Valonia ventricosa* (as described in Chapter 3).

Figure 2.20 *Hermann Mark's research group at the Kaiser Wilhelm Institute, Dahlem, Berlin, in the 1920s. Max Perutz, who was taught by Mark as an undergraduate in Vienna, always used to say that one of the secrets of a successful laboratory was a happy atmosphere among its staff.*

(By kind permission of the archives of the Max Planck Institute and Prof Gerhard Ertl)

It is appropriate to close this chapter by recalling the somewhat unlikely fact that it was at the Department of Mineralogy and Petrography in the University of Cambridge that Bernal, Hodgkin, Perutz, and Fankuchen[46–48] mounted major investigations of the 'living' molecules mentioned above (see Section 3.5). Indeed, it was the structure and chemistry of the sterols that constituted Dorothy Crowfoot's PhD thesis.

APPENDIX

Calculating the Density of a Solid from its Crystal Structure, Illustrated by the Case of Zinc Sulfide (ZnS)

Knowledge of the crystal structure of a solid provides direct information on the contents of the repeating unit (the 'unit cell') and the volume of the unit cell, from which the density (ρ) of the solid can be readily calculated as follows:

$$\rho = \text{(mass of the contents of the unit cell)} / \text{(volume of the unit cell)} \qquad [1]$$

which can be expressed in terms of molar quantities as:

$$\rho = \text{(mass of the contents of one mole of unit cells)} / \text{(volume of one mole of unit cells)} \qquad [2]$$

In the case of zinc sulfide (see Figure), the contents of the unit cell (Zn_4S_4) are *four* zinc cations (blue) and *four* sulfide anions (yellow). The unit cell is cubic and the length of the unit cell edge (at ambient temperature) is: $a = 5.406$ Å $= 5.40 \times 10^{-10}$ m.

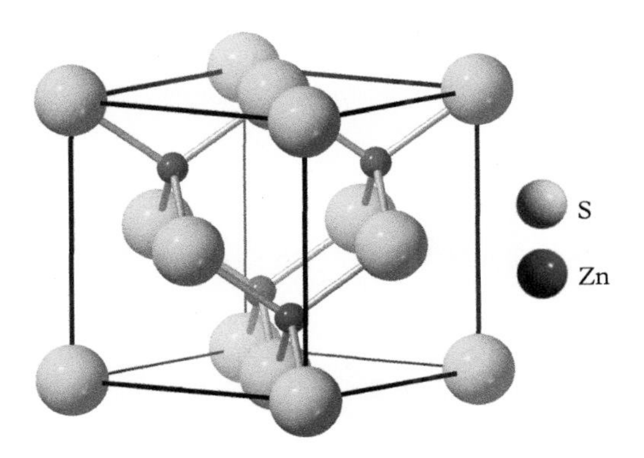

The molar mass of the formula unit (ZnS) is:

$$M_m = \text{(molar mass of } Zn) + \text{(molar mass of } S)$$
$$M_m = 65.39\ \text{g mol}^{-1} + 32.07\ \text{g mol}^{-1}$$
$$M_m = 97.46\ \text{g mol}^{-1}$$

and the molar mass of the unit cell contents (Zn_4S_4) is therefore 4 × M_m.

For a cubic system, the volume (V) of the unit cell is given by:

$$V = a^3$$

and the volume of one mole of unit cells is $N_A \times V$ where N_A denotes Avogadro's number.

From equation [2],

$$\rho = \text{(mass of the contents of one mole of unit cells)} / \text{(volume of one mole of unit cells)}$$

It follows that the density of zinc sulphide is given by:

$$\rho = (4 \times M_m)/(N_A \times V)$$

and hence:

$$\rho = (4 \times 97.46\ \text{g mol}^{-1})/[6.022 \times 10^{23}\ \text{mol}^{-1} \times (5.40 \times 10^{-10}\ \text{m})^3]$$
$$\rho = 4.097 \times 10^6\ \text{g m}^{-3}$$
$$\rho = \mathbf{4.097\ g\ cm^{-3}}$$

This illustrative exercise also reveals that, provided the density of a crystal is known (measured), then the molar mass (i.e. molecular weight) may be determined from a knowledge of the volume of the unit cell. This fact is amplified in Section 3.5.1.

REFERENCES

1. W. L. Bragg, *Scientific American*, **1968**, *219*, 58.
2. W. L. Bragg, *Proc. Roy. Soc.*, **1913,** *A89*, 248.
3. W. L. Bragg and W. H. Bragg, *Proc. Roy. Soc.*, **1913,** *A89*, 277.
4. J. M. Thomas, *Nature*, **2012**, *491*, 186.
5. J. M. Thomas, *Angew. Chem. Int. Ed.*, **2012**, *51*, 12946.
6. J. M. Thomas, *The Times (London)*, 10 November **2012**, p. 89.

7. W. Friedrich, P. Knipping, and M. Laue, *Bayerische Akademie der Wissenschaften*, **1912**, *303*.
8. L. Vegard—see C. H. Schwalbe, *Crystallogr. Rev.*, **2014**, *20*, 9.
9. W. W. Schmidt and W. Steiner, *Acta Crystallogr., Sec A*, **2012**, *68*, 1.
10. I. Olavsson, P. Liljas, and S. Lidin, '*From a Grain of Salt to the Ribosome: The History of Crystallography as Seen Through the Lens of the Nobel Prize*', World Scientific Publishing, Singapore, **2015**, pp. 3–32.
11. D. C. Phillips, *Biographical Memoirs of Fellows of the Royal Society*, **1979**, *25*, 1.
12. D. C. Hodgkin, *Ann. N. Y. Acad. Sci.*, **1979**, 66.
13. W. H. Bragg, '*An Introduction to Crystal Analysis*', G. Bell and Sons, London, **1928**.
14. W. L. Bragg, *Proc. Camb. Phil. Soc.*, **1912**, *17*, 43.
15. P. P. Ewald in '*Fifty Years of X-ray Diffraction*', International Union of Crystallography, Utrecht, **1962**, Sections I and II.
16. W. L. Bragg. The words used by the author of this paper, referring to ref [14] above, are '…the crystal actually manufactures the light of definite wavelength, much as, according to Schuster, a diffraction grating does. The difference in this case lies in the extremely short wavelength of the waves.'
17. Of C. T. R. Wilson, the Nobel Laureate P. M. S. Blackett, former President of the Royal Society and a contemporary of Wilson's at the Cavendish Laboratory, said he '*was the most gentle and serene, and the most indifferent to prestige and honours*'.
18. W. L. Bragg, *Nature*, **1912**, *90*, 410.
19. J. D. Bernal, *Proc. Roy. Soc.*, **1924**, *A106*, 749.
20. J. M. Thomas, *Nature*, **1993**, *364*, 478.
21. T. Svedberg, see S. Classon, K. O. Pedersen, *Biographical Memoirs of Fellows of the Royal Society*, **1972**, *18*, 594.
22. W. H. Bragg, *Phil. Trans. Roy. Soc.*, **1915**, *A215*, 253.
23. F. H. C. Crick and J. C. Kendrew, *Adv. Protein Chem.*, **1957**, *12*, 2005.
24. R. E. Dickerson, '*Present at the Flood: How Structural Molecular Biology Came About*', Sinauer Associates, Sunderland, MA, **2005**.
25. G. Harburn, C. A. Taylor, and T. R. Welberry, '*Atlas of Optical Transforms*', Bell and Hyman, London, **1975**.
26. R. J. Havinghurst, *Phys. Rev.*, **1927**, *29*, 1.
27. G. C. Darwin, *Phil. Mag.*, **1914**, *27*, 315.
28. W. L. Bragg and B. E. Warren, *Z. Kristallogr.*, **1928**, *69*, 118.
29. M. F. Perutz, *Scientific American*, **1964**, *208*, 65.
30. J. M. Bijvoet, J. D. Bernal, and A. L. Patterson, *Nature*, **1952**, *169*, 948.
31. Z. Dauter, M. Dauter, and E. Dodson, *Acta Cryst.*, **2002**, *D58*, 494.
32. E. Dodson, *Acta Cryst.*, **2003**, *D59*, 1958.
33. T. L. Blundell and L. N. Johnson, '*Protein Crystallography*', Academic Press, London, **1976**.
34. J. C. Phillips and K. O. Hodgson, *Acta Cryst.*, **1980**, *A36*, 856.
35. W. A. Hendrickson, *Science*, **1991**, *254*, 51.
36. V. Ramakrishnan, '*Gene Machine: The Race to Decipher the Secrets of the Ribosome*', One World, London, **2018**.
37. For example, W. L. Bragg, '*The Structure of Silicates*', Akad Verlag, Leipzig, **1930**.
38. W. L. Bragg and W. H. Bragg (eds.), '*The Crystalline State, a General Survey*', G. Bell and Sons, London, **1933**.
39. W. L. Bragg, '*Atomic Structure of Minerals*', Cornell University Press, Ithaca, NY, **1937**.

40. W. L. Bragg and C. F. Claringbull, '*Crystal Structure of Minerals*', G. Bell and Sons, London, **1965**.
41. W. L. Bragg and J. West, *Proc. Roy. Soc. A*, **1926**, *111*, 691.
42. L. Pauling, *J. Am. Chem. Soc.*, **1927**, *49*, 765. See also L. Pauling, *J. Am. Chem. Soc.*, **1929**, *51*, 1010.
43. L. Pauling, *Proc. Natl. Acad. Sci. U. S. A.*, **1930**, *16*, 123.
44. L. Pauling, *J. Am. Chem. Soc.*, **1929**, *51*, 1010.
45. L. Pauling, '*The Nature of the Chemical Bond*', Cornell University Press, Ithaca, NY, **1938**.
46. J. D. Bernal, *Z. Kristallogr.*, **1931**, *78*, 363.
47. J. D. Bernal, *Nature*, **1932**, *129*, 277.
48. J. D. Bernal, I. Fankuchen, and M. F. Perutz, *Nature*, **1938**, *141*, 523.

3

W. H. Bragg and His Creation of a World-Famous Centre for X-ray Crystallography at the Davy-Faraday Research Laboratory

3.1 Introduction

In 1923, Sir James Dewar (see Figure 3.1), renowned for his numerous studies in low-temperature physics, passed away. Dewar had invented the thermos flask (also known as the Dewar vessel), as well as made significant contributions to spectroscopy and to the development of the explosive known as cordite. With his death, the Fullerian Professorship of Chemistry at the Royal Institution (RI) and the Directorship of the Davy-Faraday Research Laboratory (DFRL), housed within the RI at Mayfair, London, became vacant, and W. H. Bragg was appointed his successor to these two posts.[1]

The Fullerian Professorship of Chemistry at the RI was created for Michael Faraday in 1833, as a result of a generous bequest by an eccentric English Member of Parliament John Fuller (see Figure 3.2), who was also responsible for a chair of physiology, the first occupant of which was Peter Mark Roget (1779–1869), of Roget's Thesaurus fame.[2] It was said of Mr Fuller MP: '*...the feebleness of whose constitution denied him at all other times and places the rest necessary for health could always find repose and even quiet slumber amid the murmuring lectures of the Royal Institution and that, in gratitude for the peaceful hours thus snatched from an otherwise restless life, he bequeathed to the Royal Institution the magnificent sum of £10,000.*'

Figure 3.1 *Sir James Dewar at his desk at the Davy-Faraday Research Laboratory.*
(Courtesy of JET Photographic and the Master and Fellows of Peterhouse, Cambridge)

Figure 3.2 *John Fuller, MP, Founder of the Fullerian Professorship of Chemistry and Physiology at the Royal Institution.*
(Courtesy of the Royal Institution, London)

3.2 Origins of the Davy-Faraday Research Laboratory

The DFRL was inaugurated towards the end of the nineteenth century (1896) within the precincts of the RI, a private research establishment and showcase of science, through a bequest from Ludwig Mond, the German-born industrial chemist and philanthropist. The RI itself was founded in 1799 by the picaresque American-born Count Rumford of the Holy Roman Empire, also known as Sir Benjamin Thompson.[3]

Rumford was a spectacularly successful judge of extraordinary talent. In the early 1800s, he recruited to the RI two gifted West Country Englishmen—Humphry Davy and Thomas Young, each of whose names will forever be remembered in the history of natural science. Davy, the more charismatic of the two (and a coruscatingly brilliant lecturer), made many important discoveries in electrochemistry at the RI. He was the first to isolate the elements sodium, potassium, calcium, strontium, and magnesium, and to establish the elemental nature of both boron and chlorine. As a young man, he discovered the anaesthetic properties of nitrous oxide (laughing gas), which present-day dentists and surgeons still use. He also invented the miners' safety lamp (assisted by the young Michael Faraday) and the technique of cathodic protection—a phenomenon still used in all sea- and ocean-going ships. It was Davy, with his work on hydrochloric acid, who was the first to disprove Lavoisier's assertion that all acids contain oxygen.

Thomas Young carried out his famous two-slit experiment in 1801 at the RI, thereby resuscitating the wave theory of light. His work on capillarity, surface energy, and elasticity and on the physiology of vision and as an Egyptologist is of everlasting renown. He presciently inferred that human colour vision arises from three kinds of receptors. And, as an Egyptologist, he was the first to make a breakthrough, ahead even of Champollion, in deciphering the Rosetta Stone. He identified the cartouche of Cleopatra.

Michael Faraday, who occupied the Fullerian Professorship of Chemistry from its inception for over thirty years, carried out essentially all his exceptionally wide-ranging and profound researches as a physicist and chemist at the RI.[2–4] Apart from his discovery of electromagnetic induction, he established several new branches of science, including magnetochemistry and opto-electronics. Indeed, when Linus Pauling and his co-worker Coryell in 1936 reported[5] their seminal studies on the interaction between haemoglobin and a range of gases, such as oxygen and carbon monoxide, they began their paper in the *Proceedings of the National Academy of Sciences of the United States of America* with the following remark: '*Over ninety years ago, on November 8, 1845, Michael Faraday investigated the magnetic properties of dry blood and made a note: "must try recent fluid blood". If he had determined the magnetic susceptibilities of arterial and venous blood, he would have found them to differ by a large amount.*'

Neither Davy nor Faraday—apart from the study reported above by Pauling, and a few attempts to record the magnetic behaviour of human flesh[2,4]—nor the first and subsequent Directors of the DFRL carried out work of a biological nature. (The first joint Directors of the DFRL when it was inaugurated in 1896 were the Third Lord Rayleigh and Sir James Dewar.)

3.3 Popularization of Science at the Royal Institution and the DFRL

From 1800 onwards, a dazzling array of eminent scientists have been involved in presenting their subjects to non-specialist audiences at the theatre of the RI, itself the oldest, continuously used theatre in London. The young Humphry Davy, in the early decades of the 1800s, attracted enormous audiences to the RI, as a result of his coruscatingly brilliant lecturing skills, which combined remarkable fluency of expression with an astonishing range of experimental demonstrations. So popular were his public performances that Albemarle Street became the first one-way street in London. It was while listening, transfixed, to Davy deliver four lectures—his last as Director of the RI—in 1812 that the young Michael Faraday (then a bookbinder's apprentice) resolved to request an opening as a laboratory assistant at the RI—see ref [2] for fuller details. Ever since Michael Faraday initiated the still-continuing programme of Friday Evening Discourses in 1826, audiences at the RI have witnessed numerous events organized specifically with the intention of bridging different cultures, especially the arts and the sciences. All subsequent Heads of the RI have maintained that tradition, the irascible[6] Sir James Dewar being exceptionally successful in so doing. Dewar was particularly good in inviting eminent scientists to lecture at the RI many years before they were awarded the Nobel Prize. As Table 3.1 shows, Dewar had the uncanny skill of identifying work of Nobel Prize-winning quality several years (sometimes as much as a quarter of a century) ahead of the decisions reached in Stockholm.

The appearance of Marconi as both a Discourse speaker and a Nobel Prize winner merits elaboration. Marconi was born the son of an aristocrat in Bologna, Italy, in 1874. He was taught at home, and he never pursued higher education but read avidly about chemistry, mathematics, and physics, and was greatly impressed by what he learned about Faraday and Maxwell, and later the German physicist Heinrich Hertz. He moved to London, aged 22, and quickly became a member of the RI (like another, somewhat older, contemporary member Lord Kelvin, Marconi was not only a scientist, but also an entrepreneur and a businessman). As a result of a fortunate occurrence, he came into contact with Sir William Preece, a Welshman who was, at that time, the Chief Electrical Engineer of the British Post Office.

In 1878, 1879, 1880, and 1897, Preece gave Friday Evening Discourses on the following topics: the telephone; multiple telegraphy; the telegraphic achievements

Table 3.1 *Discourses arranged by Sir James Dewar: natural philosophers who were later, or were already, Nobel Laureates*

Date	Speaker	Title	Date of Nobel Prize
5 April 1895	Lord Rayleigh	*Argon*	1904
17 April 1896	Gabriel Lippmann	*Colour photography*	1908
2 February 1900	Guglielmo Marconi	*Wireless telegraphy*	1909
2 March 1900	Ronald Ross	*Malaria and mosquitoes*	1902
7 March 1902	A. H. Bequerel	*Sur la radio-activité de la matière*	1903
19 June 1903	Pierre Curie	*Le radium*	1903
19 February 1904	C. T. R. Wilson	*Condensation nuclei*	1927
3 June 1904	S. Arrhenius	*The development of the theory of electrolyte dissociation*	1905
19 April 1907	C. S. Sherrington	*Nerve as a mastery of muscle*	1932
3 March 1911	J. B. Perrin	*Movement Brownian et grandeurs moléculaires*	1926
5 June 1914	W. H. Bragg	*X-rays and crystalline structure*	1915
26 May 1916	C. G. Barkla	*X-rays*	1917
10 May 1918	F. Gowland Hopkins	*The scientific study of human nutrition*	1929
12 May 1922	H. H. Dale	*The search for specific remedies*	1936
16 February 1923	A. V. Hill	*Muscular exercise*	1922

of Wheatstone; and '*Signalling Through Space Without Wires*', the last of which was heard by Marconi. The relationship between Preece and Marconi blossomed, and Marconi had already made significant advances in telegraphy. In February 1900, Marconi gave the first of his Discourses at the RI on his early work on radio. These were on: wireless telegraphy; recent advances in wireless telegraphy (1905); trans-Atlantic wireless telegraphy (1908); radiotelegraphy (1911); and radio communications by means of very short electronic wires (1932).

The aura and tradition of the RI influenced Marconi, and this showed in his opening words of many of his lectures—thus, on 3 March 1905, his initial remarks were: '*The phenomena of electromagnetic induction, revealed chiefly by the memorable researches and discoveries of Faraday carried out in the Royal Institution, have long since shown how it is possible for the transmission of electrical energy to take place across a small air-space between a conductor traversed by a variable current and another conductor placed near it; and how such transmission may be detected and observed at distances greater or less, according to the more or less rapid variation of the current in one of the wires, and also according to the greater or less quantity of electricity brought into play.*

Maxwell, inspired by Faraday's work, gave to the world in 1873 his wonderful mathematical theory of electricity and magnetism, demonstrating on theoretical grounds the existence of electromagnetic waves, fundamentally similar to, but enormously larger than waves of light. Following up Maxwell, Hertz in 1887 furnished his great practical proof of the existence of these true electromagnetic waves.'

In a dramatic opening to his Discourse on '*Transatlantic Wireless Telegraphy*' on 18 March 1908, with Sir William Crookes (of Crookes Tube fame, and the discoverer of the element thallium) in the chair, Marconi said: '*Before I go into my subject, it might interest you to know that the invitation from the Royal Institution to deliver this lecture was sent to me by transatlantic wireless telegraphy on October 19, when I was in Canada. The following is the text of the message: "Macroni, Glace Bay... Hearty congratulations on behalf of the Royal Institution, the home of Faraday. We invite you to give first Friday Evening Discourse on January 17 next. Please reply by wireless... Sir William Crookes, Royal Institution, London". To which I replied, also by wireless: "By means of other waves across Atlantic, I thank you for honour invitation to lecture at Royal Institution. Owing uncertainty my future plans greatly obliged if you will permit me postpone acceptance until I return to London... Marconi".*'

When W. H. Bragg succeeded Dewar, he too arranged a series of impressive, world-renowned Discourse speakers, as may be gleaned from Table 3.2.

Although most of the Discourses presented at the RI since the days of Faraday are on scientific or technological topics, the arts and other important disciplines are also represented. Faraday himself invited the artist John Constable to talk about landscape painting. Over a hundred years later, W. H. Bragg invited John Maynard Keynes (1883–1946), a British economist whose ideas fundamentally changed the theory and practice of macroeconomics and the economic policies of governments. Keynes was instrumental in forming the World Bank and the International Monetary Fund (IMF). In 1929, just before he produced his '*Treatise on Money*' (published in 1930), Keynes, with a colleague, wrote a pamphlet in support of Lloyd George's (the founder of the MRC—see Appendix 1 to Chapter 1 and Section 8.3.1) contention that '*We Can Conquer Unemployment*'. On 6 February 1931, Keynes summarized the content of his *Treatise* with a Discourse entitled '*The Internal Mechanisms of the Trade Slump*'.

Among the many lucid remarks and analogies that he discussed in his Discourse were the following: '*We are now in the midst of one of the most violent slumps which have ever occurred. The most parallel case is that of the 1890s after the great age of railway expansion had passed its point of greatest activity and when Great Britain was disinclined for various reasons to lend abroad, just as France and the United States are disinclined today. In that case it took six years before recovery came.*' He continued his diagnosis thus: '*This reduced expenditure by entrepreneurs was not balanced by an increased expenditure on the part of consumers. For, unluckily, each individual is impelled by his paper losses or profits in the general interest; that is to say, he saves too little in the boom, and too much in the slump.*'

Keynes, who was Bursar of King's College, Cambridge, and highly influential as a collector and supporter of the arts and a member of the egregious Bloomsbury

Table 3.2 *A selection of the Discourses arranged by Sir W. H. Bragg*

Date	Speaker	Title
15 February 1924	Sir James Jeans	*The origin of the solar system*
13 March 1925	Gilbert Murray	*The beginnings of the science of language*
1 May 1925	W. L. Bragg	*The crystalline structure of inorganic salts*
8 May 1925	Sir Henry Dale	*The circulation of blood in capillary vessels*
5 June 1925	Howard Carter	*The tomb of Tutankhamun from ante-room to burial chamber*
25 February 1927	D'Arcy Thompson	*The solids of Plato and Archimedes*
30 April 1927	Sir Edward Appleton	*Wireless transmission and the upper atmosphere*
10 June 1927	Leonard Woolley	*The excavations at Ur*
27 January 1928	Dorothy Garrod	*Prehistoric Cave A*
8 June 1928	G. P. Thomson	*The waves of an electron*
15 February 1929	E. K. Rideal	*Chemiluminescence*
15 March 1929	V. M. Goldschmidt	*The distribution of the chemical elements*
21 February 1930	J. B. S. Haldane	*The principles of plant breeding, illustrated by the Chinese primrose*
6 February 1931	J. M. Keynes	*The internal mechanics of the trade slump*
26 May 1933	Sir Walford Davies	*Pure music and applied*
24 November 1933	W. H. Bragg	*Liquid crystals*
23 February 1934	Charles Morgan	*A defence of story telling*
20 April 1934	P. M. S. Blackett	*Cosmic radiation*
1 February 1935	Sir Francis Simon	*The approach to the absolute zero of temperature*
1 March 1935	Sir Arthur Bryant	*Samuel Pepys*
20 November 1936	H. G. Wells	*World encyclopaedia*
30 April 1937	J. Dover Wilson	*Shakespeare's universe*
11 February 1938	William Temple	*Truth in science, poetry, and religion*
20 May 1938	Kenneth Clark	*The aesthetics of Restoration*
9 December 1938	Irving Langmuir	*The properties and structure of protein helices*
27 January 1939	J. D. Bernal	*The structure of proteins*
10 March 1939	E. D. Adrian	*The development of the sense of hearing*
31 March 1939	Mary Somerville	*The broadcasts for schools*
3 December 1940	J. D. Bernal	*The physics of air raids*
5 March 1942	E. M. Forster	*Virginia Woolf*

set, was greatly admired in his day. But he has since been criticized by economists, notably Friedman (1912–2006), from the 1950s onwards. In the 1929 General Election in the UK, Keynes, like Lloyd George, was becoming a strong public advocate of capital development through deficit spending as a measure to alleviate unemployment. Winston Churchill, the Conservative Chancellor of the Exchequer at that time, took the opposite view.

In addition to inviting leading men and women to present Friday Evening Discourses (see Table 3.3), W. H. Bragg also initiated a series of impressive public lectures, held usually on Thursdays or Saturdays at more convenient times for the general public. For example, in the month of February 1933, three Nobel Laureates gave scientifically oriented talks, and there were also excursions into the use of the English Language, the plot of Shakespeare's *Hamlet*, and oriental paintings.

The Fullerian Professors of Physiology, who, unlike the Fullerian Professors of Chemistry, were not resident at the RI, but they too contributed greatly to its intellectual reputation. Some of the UK's most eminent physiologists occupied this post and frequently lectured to lay audiences: T. H. Huxley (1825–1895) and, later, his two grandsons Julian Huxley (1887–1975) and Sir Andrew Huxley (1917–2012), J. B. S. Haldane (1892–1964), William Bateson (1861–1926), (who coined the word 'genetics'), Sir Charles Sherrington (1857–1952) (who coined

Table 3.3 *Part of the Royal Institution lecture calendar, February 1933*

Date	Day	Speaker and title
2	Thursday	J. B. S. Haldane: *Recent advances in genetics*
3	Friday	Cyril Norwood: *Use of the English language*
4	Saturday	L. Binyon: *Oriental painting*
7	Tuesday	J. McLennan: *Low temperatures*
9	Thursday	J. B. S. Haldane: *Recent advances in genetics*
10	Friday	A. V. Hill: *Physical nature of the nerve impulse*
11	Saturday	L. Binyon: *Oriental painting*
14	Tuesday	Sir William Bragg: *Analysis of crystal structure by X-rays*
16	Thursday	A. R. Hinks: *Geography in the public service*
17	Friday	J. Dover Wilson: *Plot of Hamlet*
18	Saturday	Lord Rutherford: *Detection and production of swift particles*
21	Tuesday	Sir William Bragg: *Analysis of crystal structure by X-rays*
23	Thursday	A. R. Hinks: *Geography in the public service*
24	Friday	W. A. Bone: *Photographic analysis of explosion flames*
25	Saturday	Lord Rutherford: *Detection and production of swift particles*
28	Tuesday	Sir William Bragg: *Analysis of crystal structure by X-rays*

the word 'synapse'), Sir Peter Medawar, Max Perutz, Sir John Gurdon, Francis Crick, Lord David C. Phillips, Sydney Brenner, and Dame Anne McLaren (1927–2007). Other eminent scientists that occupied visiting professorships at the RI included Christopher Zeeman (of catastrophe theory fame), the Nobel Prize-winning astronomer Antony Hewish, Sir Brian Pippard, Sir J. J. Thomson, Lord Rutherford, and James Clerk Maxwell, the last four of whom occupied the Cavendish Chair of Physics at Cambridge.

It is of interest to recall that at one of the Friday Evening Discourses given by the Austrian–British physicist Otto Frisch (1904–1979), his aunt Lise Meitner (1878–1968), who, together, were the first to publish an explanation of nuclear fission in uranium, was a member of the audience. Figure 3.3 shows a photograph of Meitner, Frisch, and the US Nobel Laureate Glenn Seaborg (1912–1999) taken in Max Perutz's home in Cambridge in 1966.[7]

In 1948, the 47-year-old Linus Pauling presented[8] a memorable Friday Evening Discourse at the RI entitled '*Nature of Forces Between Large Molecules of*

Figure 3.3 *Photograph of Lise Meitner and her nephew (Otto Frisch) on the occasion of her being awarded the Enrico Fermi Award by Glenn Seaborg in the home of Max Perutz in Cambridge in 1966.*

(Property of the late Max Perutz. Photographer unknown)

Biological Interest'. During the course of it, Pauling pointed out that, at that time, no one knew the three-dimensional structure of a single enzyme. As we shall see in Chapter 6, it was at the DFRL, in 1965, that David C. Phillips (later Lord Phillips of Ellesmere) and his team, consisting of Louise (later Dame Louise) Johnson, Professor Tony North, Dr Colin Blake, and others, announced the determination of the folded structure of lysozyme, the first enzyme to be characterized and its mode of action as a catalyst interpreted in atomic detail. Equally important, it was the first polypeptide structure in which Linus Pauling's predicted β-sheet motif (see Section 6.5) was identified.

3.3.1 W. H. Bragg and his acolytes at the DFRL

Readers who are familiar with the compilation[9] by P. P. Ewald '*Fifty Years of X-ray Diffraction*', published in 1962, will know what two of W. H. Bragg's eminent scientists Dame Kathleen Lonsdale and J. D. Bernal have said about the atmosphere and achievements that pervaded the DFRL in their days there. Lonsdale, in particular, gives an evocative description of scientific life under the aegis of W. H. Bragg at the DFRL: '*. . . the triple appeal of the laboratory, library and lectures was an inspiration. My main impressions were of the happy family atmosphere with formality in the background; the casual way world figures appeared at tea-breaks; the loose organisation . . .; the dearth of mathematical texts marking the emphasis on experimental science . . . Tea-time at 4pm was not to be missed. To begin with, W. H. Bragg was nearly always there and there were generally Bourbon biscuits too. And all sorts of interesting visitors turned up. Some of them were Friday Evening lecturers come to prepare the experiments for their discourses. It might be Sir Ernest (afterwards, Lord) Rutherford, about to talk on the "Life History of an α-particle from Radium", or on the Nature of the atom; J. H. Jeans on "The Origin of the Solar System" or Lord Rayleigh on "The Glow of Phosphorous" . . . The Royal Institution library was well-stocked with books and especially with periodicals. It was thrilling occasionally to open very early back numbers of, say, the Phil Trans Roy Soc., and find that Faraday had made some comment in the margin, and even more thrilling to meet an aged member of the RI (Mr William Stone)*[10] *who remembered, as a small boy, sitting next to Faraday and talking to him, in the gallery of the RI Lecture Theatre, during some Christmas Lectures.*'

J. D. Bernal's survey of the British and Commonwealth Schools of Crystallography in Ewald's book[11] makes it clear how influential W. H. Bragg's centre at the DFRL was: '*The position of the British schools in the history of the development of our subject is necessarily quite a special one. Not only did Sir William and Sir Lawrence Bragg effectively start the study of crystalline structures by means of X-ray diffraction, but for many years their respective schools at the Royal Institution and in Manchester were the centres of world study in those fields.*'

W. H. Bragg assembled at the DFRL a galaxy of young researchers who were to make transformative contributions to X-ray crystallography and who facilitated the genesis of structural biology. A modern reflection of the vitally important contribution made by W. H. Bragg to the science of macromolecular crystallography is depicted in Figure 3.4, taken from the review by Jaskolski *et al.*[12]

The most significant individuals to be recruited as co-workers by W. H. Bragg at the DFRL were: W. T. Astbury, J. D. Bernal, Kathleen Lonsdale (Yardley, earlier), W. G. Burgers, E. G. Cox, A. L. Patterson, and J. M. Robertson (see Figure 3.5).

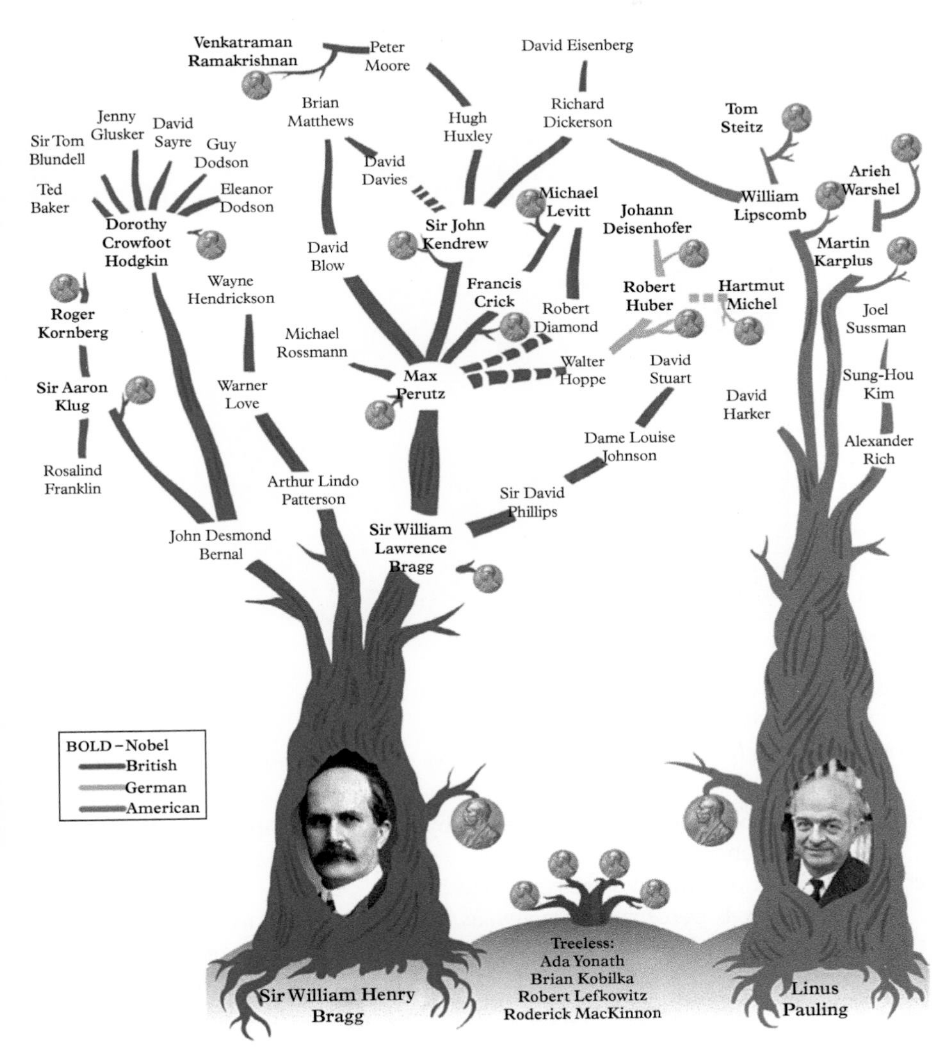

Figure 3.4 *Crystallographic ancestry from W. H. Bragg and L. Pauling.*[12] *There are some errors in this ancestral picture. Thus, Venkatraman Ramakrishnan did not learn crystallography from Peter Moore.*

K. Lonsdale W. T. Astbury J. D. Bernal

J. M. Robertson E. G. Cox

Figure 3.5 *Some of W. H. Bragg's acolytes at the Davy-Faraday Research Laboratory, Royal Institution, London, from 1923 to 1940.*

(Courtesy of The Royal Society, The Royal Institution, Copyright Godfrey Argent, and Oregon State University)

Brief descriptions of the work carried out by these formidable crystallographers and solid-state chemists have been given elsewhere,[13] but a few of their major contributions and influences are presented below.

W. T. Astbury is perceived by many, as described earlier, as the progenitor of structural molecular biology—as distinct from its phage origins (pioneered by Delbrück),[14] where the emphasis is not so much on the three-dimensional structure, but on genetically governed information transmission. The debate between these two perceptions of what constitutes modern molecular biology is admirably summarized both in Kendrew's book review[15] of the volume produced by Cairns, Stent, and Watson[16] in their celebratory volume for Delbrück's sixtieth birthday, and also in the authoritative article of Gunther S. Stent.[17] (See also Appendix 2 to Chapter 1.)

We begin by focusing on Astbury's contributions and the manner in which W. H. Bragg influenced him. Astbury found the general direction of the work for the rest of his life almost by accident. One of W. H. Bragg's most subtle ways of directing research—because (in the words of Bernal)[18] '*less of a Director you could hardly imagine*'—was casually to ask one of the research workers to help him in preparing some photographs or material for a lecture. Bernal states:[18] '*we did not like this too much at the DFRL because it took time off our work, nevertheless, though we did not know it, it often proved the most valuable and instructive part of our research training.*'

W. H. Bragg had the idea of going beyond well-formed crystals and giving a lecture on '*The Imperfect Crystallisation of Common Things*', which included, among others, fibres. Work on fibres had already started, mostly at the Kaiser Wilhelm Institute in Dahlem, Berlin, where Meyer and Mark had already studied X-ray patterns given by simple textile materials like cotton and silk. W. H. Bragg asked Astbury in 1926 to help in getting photographs of fibres like wool. Astbury took this task to heart, and it gradually became his life's work.

To understand more fully the significance of Astbury's work (after he moved to the University of Leeds in 1928 to take up biomolecular structures in the Department of Textile Physics), it is necessary to acquaint ourselves with the essence of protein structures, which is described below. But before and after Astbury left the DFRL, W. H. Bragg frequently returned to the theme of 'living molecules' in his popular talks. In his Huxley Lecture,[19] delivered at Charing Cross Hospital in November 1929, he first expressed the hope that his team's work on crystal structures of organic substance would, one day, connect up with the medical care of the human body. He then alluded to the 'order of the structure' of such substances as fats and muscles, hair and wool, cotton and silk, and bone and shell. He then mentioned the list of constituents obtained (by Astbury) from the hydrolysis of wool (to which we shall return below—see Section 3.3.2).

In an influential Friday Evening Discourse[20] delivered at the RI in January 1933 entitled '*Crystals of the Living Body*', W. H. Bragg first reminded his audience that '*A hair, for example, is largely composed of a species of the proteins known as the keratins.*' He then moved on to other proteins, nerves, muscles, and tendons, that all possessed arrangements of atoms similar to those of keratin. And then he made the important point that the new method of analysis involving X-ray diffraction '*provide us with a means of examination of structure which are of much greater power than any that we possessed previously.*' He continued thus: '*Our chemical methods, it must be pointed out, do not reveal the nature and details of molecular arrangements. When we employ them for the analysis of a material, we begin by pulling the material to pieces and so destroying that very arrangement of molecules which we should be glad to examine. We knock the house down, and discover the numbers and natures of its components; so many bricks, so many slates, so many planks and so on; but we have lost the plan of the house. We must differentiate between the arrangements of atoms in the molecule and of the molecules with respect to one another. The former has long been the study of the chemist, and especially of the organic chemist. On the other hand, the mutual arrangement of the molecules in the solid is fundamentally conserved in those directive properties which are characteristic of the solid; it is this arrangement which is now open to our examination.'*

These last few sentences of W. H. Bragg's 1933 paper are echoed almost exactly by John Kendrew in his 1966 monograph '*The Thread of Life*', the relevant passage being:[21] '*The biochemist started off using the traditional techniques of chemistry. These techniques which can readily handle small molecules, containing, say, twenty or thirty atoms. The limitations of this approach derived from the fact that most of the really important molecules in living organisms are very much larger than this. They contain thousands of atoms in each molecule instead of tens. At this point the traditional chemical techniques, though developed to a very high point of sophistication, in effect reached their limit. It was here that molecular biology, with its new techniques for studying very large molecules, began to make a significant impact on the subject.*'

It was the nature of that impact—that of a purely physical tool (X-ray crystallography)—that Kendrew described in his '*Thread of Life*' series of broadcast lectures. Nowadays, with the astonishing advances made in several aspects of mass spectrometry—in which unfragmented molecules of mega Dalton mass can be manipulated and identified in atomic detail (see the work of C. A. Robinson,[22] for example) and also in cryoelectron microscopy recognized in the award of the Nobel Prize in Chemistry 2017 to Dubochet, Frank, and Henderson—there is less validity in the arguments of W. H. Bragg and J. C. Kendrew outlined above. (See Chapters 9 and 10, especially the latter for an outline of the importance of low-temperature electron microscopy in present-day structural biology.)

3.3.2 A guide to proteinaceous and a few other living molecules

To appreciate the significance of the structural molecular biological work started by Astbury, Bernal, Robertson, Cox, and Lonsdale at the DFRL, it is necessary to acquaint ourselves with some basic facts pertaining to the materials they and their scientific progeny (especially Dorothy Hodgkin, Perutz, and Kendrew) investigated. Keratin, which has been mentioned previously, is a protein, but it is also described by the synonym polypeptide. A dipeptide is formed when two amino acids combine, with a loss of a water molecule, as shown in Figure 3.6.

When three amino acids combine, likewise a tripeptide forms, the example shown in Figure 3.7 being that of the well-known antioxidant (that occurs in all plants and animals) glutathione, which is formed from three familiar amino acids.

When large numbers of amino acids combine to eliminate water, a polypeptide (protein) is formed. Proteins are macromolecules consisting of one or more chains of polypeptides. They perform a vast array of functions in the living world. Keratin is but one of a family of fibrous proteins, and it is the key structural material of hair, wool, horns, claws, hooves, and the outer layers of human skin. The so-called S-layer proteins assemble to form planar sheets on the surface of living cells.

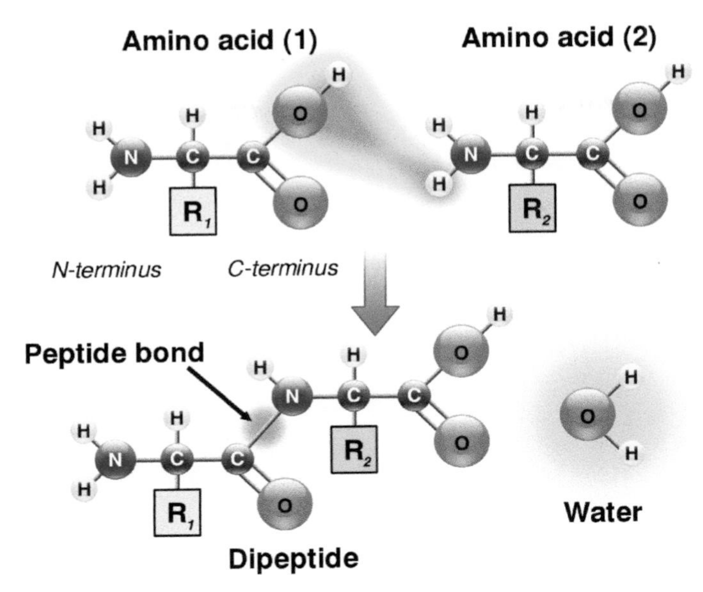

Figure 3.6 *Two amino acids, when combined, with loss of water, form a peptide, the peptide link being planar, in the sense that the groups attached to the atoms of the link must lie on a plane. Those carbon atoms of the peptide to which are attached R groups are called alpha- (α-) carbon atoms.*

(From Wikipedia)

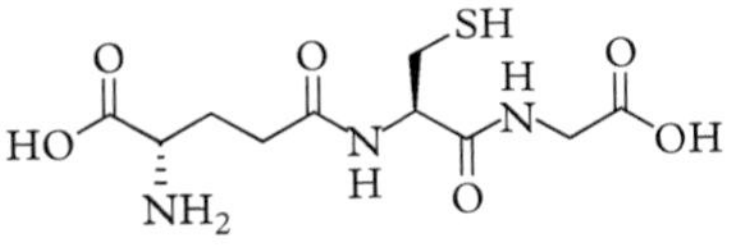

Figure 3.7 *The antioxidant known as glutathione is an example of a tripeptide.*

S-layers act as the outermost permeability barrier to protect cells from extracellular attack. The skin, muscles, and tendons are all proteins.

It is estimated that the human body has the ability to generate two million different types of proteins which, it is now known, are coded by some 20,000 genes. Almost 500 amino acids occur naturally, but only twenty of them occur in the living world. Their names (which consist of three-letter symbols) and structures are shown in Figure 3.8, which also describes whether these naturally occurring amino acids have basic or acidic properties and whether they exhibit aromatic (i.e. like benzene) or aliphatic (like long-chain hydrocarbons) characters.

The hormone insulin is a protein that consists of two polypeptide chains, as depicted in Figure 3.9.

These two chains are also connected by S–S bridges. In all, there are fifty-one amino acid residues that constitute insulin. The sequence of residues in the two-dimensional depiction of insulin was determined by Fred Sanger,[23] for which

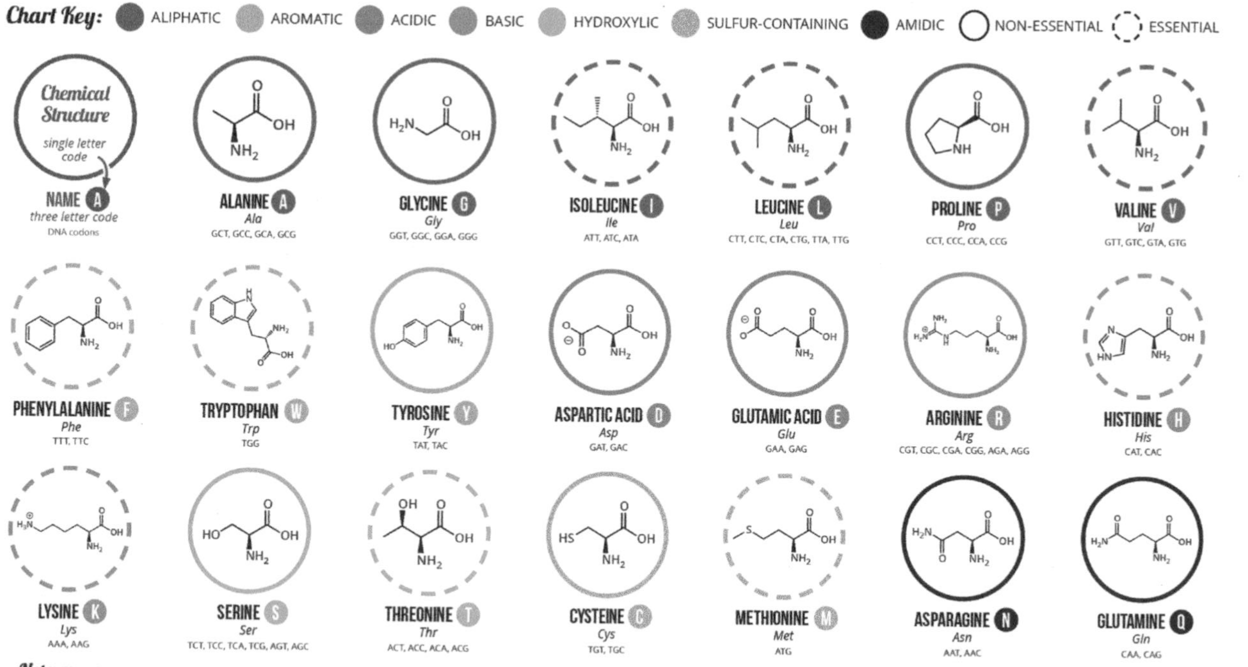

Figure 3.8 *These are the twenty 'essential' amino acids which the human genetic code directly encodes. (Selenocysteine is regarded as the twenty-first amino acid, and it is encoded in a special manner.)*

(Reproduced from https://www.compoundchem.com/2014/09/16/aminoacids/)

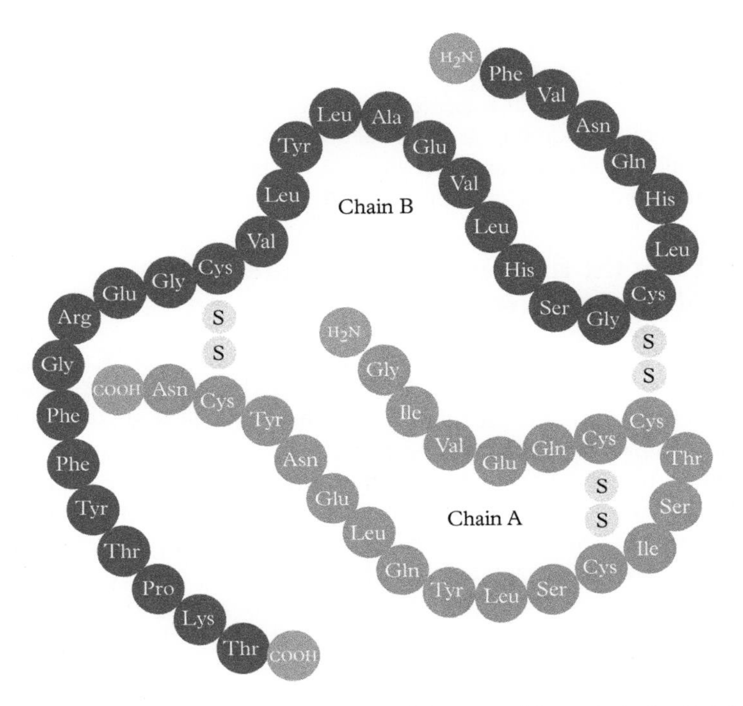

Figure 3.9 *Depiction of the two-chain nature, with intra- and inter-chain disulfide bridges, of human insulin.*

(Courtesy Dr D. N. Johnstone)

he was awarded the Nobel Prize in Chemistry in 1958. When Sanger proved the structure of insulin, it opened up the possibility of a synthetic method for its production, thus obviating the need thereafter to use animal sources. (These days, biotechnological methods are used to produce insulin for sufferers of diabetes.)

Other examples of protein, mentioned earlier, are haemoglobin, myoglobin, and the antibiotic known as gramicidin-S. Single red blood cells contain about 280 million molecules of haemoglobin. Each molecule has 64,500 times the mass of a hydrogen atom and is made up of hydrogen, carbon, nitrogen, oxygen, and sulfur, plus four atoms of iron, which are more important than the rest. As we shall see in Chapter 5, there are four polypeptide chains in a molecule of haemoglobin, consisting of a total of 574 amino acids.

All enzymes, which are the biological catalysts that facilitate essentially all of the chemical conversions of the living world, are proteins. These remarkably effective agents may often be hundreds of times as large as the molecules that they convert. And each individual living process has its own (evolved) enzyme. Thus, the simple conversion of glucose to alcohol is now known to involve twelve consecutive steps, and each step uses a different enzyme. Moreover, these enzymes are, by comparison with man-made inorganic catalysts (like the iron compound used to convert hydrogen and nitrogen to ammonia), extraordinarily efficient. They can often proceed to 'turn over' one substance to another in a living system more than a million times per second at room temperature. Lysozyme, discovered in his own tears by Alexander Fleming, as we shall see in Chapter 6, yielded its secrets to X-ray crystallography (at the DFRL) in 1965. It is the antibiotic enzyme that disrupts the membranous sac enveloping a cell, as exemplified further in Chapter 6.

3.3.3 How the amino acid constituents of a protein were determined

It was at the Wool Research Institute in Leeds that two former Cambridge scientists A. J. P. Martin and R. L. Synge[24–26] first evolved a method of determining the components of wool. This they did by applying their revolutionary technique of partition chromatography. They first hydrolysed the wool to break it down to its constituent amino acids, which were then identified in their (paper) partition chromatograph, work that earned them the Nobel Prize for Chemistry in 1952.

Sanger[23] extended and improved their method in his work on the precise sequencing of the various amino acids that formed part of the bovine insulin that he investigated. He used not only acids to hydrolyse the protein, but also other enzymes such as trypsin and chymotrypsin (which are themselves proteins) to break up the parent protein at specific regions in the polypeptide chain. An illustration of how he chopped the chain in different places and produced a different set of fragments overlapping the original set is shown in Figure 3.10.

The amino acid sequence of sperm whale myoglobin was determined at the Laboratory of Molecular Biology (LMB), Cambridge, by Edmundsen,[27] and he and others later worked out the amino acid sequence of human haemoglobin, a protein that we shall be discussing more fully in Chapter 5. The molecule of

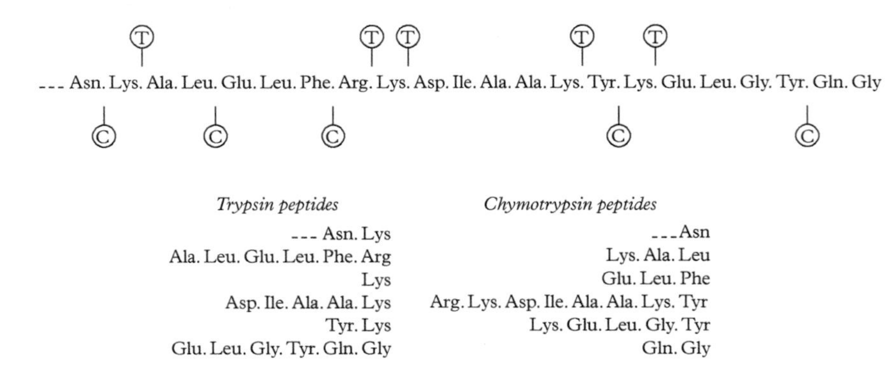

Figure 3.10 *The last twenty-two residues in the amino acid sequence of myoglobin (of the sperm whale), showing the points at which the chain is split by trypsin (T) and chymotrypsin (C), and also the peptides resulting from tryptic and chymotryptic splitting.*

(Courtesy J. C. Kendrew/MRC/LMB)

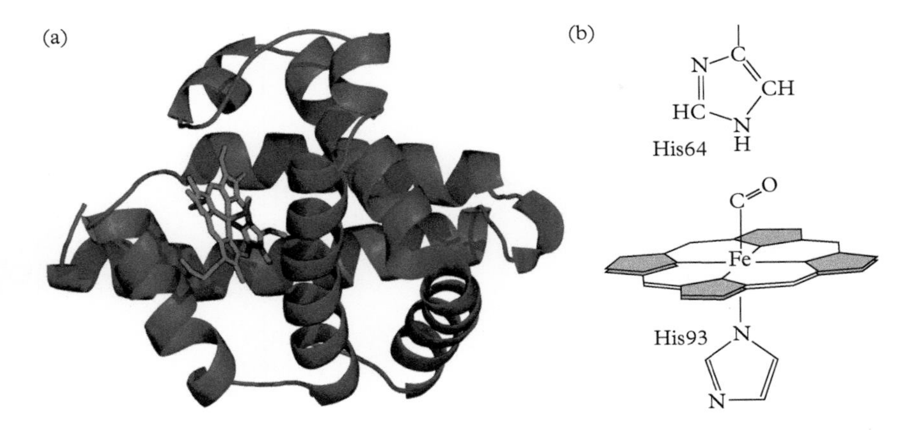

Figure 3.11 *Depiction of the chains (a) and the haem group (b) in the myoglobin structure. (Reproduced with permission from The Company of Biologists Ltd.)*

haemoglobin contains four polypeptide chains of two different kinds—two so-called α-chains and two β-chains (see Chapter 5). Figure 3.11 gives the amino acid sequences of the chains and shows just how complicated the chemical make-up of a typical protein is.

Myoglobin consists of a backbone and haem-binding domain (see Figure 3.11, panel (a)). Myoglobin was the first protein to be subjected to X-ray crystallography. The backbone of myoglobin consists of eight α-helices (blue) that wrap around a central pocket containing a haem group (red), which is capable of binding various ligands, including oxygen, carbon monoxide, and nitric oxide (see Figure 3.11, panel (b)). The protohaem group is bracketed or stabilized by histidine residues above (His64) and below (His93) (see Figure 3.11, panel (b)).

For some proteins, including the haemoglobins, the amino acid sequence has been determined for a number of different species, and one finds that if, for example, one compares the amino acid sequence of human and horse haemoglobin, for the most part, the sequences are the same, but every now and then, there comes a difference. Broadly speaking, the further apart the two animals lie in the evolutionary tree, the more differences there are to be found. This phenomenon is mentioned again in Chapter 7, in connection with Linus Pauling's work on the use of natural proteins as an evolutionary clock.

3.4 Astbury's Pioneering Work on Keratin

Astbury's studies of fibrous proteins, represented by keratin, the chief constituent of wool, hair, horns, etc., as described in Section 3.3.2, soon revealed that the structures of the unstretched and stretched forms were significantly different (see Figure 3.12).

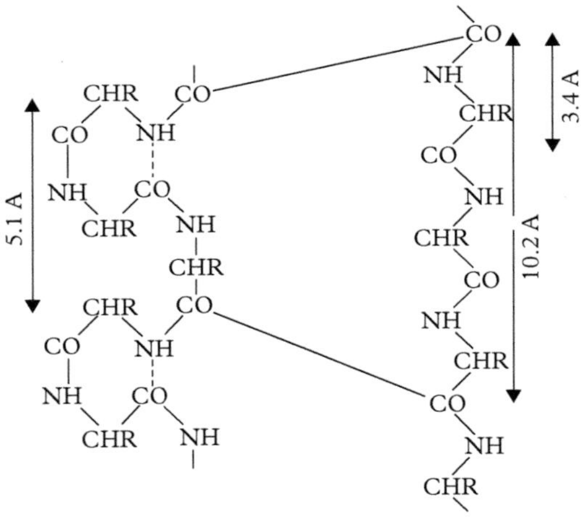

Figure 3.12 *Structures proposed by Astbury for both α-keratin (left) and β-keratin (right). (Courtesy of The Royal Society of Chemistry)*

For the so-called α-keratin (unstretched), he formulated the structure shown on the left of Figure 3.12. The β-keratin, stretched form, he depicted on the right of this figure (where the substituent R varies, according to the source of the keratin).

In a memorable article, entitled '*Fibrinogen and Fibrin as Members of the Keratin-Myosin Group*',[28] Astbury and his colleagues concluded their remarks with the following important statement that reflects his deep insight into recurring structural motifs: '*When we consider that the fibrous proteins of the epidermis, the keratinous tissues, the chief muscle protein, myosin, and now the fibrinogen of the blood, all spring from the same peculiar shape of molecule and are therefore probably all adaptations of a single root idea, we seem to glimpse one of the great co-ordinating facts of the lineage of biological molecules.*'

It is still universally recognized that his picture of β-keratin is valid, but that of α-keratin is erroneous. (We shall see in Chapter 4 that Linus Pauling's famous α-helix structure, proposed in 1949–1951, is what dominates the vast majority of protein structures—not just keratin, but in enzymes and vast ranges of other proteins.)

Astbury and Bernal continued to interact on their structural elucidation of '*living molecules*' to which they had been introduced at the DFRL, even after they each had migrated to Leeds in 1928 and Cambridge in 1927, respectively. In 1932, for example, they published together a paper on the walls of the algae *Valonia ventricosa*.[29] It was W. H. Bragg who stimulated their continuing collaboration and who communicated their results to the *Proceedings of the Royal Society*. It was W. H. Bragg who also encouraged Bernal to determine the structure of graphite, described in Chapter 2.

3.5 Bernal's Work at the DFRL and Shortly After His Return to Cambridge

In her comprehensive analysis of Bernal's life and work,[30] Dorothy Hodgkin not only praises the definitive work on the structure of graphite, but also highlights what Bernal achieved—for the benefit of the entire X-ray crystallographic community—in designing a universal photogoniometer. This was a versatile instrument designed to take single crystal rotation and oscillation X-ray photographs and Laue and powder photographs, and also, if desired, to be used as an optical goniometer for spectrographic work. According to Hodgkin, with this work, Bernal[31] provided a manual to '*handle, mount, and set all kinds of crystals, clear or opaque, with or without reflecting planes, unstable under ordinary conditions, at high or low temperature*'. He prepared '*the rest of the profession to take on any problem that might appear*'.[30]

3.5.1 Bernal and the structures of living molecules

Upon his return to Cambridge (as a member of the Department of Mineralogy and Petrography), Bernal broadened his interests considerably, partly stimulated by conversations with his contemporaries there: J. B. S. Haldane, Joseph and Dorothy Needham, N. W. Pirie, C. H. Waddington, R. L. Synge, and A. J. P. Martin. Pirie, for example, gave him a sample of the antioxidant glutathione (see Figure 3.7). Others, including many eminent organic chemists, like Diels and Butenant in Germany, Ruzicka in Switzerland, and R. D. Howarth and J. W. Cook in the UK, provided him with amino acids, vitamin D hydrochloride, β-carotene, some steroid molecules that were sex hormones, and a total of 23 hydrocarbons related to sterols.[32]

Table 3.4 enumerates some of the major studies that Bernal *et al.* carried out in Cambridge up to the time of the Second World War. By now, Dorothy Crowfoot (later Hodgkin) had joined him from Oxford in 1934, Max Perutz from Vienna in 1936, and Isidore Fankuchen from the United States via Manchester in 1935. His work in that period was particularly fruitful in four distinct areas:

Sterols and sex hormones;

Proteins;

Viruses and liquid crystals; and

The structure of liquids.

We shall deal here with some of these adventures, but first it is important to emphasize that, so far as protein crystallography was concerned, his most influential paper was one he published with Crowfoot entitled '*X-ray Photographs of Crystalline Pepsin*',[33] the digestive enzyme. It is this X-ray diffraction pattern, which is extraordinarily rich in detail (with diffraction spots going out to large distances in reciprocal space—see Chapter 2) that made it clear to the authors and all those interested in macromolecular protein species that the detailed atomic structure of the latter could, in principle, be retrieved from the methods of X-ray crystallography. Equally important is that this work showed that proteins were not random polymers but were composed of well-defined three-dimensional structures.

It is noteworthy that, when this paper by Bernal and Crowfoot was published, not only did it constitute a turning point in the history of protein science and crystallography, but it also revealed to non-crystallographers the ability that simple measurements of the density ρ and the volume of the unit cell V of a crystalline protein (or any—soluble or insoluble) yielded the molecular weight M of the molecule under study, as described in Section 2.3. The ultra-centrifuge community in Sweden, who were the experts in determining molecular weights of macromolecules, were impressed when Bernal told them that his independent method of determining molecular weights agreed precisely with theirs:

$$\rho = nM/V \qquad (1)$$

Bernal and his associates, particularly Fankuchen, were among the first scientists to explore the mysteries of viruses—see, for example, his paper with Riley and Fankuchen in *Nature*, 1938, as given in Table 3.4. His colleagues in the Biochemistry

Table 3.4 *List of papers on structural molecular biology published by J. D. Bernal and his co-workers up to 1940*

1931	The crystal structure of natural amino acids, *Z. Kristallogr. Kristallgeom.*, **78**, 363
1932	Crystal structure of vitamin D and related compounds, *Nature, Lond.*, **129**, 277–8
	A crystallographic examination of oestrin, *J. Soc. Chem. Ind., Lond.*, **51**, 259 (Biochem. Soc.)
	Carbon skeleton of the sterols, *J. Soc. Chem. Ind., Lond.*, **51**, 466
	Properties and structures of crystalline vitamins, *Nature, Lond.*, **129**, 721
	Crystal structure of complex organic compounds, Lecture at Manchester University, 13 July (printed by Metropolitan Vickers)
	(With W. T. Astbury and T. C. Marwick) X-ray analysis of the structure of the wall of *Valonia ventricosa*, *Proc. R. Soc. Lond.*, **B 109**, 443–50
1933	Contribution on the chemical constitution of oestrin, *J. Soc. Chem. Ind.*, **52**, 288 (Disc. Chem. Soc.)
	(With Dorothy Crowfoot) Crystal structure of vitamin B_1 and of adenine hydrochloride, *Nature, Lond.*, **131**, 911–12
	Liquid crystals. (Account of conference of 'Liquid Crystals and Anisotropic Melts', organized by the Faraday Society, 24–25 April), *Nature, Lond.*, **132**, 86–9
	(With Dorothy Crowfoot) Crystalline phases of some substances studied as liquid crystals, *Trans. Faraday Soc.*, **29**, 1032–49

1934 Application of X-ray methods in the food industry, *Chemy. Ind.*, **12**, 1075–7

(With Dorothy Crowfoot) X-ray photographs of crystalline pepsin, *Nature, Lond.*, **133**, 794

(With Dorothy Crowfoot) X-ray crystallographic measurements on some derivatives of cardiac aglucones, *J. Soc. Chem. Ind., Lond.*, **53**, 953–6

1935 (With Dorothy Crowfoot) The use of the centrifuge in determining the density of small crystals, *Nature, Lond.*, **135**, 305

(With Dorothy Crowfoot) The structure of some hydrocarbons related to the sterols, *J. Chem. Soc. Trans.*, part 1, 93–100

1936 (With Dorothy Crowfoot) X-ray crystallographic data on the sex hormones, *Z. Kristallogr. Kristallgeom.*, **93**, 464–80

1937 X-ray studies and the structure of proteins (Contribution to a Discussion), *Trans. Faraday. Soc.*, **33**, 1143

Structure types of protein 'crystals' from virus infected plants, *Nature, Lond.*, **139**, 923–4

1938 (With I. Fankuchen and D. Riley) Structure of the crystals of tomato bushy stunt virus preparations, *Nature, Lond.*, **142**, 1075

Rayons-X et structure des proteins, *J. Chim. Phys.*, **35**, 179–84

Virus proteins—structure of the particles (Contribution to a Discussion), *Proc. R. Soc. Lond.*, **B 125**, 299–301

A speculation on muscle, *Perspectives in Biochemistry* (Presentation Volume to Sir Frederick Gowland Hopkins), pp. 45–6, Cambridge University Press

(With I. Fankuchen and M. F. Perutz) An X-ray study of chymotrypsin and haemoglobin, *Nature, Lond.*, **141**, 523–4

Molecular architecture of biological systems (Four lectures to the Royal Institution, 25 January; 1, 8, and 15 February. Summaries published)

1939 X-ray evidence for the structure of the protein molecule, *Proc. R. Soc. Lond.*, **A 170**, 75-8, **B 127**, 36–9

The structure of proteins, *Proc. R. Instn. Gt. Br.*, **30**, part 3, 541–57 (Also *Nature, Lond.*, **143**, 663–7)

(With I. Fankuchen and D. Riley) X-rays and the cyclol hypothesis, *Nature, Lond.*, **143**, 897

1940 Structural units in cellular physiology, *The cell and protoplasm* (Proc. 10th Symp. Of the Am. Ass. Adv. Sci. On the Cell, 1939) *Am.Ass.Adv. Sc.* (Washington), no. 14, pp. 199–205

(With Dorothy Crowfoot and I. Fankuchen), X-ray crystallography and the chemistry of the sterols, *Phil. Trans. R. Soc. Lond.*, **A 239**, 135–82

X-ray evidence on size and structure of plant virus preparations, *Rep. Third Int. Congr. Of Microbiology* (1939); section on 'The nature and characteristics of filterable viruses', (ed. R. St. John-Brooke)

Department in Cambridge were the ones that discovered new types of viruses, as distinct from the cylindrical tobacco mosaic ones that had, by then, been well recognized. (It was in Bernal's Department of Physics at Birkbeck College, as described in Chapter 9, that Rosalind Franklin and Aaron Klug and their associates Don Caspar, John Finch, and Ken Holmes did their brilliant work. Many years later, through his efforts with Crowther, in particular at the LMB, Klug earned his Nobel Prize in Chemistry in 1982.)

In regard to Bernal's pioneering work (summarized in this section) in structural biology, it is interesting to recall that one of his great admirers Linus Pauling, in his famous RI Discourse,[8] said the following: '*What is skin, fingernail? How do fingernails grow?*'

It is noteworthy to report that in this exceptionally busy scientific period of Bernal's life, he became increasingly involved in political activity. He visited the USSR several times. He interacted with several other left-wing scientists like P. M. S. Blackett, Joseph Needham, J. B. S. Haldane, and P. Langevin, as well as with Aldous Huxley.

Of this work Dorothy Hodgkin, who was his brilliant PhD student in Cambridge, as from 1934, has written evocatively about it:[30] '*As with the discovery of X-ray diffraction itself, there were both accidents and purposes contributing to the taking of the first X-ray photograph of protein crystals by J. D. Bernal*[33–35] *in the spring of 1934. John Philpot was working at that time on the purification of pepsin at Uppsala; a preparation he left standing in a fridge while he was away skiing produced very large – 2mm long – and beautiful crystals of the hexagonal bipyramidal form obtained earlier by Northrop and Kunitz. They were seen by Glenn Millikan, a passing visitor and friend of Bernal's who knew of his interests in proteins and his appeal for crystals. Philpot gave Millikan a tube of crystals in their mother liquor to take back to Cambridge.*'

3.5.2 The structure of sterols

Nowadays, members of the general public are well aware of the existence, and sometimes the nature, of steroids. They are: (i) sex hormones, which influence sex differences and support reproduction—they include oestrogens, androgens, and progesterone; (ii) corticosteroids, which include most synthetic drugs; and (iii) anabolic steroids, which interact with androgen receptors to increase muscle and bone synthesis. Testosterone is the principal male sex hormone; cholic acid is a bile acid, and progesterone is a steroid hormone involved in the female menstrual cycle.

It was in the late 1920s that two eminent (Nobel Prize-winning) German organic chemists first proposed the chemical structure of the family of steroidal compounds. Straightforward chemical tests had shown that a typical steroid contained 27 carbon atoms. In Heinrich Wieland's Nobel Lecture (1929),[36] he presented the structural formula shown in Figure 3.13, which was also in full agreement with the structure proposed by his fellow Laureate Adolf Windaus.

Although some contemporary organic chemical research, at that time, cast some doubt on the validity of this structure, it was Bernal's work, using straightforward unit cell measurements by X-rays, together with the observed space group of a sterol crystal and its individual lattice parameters, that led him to point out that his measurements, especially a 7 Å dimension found in many crystal forms, were '*difficult to reconcile with the usually accepted sterol formula*'. Bernal showed unequivocally that the sterol was flat and lath-like.[37] The accepted formula for cholesterol is also shown in Figure 3.13 (marked II and III).

Dorothy Hodgkin carried out extensive work on the sterols—it was the subject of her PhD thesis[38]—with Bernal and his associates. A comprehensive paper, entitled '*X-ray Crystallography and the Chemistry of the Sterols*', by Bernal, Crowfoot, and Fankuchen appeared in the *Philosophical Transactions of the Royal Society* in 1940,[39] in which they reported the results of their investigations of over eighty distinct kinds of steroids.

Dorothy Hodgkin retained a strong interest in steroids after her return to the Oxford Department of Mineralogy and Crystallography. And her later work with Carlisle[40] (on cholesterol iodide), as well as that with Dunitz[41] (on calciferol) (see Chapter 1), are landmarks in the history of X-ray structure determination.

For the benefit of the general reader, the framework (carbon) structure of the steroid ring system, as well as the International Union of the Pure and Applied Chemistry (IUPAC)-approved ring lettering and atom numbering, is shown in Figure 3.14 (compare with structures II and III of Figure 3.13).

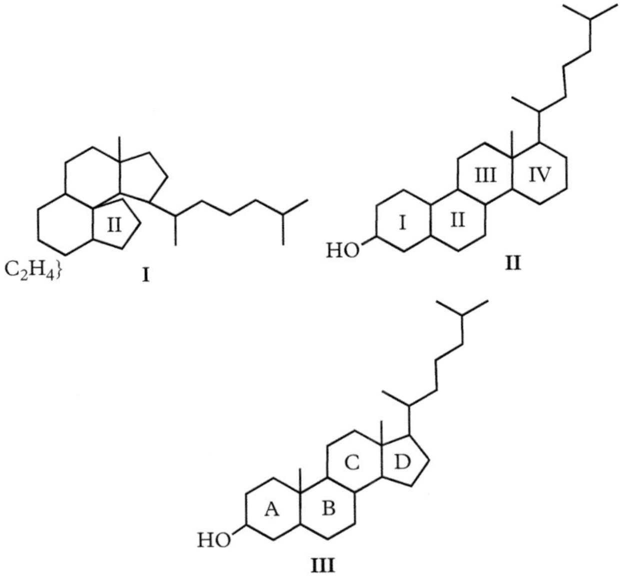

Figure 3.13 *(I) The Wieland–Windaus version of the structure of the steroid framework. (II) and (III) The correct framework structure, deduced, in part, from Bernal et al.'s X-ray diffraction studies (after Hodgkin[30]).*

(Courtesy of The Royal Society)

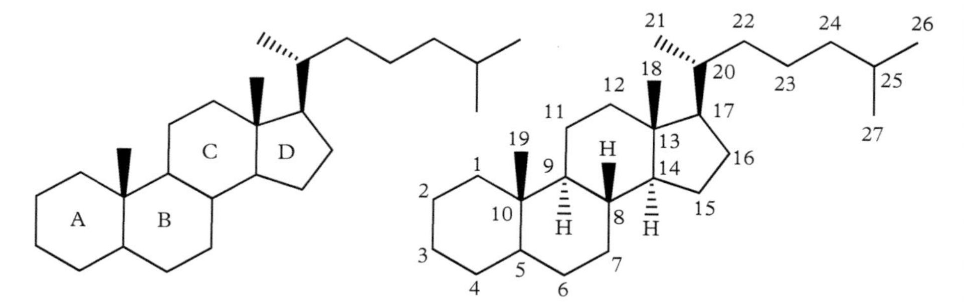

Figure 3.14 *Framework structure, with IUPAC-approved numbering of the atoms, of the steroid ring system. (IUPAC = International Union of Pure and Applied Chemistry)*

3.5.3 W. H. Bragg's indirect influence on Dorothy Hodgkin

As a teenager, thanks to the fact that her mother had drawn Dorothy Crowfoot's attention to the books that W. H. Bragg had published, based on his RI Christmas Lectures—in 1923 and 1925—entitled '*Concerning the Nature of Things* and *Old Trades and New Knowledge*', her interest in experimental science was kindled. Later, as an undergraduate at Somerville College, Oxford, having been aroused by the possibility of seeing the internal atomic structure of crystals with X-rays, she began to read the work that emanated from the DFRL carried out by J. D. Bernal, W. T. Astbury, and Kathleen Lonsdale. Her essays revealed that she was acquainted with some of the important results published by W. H. Bragg's acolytes. For example, she knew that the benzene molecule is flat, as she had read Kathleen Lonsdale's important studies of the structure of the hexachloro and hexamethyl derivatives of benzene.

Her extensive work with Bernal, described above, drew her closer to W. H. Bragg as a person. And in the spring of 1937, Sir William invited her to spend a week at the DFRL so that she could use the more powerful X-ray sources that were available there.

Scientifically, she did not benefit much from this week's visit to London. But, on a personal basis, it was transformative. While staying in the home of Margery Fry during the Easter vacation, she met her future husband Thomas Lionel Hodgkin.

Having been taught by Bernal on how to determine molecular weights of materials from X-ray and density measurements, Dorothy Hodgkin was often in demand by eminent organic chemists (notably Sir Ian Heilbron at Imperial College) to lecture to them on the merit and power of this simple method. This further enhanced her reputation as an able chemist. Her stature at Oxford grew as a formidable and penetrating X-ray crystallographer, and she was soon investigating major challenges such as the determination of the structures of insulin, lacto-globin, and penicillin. Her crowning glory, as described in Chapter 8, was the determination of the structure of vitamin B_{12}.

3.5.4 The legacy of W. H. Bragg's other acolytes: J. M. Robertson, E. G. Cox, A. L. Patterson, and Kathleen Lonsdale

Robertson's reputation as an X-ray crystallographer rests on three main achievements: first, his pioneering work on the electron density distribution and the structures of aromatic hydrocarbons (see Figure 1.6 of Chapter 1); second, his introduction, under W. H. Bragg's aegis at the DFRL, of the so-called '*heavy atom technique*'; and third, his creation at Glasgow University from the 1940s onwards of one of the world's leading X-ray crystallographic centres, where such distinguished individuals as Jack Dunitz and particularly Michael Rossmann, who did crucial work on the globins for Perutz and Kendrew (see Chapter 5), were launched on their careers.

The classical example for the application of the heavy atom method is Robertson's determination of the structures of the phthalocyanines of nickel (Ni), copper (Cu), and the platinum (Pt) derivatives.[42] These are particularly favourable materials because it is possible to compare the diffracted intensities of the pure organic molecule $C_{32}N_8H_{18}$ with those obtained after insertion of a Ni, or Pt, or Cu atom at the centre of the organic group. The contribution of the heavy atom outweighs that of the other parts in determining the phases, and the Fourier synthesis can proceed without any previous model. Even the assumption of the existence of atoms need not be made! And the resulting structure is so meaningful that, from the electron densities, one may read off the precise location of all the constituent atoms, as seen in Figure 3.15, which shows the projections of the electron densities (along the *b* crystal axes).[43,44]

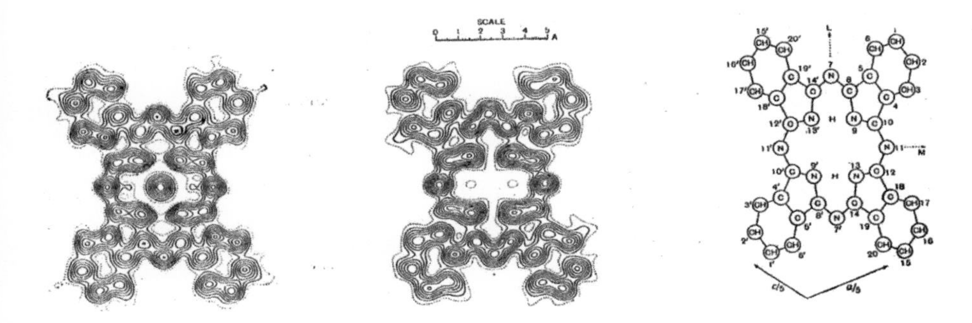

Figure 3.15 *The electron density of phthalocyanine determined by J. M. Robertson.*[44] *(Courtesy The Royal Society of Chemistry)*

It was this technique, in a refined, advanced form, that enabled Perutz and Kendrew to solve the structures of haemoglobin and myoglobin, as is described in Chapter 5.

It is to be emphasized, however, that even when no heavy atom substitution is employed, it is still possible, in favourable circumstances, to determine the detailed molecular structure of a new substance from the X-ray diffraction pattern alone. This was done by Dorothy Hodgkin when she solved the structure of penicillin in 1944. The eminent organic chemists who had provided her with the specimens of the penicillin—Sir Robert Robinson and Sir John Cornforth (each to become Nobel Laureates in Chemistry)—were unconvinced by the structure that she had determined (see Figure 3.16). They were perplexed to see that it contained a so-called β-lactam ring. It is on record that Cornforth said that if Hodgkin's structure of penicillin was right and it did indeed contain a β-lactam ring, he would give up organic chemistry and grow mushrooms. According to the organic chemists, its occurrence in penicillin was so much at variance with their chemical intuition that they suggested that the nature of the penicillin was modified by exposure to X-rays during the course of the diffraction study. But they were proved wrong by the X-ray crystallography.

3.5.4.1 E. G. Cox (1906–1996)

A graduate of A. M. Tyndall's famous Department of Physics at the University of Bristol, Gordon Cox's first task, assigned to him by W. H. Bragg, was to try to locate the carbon atoms in molecules of benzene (a substance that Faraday had discovered and characterized at the RI in 1825). W. H. Bragg then persuaded him to join Sir Norman Howarth, the Nobel Prize winner, for his work on the synthesis of ascorbic acid (vitamin C).

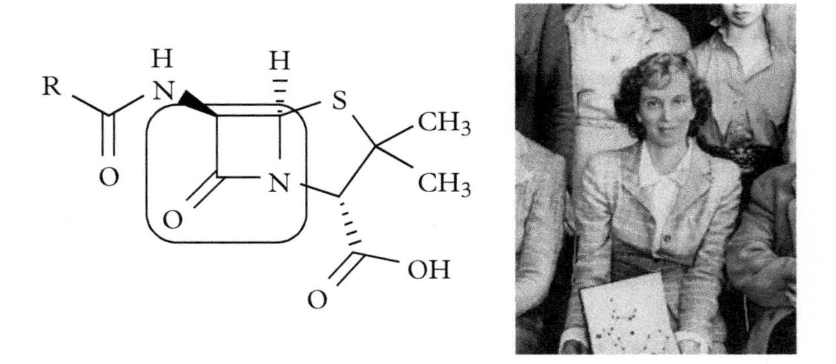

Figure 3.16 *The structure of penicillin, determined by Dorothy Hodgkin, with the* β*-lactam ring highlighted.*

(Structure copyright Dr D. N. Johnstone; photo copyright Prof E. Dodson)

3.5.4.2 A. Lindo Patterson (1902–1966)

Ever since Arthur Lindo Patterson introduced, in 1934, the Patterson method, to this day, it has helped crystallographers solve their problems. His work remains invaluable—it is used extensively to assist in the solving of the phase problem, and the work of both Dorothy Hodgkin and Rosalind Franklin was greatly facilitated by the use of Patterson's method.

In essence, Patterson's contribution was to show that from the intensities of the various diffractions, it is possible to construct a 'vector map' that gives the distances between pairs of atoms. We may describe a 'vector map', following Judson,[45] using the analogy of a party. Everyone has their shoes nailed to the floor and everyone wants to meet everyone else! The vectors are the directions each person has to turn to shake the hand of another guest and how far that person has to extend his or her arms. The strength of the handshake is analogous to the atomic numbers of the two atoms at the end of the guests at the party. It is as if the only information one possesses about the party is the distance and angle of the handshake of each person when he or she met everyone else. Equipped with this complicated knowledge, it should then be possible to find where everyone stood in the room relative to each other.

Patterson's key paper was written at the Massachusetts Institute of Technology where he was an unpaid guest of B. E. Warren, an eminent crystallographer, who had earlier worked with W. L. Bragg. Patterson, a Canadian, obtained a rather inferior initial degree at McGill University, Montreal, but W. H. Bragg took him on as a researcher because of his mathematical potential. At the DFRL, in 1924–1926, he determined the unit cell and space groups of various phenylaliphatic acids.

3.5.4.3 Kathleen Lonsdale (1903–1971)

Dorothy Hodgkin, in her *Biographical Memoir*[46] of Kathleen Lonsdale, described her as a '*force of nature*'. She was a phenomenally able and well-organized scientist—and something of a workaholic—but endowed with a great sense of humour. Of all W. H. Bragg's acolytes, she spent the largest proportion of her time at the DFRL during Bragg's Directorship of it (over twenty years).

In Section 3.3.1 above, Kathleen Yardley (as she was before marriage) gave a vivid description of the exciting scientific and personal experiences she savoured at the DFRL in W. H. Bragg's day as Director. During the course of a two-day visit she paid to my laboratory in Aberystwyth at the University College of Wales in 1970, she described to me the joy and excitement that pervaded the laboratories, tea-room, and table-tennis room of the house in Albemarle Street, Mayfair, London, which they occupied.

Dorothy Hodgkin, in her characteristic warm, sympathetic, authoritative way, has described at length in her charming biographical memoir[46] the quintessential character and enormous contribution made by Kathleen Lonsdale, the tenth child of a postmaster from Newbridge, Ireland. Anyone who wants to be further reminded of the truth of Perutz's assertion (see Figure 2.19) that a happy atmosphere is conducive of good science should consult Lonsdale's chapter in Ewald's book.[9] Figure 3.17, taken from Dorothy Hodgkin's memoir, conveys the happiness that was such a feature of the pioneering days when Bernal, Robertson, Lonsdale, and so many other distinguished crystallographers, including the Dutch scientist Burgers and the Russian Orelkin, worked at the DFRL.

The charm of Hodgkin's memoir on Lonsdale (also seen in her equally impressive one on Bernal[30]) is attributable to her skill in mingling the strictly scientific with personal, even forensic, details. In commenting on the atmosphere among co-workers at the DFRL, she says: '*They shared and criticized each other's ideas, talked politics and religion, as well as science and played table-tennis internationals in the basement.*' Of W. H. Bragg, Hodgkin quotes Kathleen Lonsdale: '*He inspired me with his own love of pure science and with his enthusiastic spirit of enquiry and at the same time left me entirely free to follow my own line of research.*'

Apart from her work, outlined earlier, on benzene—and, in particular, the determination of the 1.42 Å bond length of the C–C bonds in the benzene ring—she carried out outstanding work on the magnetic anisotropy of molecular crystals. We have also mentioned that, along with her colleague W. T. Astbury, she was

Figure 3.17 *From left to right: front row, W. T. Astbury, Kathleen Yardley, W. G. Burgers, J. M. Robertson, R. E. Gibbs; back row, Eric Holmes, Boris Orelkin.*

(Courtesy Royal Society)

responsible for the birth of the '*International Tables for X-ray Crystallography*', a major aid (as it still is in its present enlarged issue) to all X-ray crystallographers. Later, as Head and Professor at University College, London, she broadened her interests considerably; she pioneered the use of neutrons for structural elucidation, and she initiated studies in organic solid-state photochemistry, in the physics of diamond, and in the role of epitaxy in biomineralogy—one of her last papers dealt with the formation of urinary calculi and bladder stones.[47]

As a Quaker and militant pacifist, she refused to register in 1939 for government service for World War II. Shortly thereafter, she spent a month in Holloway Prison. But with the aid of Sir Henry Dale, who became Head of the RI after W. H. Bragg's death in 1942 and who was President of the Royal Society, it was arranged for her to receive scientific papers in prison where she did some of her research work.

3.6 The Return of Protein Crystallography to the DFRL

When Sir Lawrence Bragg decided to relinquish his Cavendish Professorship of Physics in Cambridge in 1953, he took up the Fullerian Professorship of Chemistry at the RI and the Directorship of the DFRL there. He quickly assembled a group of inventive crystallographers and supporting staff, assisted, in part, by Dorothy Hodgkin, who recommended two exceptionally gifted X-ray crystallographers David C. Phillips and Jack Dunitz. W. L. Bragg also recruited A. C. T. North from King's College, London, where he had worked with Sir John Randall. In addition, W. L. Bragg signed up both Max Perutz and John Kendrew as Honorary Readers at the DFRL.

A new era of protein studies was about to dawn, as we describe in Chapters 5 and 6, after first describing a clash between the Cavendish team and their rivals in California.

REFERENCES

1. J. Jenkin, '*William and Lawrence Bragg: Father and Son: The Most Extraordinary Collaboration in Science*', Oxford University Press, New York, NY, **2008**.
2. J. M. Thomas, Michael Faraday, and the Royal Institution, '*The Genius of Many and Place*', Taylor & Francis. Originally published by Institute of Physics, Bristol, **1991**.
3. J. M. Thomas, *Proc. Am. Phil. Soc.*, **1998**, *142*, 597.
4. J. M. Thomas, *Chem. Commun.*, **2017**, *53*, 9179.
5. L. Pauling and C. R. Coryell, *Proc. Natl. Acad. Sci. U. S. A.*, **1936**, *22*, 210.
6. J. S. Rowlinson, '*Sir James Dewar, 1842–1923: A Ruthless Chemist*', Ashgate, Aldershot, **2012**.
7. Max Perutz nominated Lise Meitner for the Fermi prize, partly because she was not included in the Nobel Prize awarded for the discovery of nuclear fission. The Chairman

of the Fermi Award Committee Glenn Seaborg travelled to Cambridge to present her her medal, as she herself was too frail to journey to the United States.
8. L. Pauling, *Nature*, **1948**, *161*, 707.
9. P. P. Ewald (ed.), '*Fifty Years of X-ray Diffraction*', International Union of Crystallography, Utrecht, **1962**.
10. William Stone (1857–1958) studied at Peterhouse, Cambridge, and later became one of its greatest benefactors.
11. J. D. Bernal, ref [9], pp. 374–404.
12. M. Jaskolski, Z. Dauter, and A. Wlodawer in '*A brief history of macromolecular crystallography, illustrated by a family tree and its Nobel fruits*', *FEBS J.*, **2014**, 1–2. See also I. Olovson, A. Liljas, and S. Lidin, '*From a Grain of Salt to the Ribosome: The History of Crystallography as Seen Through the Lens of the Nobel Prize*', World Scientific Publishing, Singapore, **2015**, p. VII.
13. J. M. Thomas, *Notes and Records of the Royal Society*, **2011**, *65*, 163.
14. M. Delbrück, *Trans. Conn. Acad. Arts Sci.*, **1949**, *38*, 173.
15. J. C. Kendrew, *Scientific American*, **1967**, *216*, 141.
16. J. Cairns, G. S. Stent, and J. D. Watson (eds.), '*Phage and the Origins of Molecular Biology*', Cold Spring Harbor Press, New York, NY, **1966**.
17. G. S. Stent, *Science*, **1968**, *160*, 390.
18. J. D. Bernal, *Biographical Memoirs of Fellows of the Royal Society, W. T. Astbury*, **1963**, *9*, 1.
19. W. H. Bragg, *Lancet*, 21 December **1929**.
20. W. H. Bragg, *Nature*, **1933**, *132*, 11 and 50.
21. J. C. Kendrew, '*The Thread of Life*', **1966**, p. 18.
22. B. T. Ruotola, J. L. P. Bluesh, A. M. Sandercock, S. J. Hyung, and C. A. Robinson, *Nature Protocols*, **2008**, *3*, 1139.
23. F. Sanger, *Biochem. J.*, **1945**, *39*, 507.
24. H. Gordon, A. J. R. Martin, and R. L. M. Synge, *Biochem. J.*, **1943**, *37*, 19.
25. R. Cosden, A. H. Gordon, and A. J. P. Martin, *Biochem. J.*, **1944**, *38*, 224.
26. W. J. Whelan, *IUBMB Life*, **2001**, *51*, 329.
27. E. Edmundsen, *Nature*, **1965**, *205*, 883.
28. K. Bailey, W. T. Astbury, and K. M. Rudall, *Nature*, **1943**, *157*, 716.
29. W. T. Astbury, T. C. Marwick, and J. D. Bernal, *Proc. Roy. Soc. B*, **1932**, *109*, 443.
30. D. M. Hodgkin, *Biographical Memoirs of Fellows of the Royal Society, J. D. Bernal*, **1980**, *26*, 16.
31. J. D. Bernal, '*A universal X-ray photogoniometer*'. A series of papers in *J. Sci. Instruments*, **1927**, *4*, 273; **1928**, *5*, 241; **1929**, *6*, 343.
32. J. D. Bernal and D. Crowfoot, *J. Chem. Soc.*, **1935**, 93.
33. D. Crowfoot and J. D. Bernal, *Nature*, **1934**, *138*, 794.
34. D. C. Hodgkin, *Ann. N. Y. Acad. Sci.*, **1979**, 66.
35. Although the paper describing this work has two authors, Dorothy Hodgkin has revealed that she was not even in the laboratory when Bernal made his sensational discovery. It is a measure of Bernal's generosity of spirit that he felt impelled to include the name of his research student—and it came first!—in the resulting paper in *Nature*.
36. H. Weiland, *Angew. Chem.*, **1929**, *42*, 421.
37. J. D. Bernal, *Nature*, **1932**, *129*, 277.

38. D. Crowfoot, '*X-ray Crystallography of Sterols*', PhD thesis, University of Cambridge, Cambridge, **1936**.
39. J. D. Bernal, D. Crowfoot, and I. Fankuchen, *Phil. Trans. Roy. Soc.*, **1940**, *A239*, 135.
40. H. C. Carlisle and D. M. Crowfoot, *Proc. Roy. Soc.*, **1945**, *A184*, 64.
41. J. D. Dunitz and D. C. Hodgkin, *Nature*, **1948**, *162*, 608.
42. J. M. Robertson, *J. Chem. Soc.*, **1935**, 615.
43. J. M. Robertson and I. Woodward, *J. Chem. Soc.*, **1937**, 219.
44. J. M. Robertson, '*X-ray analysis and application of Fourier series methods to molecular structures*' in *Rep. Progr. Physics*, **1937**, *4*, 332.
45. H. F. Judson, '*The Eighth Day of Creation: Makers of the Revolution in Biology*', Simon and Schuster, New York, NY, **1979**.
46. D. M. C. Hodgkin, *Biographical Memoirs of Fellows of the Royal Society, K. Lonsdale*, **1975**, *21*, 447.
47. K. Lonsdale, *Science*, **1968**, *159*, 1199.

4

A Dispute Between the Cavendish and Caltech: The Emergence and Ubiquity of the α-Helix

4.1 Introduction

One of the most elegantly composed and plausibly argued papers that Lawrence Bragg, John Kendrew, and Max Perutz ever published was that which appeared in the *Proceedings of the Royal Society* in 1950[1] entitled '*Polypeptide Chain Configurations in Crystalline Proteins*'. But in less than a year after its appearance, it was comprehensively repudiated by Linus Pauling, Corey, and Branson[2] in a devastatingly direct paper, followed by a series of seven other short papers by Pauling and Corey in the *Proceedings of the National Academy of Sciences of the United States of America*[2–8] and one in the *Journal of the American Chemical Society*.[9] These papers by Pauling *et al.* introduced into structural molecular biology the idea of both the α-helix and the so-called β-pleated sheet features as integral motifs in the structure of proteins.

It is appropriate that we dwell on the nature and introduction of the α-helix as it has, in retrospect, turned out to be one of the most important structural motifs ever proposed to understand the fundamental nature of polypeptides and proteins generally. Furthermore, even though it was introduced by Pauling and his co-workers in his acts of disapproval of the paper by Bragg, Kendrew, and Perutz,[1] ironically, it was Kendrew *et al.*[10] (and later both Perutz and Phillips,[11] as we shall see) who first demonstrated the reality of the existence of the α-helix—not Pauling. It was Phillips *et al.*[11] who first recorded the existence of Pauling's proposed β-sheet structure in proteins.

Shortly after the publication of Pauling *et al.*'s paper,[2] such was the impact of the motifs of the α-helices and β-sheets that fewer and fewer structural papers dealing with proteins resorted to the use of the previously popular, atomically resolved models, like those shown in Figure 4.1, taken from the Bragg, Kendrew, and Perutz paper.[1]

Instead, solely because of the compelling veracity of Pauling and Corey's models of α-helices and β-sheets and their subsequent discovery by Kendrew *et al.*[10] and Phillips,[11] both the cognoscenti among biologists, as well as non-specialists, are

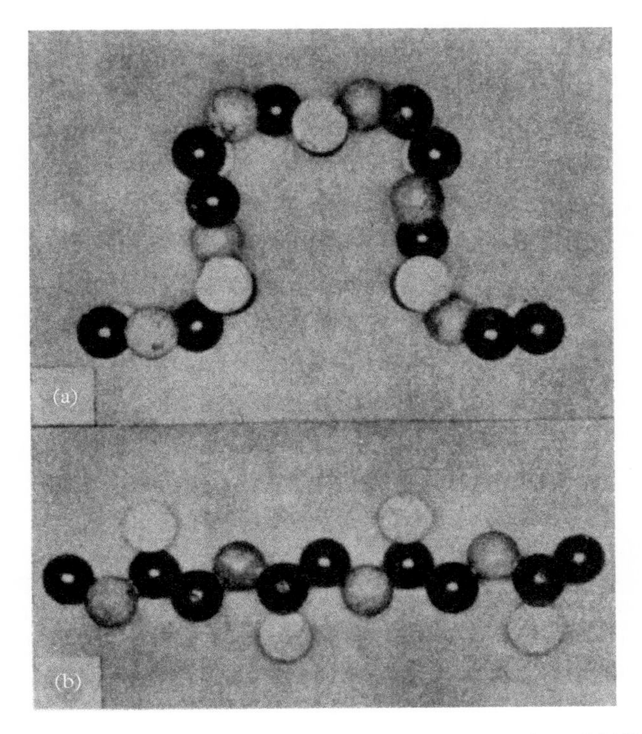

Figure 4.1 *Chain configuration proposed by Astbury for: (a) α-keratin and (b) β-keratin.*[1]
(Courtesy of The Royal Society)

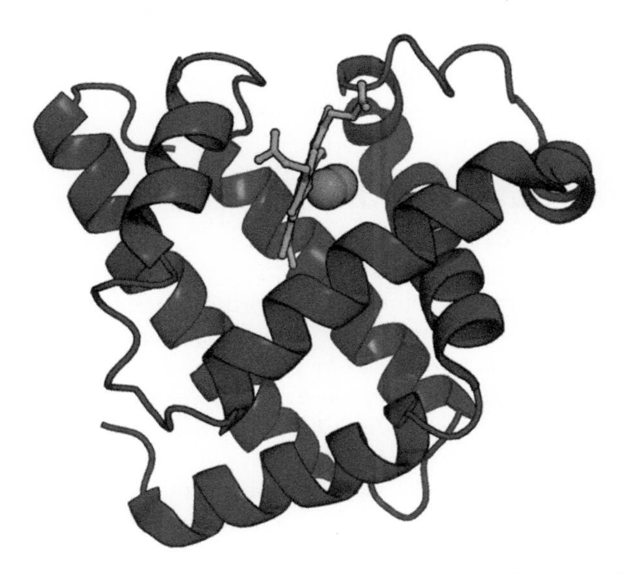

Figure 4.2 *The structure of myoglobin determined by Kendrew* et al.*, showing high prominence of the α-helix (see also Chapter 5).*
(Copyright MRC-LMB)

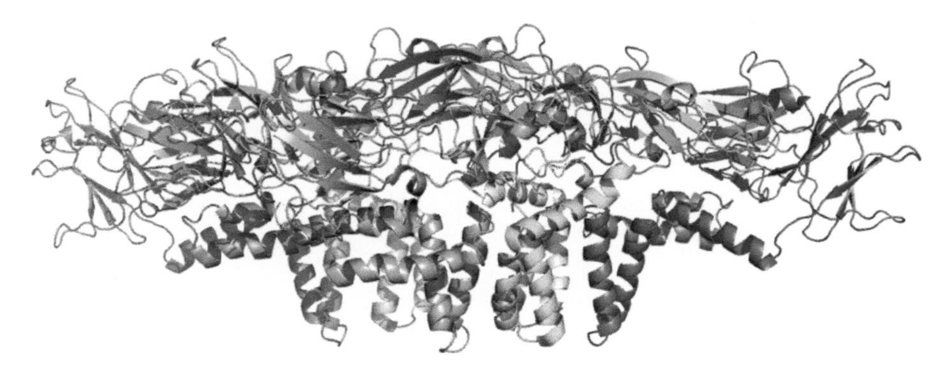

Figure 4.3 *(a) Representation of the structure of the Zika virus showing both α-helices and β-sheets. (By kind permission of Dr Balaji Santhanam*[12]*)*

presented by the authors with descriptions of protein structures in terms of the post-Pauling model. Thus, the structures of myoglobin and haemoglobin, as well as that of the enzyme lysozyme, which are discussed fully in the next chapters, are shown with their prominent α-helices. Figure 4.2 shows myoglobin, with its high preponderance of α-helices.

Later examples, cited here, are representations of the structure of insulin and of the Zika virus determined by Rossman *et al.*[12] (see Figure 4.3).

At the time (1951) that Pauling wrote these definitive, detailed articles, he was at the pinnacle of his creative work. He produced this stream of powerful papers in structural molecular biology when he was under great emotional strain, as he was fighting accusations levelled against him in the US of disloyalty to his country, owing to the pressures of the Cold War and what became known as McCarthyism.[13] We return to this matter at the end of this chapter.

4.2 A Brief Outline of the Erroneous Paper by Bragg, Kendrew, and Perutz

This paper [1] begins in the customary lucid prose associated with the writings of all these three eminent scientists: '*Astbury's studies of α-keratin, and X-ray studies of crystalline haemoglobin and myoglobin by Perutz and Kendrew, agree in indicating some form of folding polypeptide chain which has a repeat distance of about 5.1 Å, with three amino acid residues per repeat. In this paper a systematic survey has been made of chain models which conform to established bond lengths and angles, and which are held in folded form by N-H…O bonds.*' Figure 4.1 shows the picture reproduced in ref [1] that Astbury proposed for α- and β-keratins in his 1949 paper—compare with Figure 3.12 of Chapter 3. In the conclusions of their paper, Bragg, Kendrew, and Perutz[1] were quite candid (and both tentative and prophetic) about their views. They ended their abstract with the words: '*The evidence is still too slender for definite*

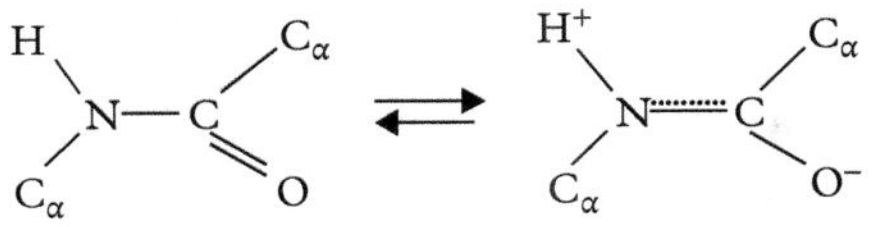

Figure 4.4 *Pauling's representation of the partial double bond joining the nitrogen and carbon atoms. This explains why the peptide link leads to planarity of the atoms attached to it.*

(L. Pauling, R. B. Corey, and H. R. Branson, Proc. Natl. Acad. Sci. U. S. A., ***1951****,* 37, 205*)*

conclusions to be drawn, but it indicates that a further study of these proteins, and in particular, of myoglobin which has promising features of simplicity, may lead to a determination of the chain structure.'

In retrospect, protein crystallographers, especially Pauling and his co-workers at the California Institute of Technology, recognized that Bragg, Kendrew, and Perutz had committed two major errors in their otherwise plausible arguments. First, they ignored the fact that the chemical groups attached to the N–C peptide link (see Figure 3.6 in Chapter 3) have to be planar because of the existence of the phenomenon of ionic–covalent resonance (see Figure 4.4). (Note that structural biologists use the 'peptide link' and the 'amide link' as synonymous.)

And, second, they assumed that, in their helical structures, integral repeats of the residues occurred along the axis of the helix. It is important to drive home this last point. Crystals of proteins are well known to exhibit both n-fold *rotation* axes and also n-fold *screw* axes. The first refers to the situation where the structure is identical after being rotated through an angle of $360°/n$ about the axis. On the other hand, a crystal has an n-fold screw axis if the structure appears identical after first rotating through an angle of $360°/n$ about the axis, and then translating it a certain distance parallel to the axis. The erroneous paper by Bragg *et al.*[1] made the assumption that a screw axis had a value of n that was an integer. As we shall see shortly, Pauling made a different assumption, based on his superior knowledge of the nature of polypeptides generally.

4.3 The Discovery of the α-Helix and β-Sheet as the Principal Motifs of the Structure of Proteins

Most of the 150,000 or so structures now deposited in the Protein Data Bank[13,14] reveal the ubiquity of the α-helix, especially, and to a lesser degree, the β-sheet formulations proposed in the landmark papers from the California Institute of Technology in 1951.[2–9] This work, largely by Pauling and Corey, had a significance for protein structure comparable to that of Watson and Crick, two years later, when they proposed the double-helix structure of DNA. Watson and Crick used the model-building approach of the Caltech workers, aided by some hard experimental data pertaining to bond lengths and bond angles for covalent bonds and for hydrogen bonds, as well as known features pertaining to their flexibility and distortability. In the opening paragraph of the Pauling, Corey, and Branson paper,[2] the authors stated: '*We have been attacking the problem of the structure of proteins in several ways. One of these ways is the complete and accurate determination of the crystal structure of amino acids, peptides, and other simple substances related to proteins, in order that information about interatomic distances, bond angles, and other configurational parameters might be obtained that would permit the reliable prediction of reasonable configurations of the polypeptide chain.*'

Pauling, Corey, and Branson[2] then asserted: '*The problem we have set ourselves is that of finding all hydrogen-bonded structures for a single polypeptide chain, in which the residues are equivalent (except for the differences in side chain R). An amino acid residue (other than glycine) has no symmetry elements. The general operation of conversion of one residue of a single chain into a second residue equivalent to the first is accordingly a rotation about an axis accompanied by translation along the axis. Hence, the only configurations for a chain compatible with our postulate of equivalence of the residues are helical configurations. For rotational angle 180° the helical configurations may degenerate to a simple chain with all of the principal atoms, C, C^1 (the carbonyl carbon), N, and O, in the same plane.*'

'*We assume*', said Pauling and Corey, '*that because of the resonance of the double bond between the carbon–oxygen and carbon–nitrogen positions the configuration of each*

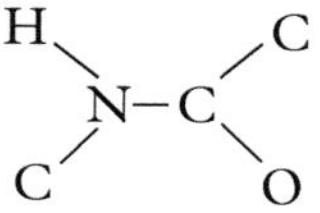

```
H       C
  \    /
   N—C
  /    \
C       O
```

residue is planar...The observed C–N distance, 1.32 Å, corresponds to nearly 50 per cent double-bond character, and we may conclude that rotation by as much as 10° from the planar configuration would result in instability by about 1 kcal mole^{-1}. The interatomic distances and bond angles within the residue are assumed to have the values shown in Figure 4.5.'

4.3.1 What prompted Pauling to think of the α-helix?

Fortunately, in 1993, Linus Pauling published an article entitled '*How My Interest in Proteins Developed*' (see Figure 4.6), so we have exactly his thought processes that led him to one of his greatest contributions to biochemistry and structural molecular biology.

As early as 1937, he had become interested in the work of Astbury, started at the DFRL of the Royal Institution and later at Leeds, in the X-ray photography of α-keratin (hair and related proteins) and β-keratin (silk and stretched hair). He was especially intrigued by the repeat distance of 5.1 Å, which would correspond to two residues in a polypeptide chain extended along the length of the hair. At that time, Pauling already knew that the carbon–nitrogen (C–N) bond in the main '*chain of the proteins would have some double-bond character (C–N) stolen from the carbonyl groups, and that as a result would consist of planar groups joined to one another at the alpha-carbon atoms*' (see Figure 4.4)...'*Also, I was sure that the N–H and*

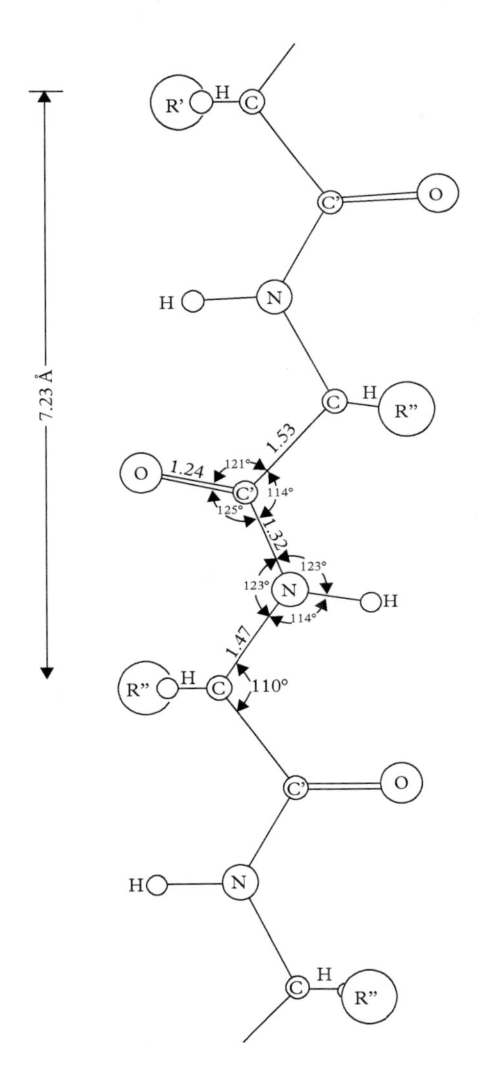

Figure 4.5 *A diagrammatic representation of a fully extended polypeptide chain, with the bond lengths and bond angles derived from crystal structures and other experimental evidence (after R. B. Corey[27]).*

(Courtesy of the Royal Society)

O=C groups would form hydrogen bonds extending in the direction of the axis of the keratin molecules. Despite my efforts to use this information to find the structure that would repeat with two amino acid residues in 5.1 Å, I was unsuccessful.'

In 1937, Pauling was joined by the expert crystallographer Robert Corey, who worked out precisely from that time onwards (to well beyond the end of World

Figure 4.6 *Linus Pauling with a model of his α-helix.*[15]
(Courtesy Ava and Linus Pauling Archives, Oregon State University, Corvallis)

War II) at Caltech the detailed structures (by X-ray studies) of amino acids like glycine, N-acetylglycine, *β*-glycylglycine, etc. So Pauling had access to a collection of precise bond lengths and bond angles involved in the atoms forming the polypeptide chain (see Figure 4.5).

But it was in March 1948, while ill in bed with flu during his tenure of the Eastman Visiting Professorship at the University of Oxford, that Pauling decided to think again about *the α-keratin* structure. In his own words:[15] *'I did not have X-ray photographs of α-keratin with me, and I decided to ignore the X-ray data and depend solely on my knowledge of structural chemistry. I decided to make the assumption that all the amino acid residues in the polypeptide chain of α-keratin are structurally equivalent, with the different side chains* [see Figure 3.12 of Chapter 3] *not exerting a significant perturbation. I remembered that as a graduate student I had heard Professor Harry Bateman state that the operation that connects an object into an equivalent object anywhere in space is a translation along an axis coupled with a rotation around the axis. I knew that to repeat this operation would give a helix. By making a drawing of a polypeptide chain on a sheet of paper and folding the paper on parallel lines passing through the alpha-carbon atoms, I tried to bring the N–H group and the O=C group into the proper orientation and distances from one another to correspond to the formation of an acceptable hydrogen bond, with the N–H...O distance about 2.8 Å. It took me a couple of hours to find this structure and to make calculations about the repeat distance. In fact, it turned out there were not two residues, but 3.6 residues in the repeat distance of the helix (the pitch of the helix) and that the pitch could be predicted to have a value close to 5.4 Å.*

This was quite satisfying to me, because this structure, which I called the alpha helix, provided a beautifully simple explanation of the properties of keratin. There was a serious difficulty, however: the X-ray photographs seemed to give a repeat distance of 5.1 Å. The discrepancy troubled me to such an extent that I did not publish anything about the alpha helix for about a year and a half, because I did not want to publish an incorrect structure and I could not explain the diffraction photographs.'

Towards the end of his article, Pauling says that after he returned to Pasadena, he noted the paper published by Bragg, Kendrew, and Perutz,[1] in which they described a number of helical structures of polypeptide chains obtained in their search for an α-keratin structure; Pauling opined: '*From my point of view, all these structures were wrong, because they did not involve planarity around the nitrogen atom. I thought it was likely, however, that in the course of time they would learn enough chemistry to see that peptide groups had a planar structure, and would discover the α-helix, so Professor Corey and I decided to publish a short description of the α-helix in the Journal of the American Chem. Soc.*[16] *We followed this publication by a more detailed description of the α-helix with Herman Branson and by some other papers including descriptions of the β-sheets, pleated sheets involving parallel extending chains of polypeptides.*'[2–9]

Before we proceed to describe the quintessential details of the α-helix and the β-sheets, it is worth emphasizing that Pauling's remark '*...they would learn enough chemistry...*' irked all of the Cambridge authors, especially Perutz and Kendrew, each of whom were chemistry graduates! (In the light of this criticism, we shall return to the subsequent action of Perutz later in this chapter.)

4.3.2 Pictorial illustrations of the α-helix and β-sheets

Figure 4.7 shows the essence of the α-helix. The two illustrations in Figure 4.7 are from a later article by Crick and Kendrew.[17] Note that here are shown both left-handed and right-handed α-helices. (Oddly, Pauling, Corey, and Branson[2] drew a left-handed helix, which is not what occurs in a polypeptide—see Dunitz[18].)

Apart from the accurate bond distances and bond angles that Pauling and Corey had accumulated for the backbone structure (see Figure 4.5) of the polypeptide, they incorporated intra-chain hydrogen bonds that gave rise to a strain-free helix. The cartoon versions of the α-helix, used universally in current structural biological publications, are depicted in Figure 4.8. In summary, the α-helix is a common motif in the structure of proteins, and it is a right-handed spiral confirmation (i.e. a helix) in which every backbone N–H group donates a hydrogen bond to the backbone C=O group of the amino acid located three or four residues earlier along the protein sequence. The α-helix is sometimes denoted as a 3.6_{13}-helix, as it signifies the average number of residues per helical turn, with thirteen atoms being involved in the ring formed by the hydrogen bond.

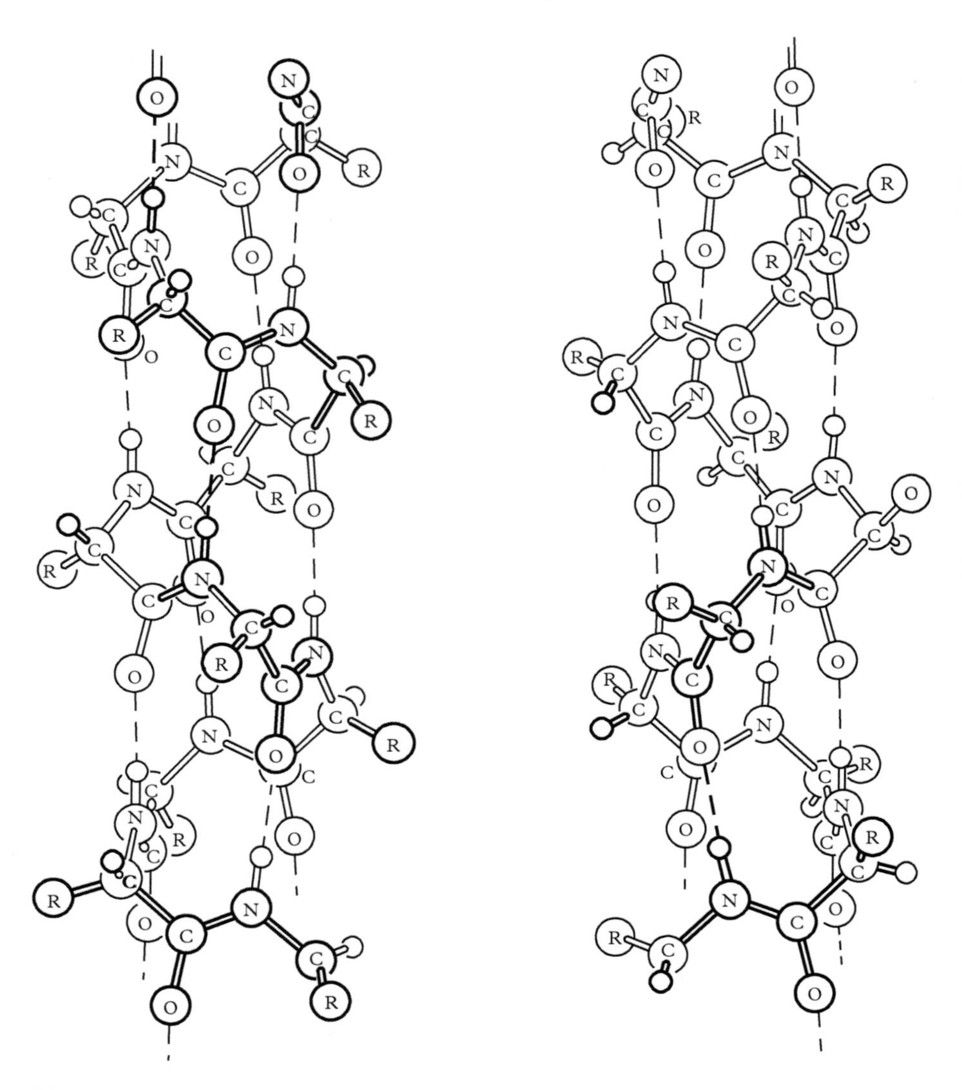

Figure 4.7 *Schematic model of the left-handed and right-handed α-helices. The R groups on the α-carbon atom are in the correct position, corresponding to the known configuration of the L-amino acids in proteins (after Crick and Kendrew*[17]*).*

(Courtesy of Elsevier)

The fundamental difference between the Pauling–Corey–Branson[2–9] α-helix structure and the earlier proposed possibilities by Bragg, Kendrew, and Perutz[1] is that there are non-integer numbers of residues per helical turn and the planarity of atoms is associated with the peptide link (the amide bond).

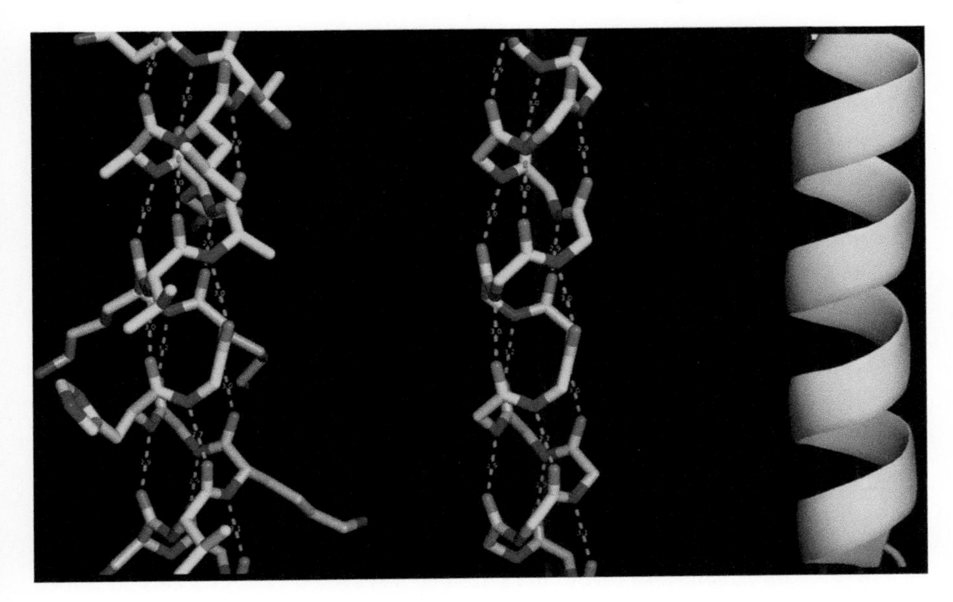

Figure 4.8 *Three representations of the α-helix. (Left) The complete α-helix, (centre) the main chain only, and (right) the cartoon version.*

(By kind permission of Dr K. Nagai)

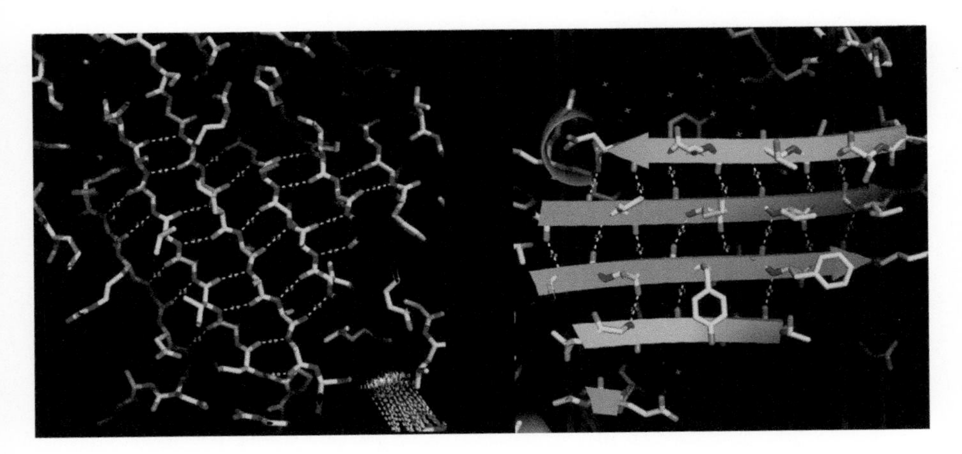

Figure 4.9 *Illustration of the β-sheets structure proposed by Pauling and Corey.*

(By kind permission of Dr K. Nagai)

The β-pleated sheets proposed by Pauling *et al.*[2–9] involving inter-chain hydrogen bonding are shown in Figure 4.9. Here we see both anti-parallel pleated sheets, as well as parallel ones. These are also described in Crick and Kendrew's paper.[17]

Professor David Eisenberg[13] has given an admirable summary of the discovery of the α-helix and β-sheet, and the reader is recommended to consult his article.

4.4 Crick's Comments on the Repeat Distance of the α-Helix and Perutz's Important Observations

The discrepancy between the observed 5.1 Å repeat, reported from the time of Astbury[19] as a feature of α-keratin, in porcupine quill, hair, horn, muscle, and related proteins and the value of 5.4 Å repeat that Pauling had arrived at drew the attention of Crick,[20,21] who pointed out that the difference in spacing is not an insuperable discrepancy, since the α-helix might be more tightly coiled. What Crick proposed was that the α-helix might be deformed into a so-called coiled-coil, as the energy involved in such a deformation was likely to be small. In Crick's view,[20,[21] '*if the side chains of the α-helix* [the R groups—see Figure 3.8 of Chapter 3) *are thought schematically as knobs on the surface of a cylinder, then it is found that the pattern on this surface consists of knobs alternating with "holes", that is, spaces into which the knobs from a neighbouring α-helix could fit.*' In his article on the packing of α-helices,[21] he shows the following diagram (see Figure 4.10).

An illustration of a coiled-coil drawn by Pauling and Corey[22] is shown in Figure 4.11.

These arrangements take care of the puzzling 5.4 Å versus 5.1 Å repeats described earlier. An idealized α-helix when packed in the coiled-coil manner yields a repeat distance of 5.1 Å.

4.4.1 Perutz's important observation concerning the α-helix

Max Perutz has described,[23] in one of the most exciting and evocative passages in the literature of structural molecular biology, how he felt on a Saturday morn-

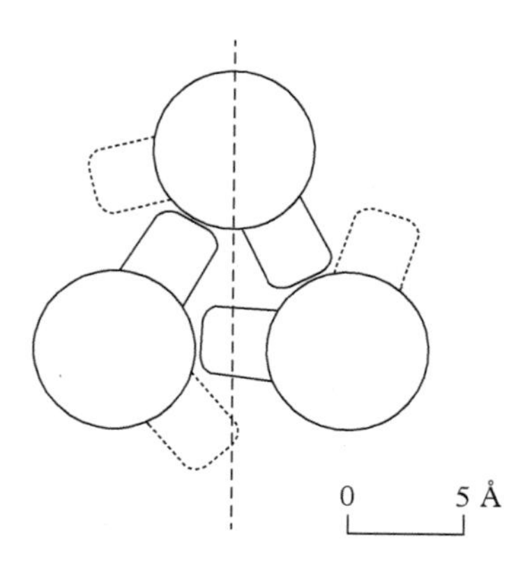

Figure 4.10 *Schematic diagram to illustrate the packing of the three-strand rope. The figure is a section perpendicular to the fibre axis. The circles represent the main polypeptide chains. The knobs represent side chains. Only the inner side chains are shown. The dotted side chains are at a slightly different level from those drawn with solid lines.*[21]

(Reproduced with permission of the International Union of Crystallographers)

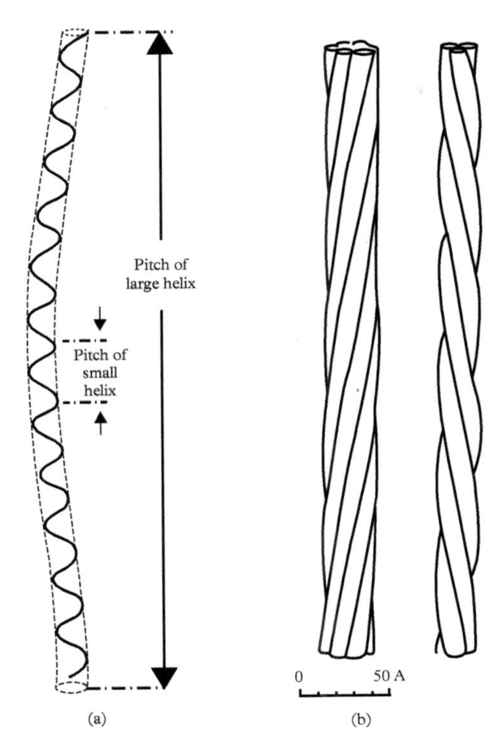

Figure 4.11 *Illustration of the general idea of a 'coiled-coil' or 'compound helix'. The figure on the left shows a single polypeptide chain. The small helix is supposed to be an α-helix, the axis of which has been distorted so that it follows a larger, more gradual helix. The figure on the right shows two possible ways of combining helices into a rope.*

(Prepared by Pauling and Corey[22] *and cited by Crick and Kendrew*[17]*) (Copyright Elsevier)*

ing shortly after his paper with Bragg and Kendrew[1] appeared in the *Proceedings of the Royal Society* in 1950. '*I went to the Cavendish Library and in the latest issue of PNAS found a series of papers by Linus Pauling together with the crystallographer R. B. Corey. In their first paper they proposed an answer to the long-standing riddle of the structure of α-keratin, suggesting that it consisted of helical polypeptide chains with a non-integral repeat of 3.6 amino acid residues per turn.*

I was thunderstruck by Pauling and Corey's paper. In contrast to Kendrew's and my helices, theirs was free of strain; all the amide groups were planar and every carbonyl group formed a perfect hydrogen bond with an amino group four residues further along the chain.

The structure looked dead right. How could I have missed it? Why had I not kept the amide groups planar? Why had I stuck blindly to Astbury's 5.1 Å repeat? On the other hand, how could Pauling and Corey's helix be right, however nice it looked, if it had the wrong repeat? My mind was in a turmoil. I cycled home to lunch and ate it oblivious of my children's chatter and unresponsive to my wife's inquiries as to what the matter was with me today.

Suddenly I had an idea. Pauling and Corey's α-helix was like a spiral staircase in which the amino acid residues formed the steps and the height of each step was 1.5 Å. According to diffraction theory, this regular repeat should give rise to a strong X-ray reflection of 1.5 Å spacing from planes perpendicular to the fiber axis. As far as I knew, such a reflection had never been reported, either from 'natural' proteins like hair and muscle or from synthetic polypeptides. Hence, I concluded, the α-helix must be wrong.

But wait! Suddenly I remembered a visit to Astbury's laboratory and realised that the geometry of his X-ray setup would have precluded observation of the 1.5 Å reflection because he oriented his fibers with their long axes perpendicular to the X-ray beam, while observation of the 1.5 Å reflection would have required inclining them at the Bragg angle of 31°. Furthermore, Astbury used a flat plate camera that was too narrow to record a reflection deflected from the incident X-ray beam by 2 × 31°.

In mad excitement, I cycled back to the lab and looked for a horse hair that I had kept tucked away in a drawer. I stuck it on a goniometer head at an angle of 31° to the incident X-ray beam; instead of Astbury's flat plate camera I put a cylindrical film around it that would catch all reflection with Bragg angles of up to 85°.

After a couple of hours, I developed the film, my heart in my mouth. As soon as I put the light on I found a strong reflection at 1.5 Å spacing, exactly as demanded by Pauling and Corey's α-helix. The reflection did not by itself prove anything, but it excluded all alternative models that had been put forward by ourselves and others and was consistent only with the α-helix.

On Monday morning I stormed into Bragg's office to show him my X-ray diffraction picture. When he asked me what made me think of this crucial experiment, I told him that the idea was sparked off by my fury over having missed building that beautiful structure myself. Bragg's prompt reply was, "I wish I had made you angry earlier!" because discovery of the 1.5 Å reflection would have led us straight to the α-helix.

The inconsistency between the 5.4 and 5.1 Å continued to worry me until one morning about two years later when Francis Crick arrived at the lab with two rubber tubes around which he had pinned corks with a helical repeat of 3.6 corks per turn and a pitch of 5.4 centimeters. He showed me that the two tubes could be wound around each other to make a double helix such that the corks neatly interlocked. This shortened the pitch of the individual chains, when projected onto the fiber axis, from 5.4 to 5.1 centimeters, as required by the X-ray pattern of α-keratin.

Such a double helix was eventually found by my colleagues A. D. McLachlan and J. Karn in the muscle protein myosin.'

4.5 Further Reflections on the Pauling–Perutz Saga Pertaining to the α-Helix

The concise paper that Max Perutz published[24] in June 1951, entitled '*New X-ray Evidence on the Configuration of Polypeptide Chains*', settled beyond doubt the essential validity of the α-helix motif in protein structures. Shortly thereafter, Perutz sent a copy of his work to Pauling, whom he thought would be pleased, '*but no, he attacked me furiously, because he could not bear the idea that someone else had thought of a test for the α-helix of which he had not thought himself. I was glad when he forgot his anger later and he became a good friend.*'

The above passage appeared in Max Perutz's Obituary of Pauling in *Nature, Structural Biology* in 1994.[25] It is a magnificent account of the work of one of the greatest scientists ever. In it, Perutz contrasts the attitude of Pauling and Einstein when their work was criticized, '*When anybody contradicted Einstein, he thought it over, and if he found he was wrong, he was delighted, because he felt he had escaped from an error, and that he now knew better than before, but Pauling would never admit that he might have been wrong.*'

There is another incident which underlines the rivalry between the Cavendish crystallographers and Pauling's group in Caltech. Shortly after Pauling had discovered (in his bed in Oxford) the essence of the essential structure of the α-helix in 1948, he visited Kendrew and Perutz in Cambridge. '*Ignorant of his Oxford experiment, I proudly showed him my three-dimensional patterns of haemoglobin which indicated that its polypeptide chain was folded in the same way as Astbury's fibres, but to my disappointment Pauling made no comment; he did not announce his discovery until the delivery of a dramatic lecture at Pasadena in the following year.*'[25]

In the elegant, pedagogically illuminating article by David Eisenberg[13] on the discovery of the α-helix and β-sheet by Pauling, Corey, and Branson, there is a host of fascinating facts that relate to the continuing progress of structural molecular biology. While Eisenberg draws attention to the fact that the bond lengths of Pauling *et al.* were not surpassed in accuracy for over forty years, he also points out that these workers did not consider the hand of the helix, as mentioned above—they drew a left-handed, rather than a right-handed, one. They also did not allow for the possibility of bent sheets. Furthermore, Pauling *et al.*, in their landmark paper, proposed structures and functions that have not been found, including the so-called δ-helix.

In Eisenberg's article, he also draws attention to the fact that, nowadays, structural molecular biologists accept without a second thought that helices do not need to have an integral number of monomer units per turn. In the 1950s, however, the crystallographic backgrounds of the Cavendish trio had saddled them with the notion of integral members of units per unit cell.

Eisenberg rightly regards the stream of definitive papers that emanated from Pauling's laboratory in 1951 as all the more remarkable when one considers the acute emotional strain (alluded to above) he was undergoing as a result of accusations from his detractors in *The House Un-American Activities Committee.* Pauling was subpoenaed to appear before various anti-communist investigation committees. He received hate mail for his work on liberal causes, and he faced cancellation of his major consulting contracts and coolness from some Caltech colleagues.[26]

On the day after Pauling and Corey submitted their seven papers for publication, *The House Un-American Activities Committee* named Pauling one of the foremost Americans involved in a 'Campaign to Disarm and Defeat the United States'. The press release read: '*His whole record indicates that Dr Linus Pauling is primarily engrossed in placing his scientific attainments at the service of a host of organisations which have in common their subservience to the Communist Party of the USA and Soviet Union.*'

As Eisenberg concludes, '*Somehow, even in the face of such false invective and multiple distractions, Pauling could maintain his focus as a top creative scientist.*'

Pauling was denied permission to travel to London for a Royal Society Discussion Meeting on Proteins in 1951, when he was scheduled to be the opening speaker. The paper was presented by Corey[27] and was entitled '*Fundamental Dimensions of Polypeptide Chains*'. In Chapter 7, we return again briefly to the emotional strain suffered by Pauling in this productive period of his work in structural molecular biology.

Two general points relating to α-helices are worthy of mention here. In answer to the question of why such structures exist, two responses are noteworthy—using post-event rationalization. First, the α-helix is a structural motif that beautifully satisfies much of the chemical requirements involved in forming a structure from a polypeptide chain. And it was the perceptive intelligence of Pauling that led to its formulation. Second, the α-helix can exhibit both hydrophobic surfaces and hydrophilic regions—in other words, it may display so-called amphipathic behaviour. Coiled-coil α-helices can produce more stable tertiary structures with hydrophobic cores. It has been frequently observed[27,28] that α-helical coiled-coils are ubiquitous protein-folding and protein interaction domains in which two or more α-helical chains come together to form bundles.[28]

Recent work on polypeptide chains rich in α-helices has uncovered the mysterious phenomenon of knotted helices, as shown in Figure 4.12, taken from the work of my Peterhouse colleague Professor Sophie Jackson.[29] At present, it is difficult to ascertain what advantages, if any, such a contorted structure confers on the protein (see also Section 9.7).

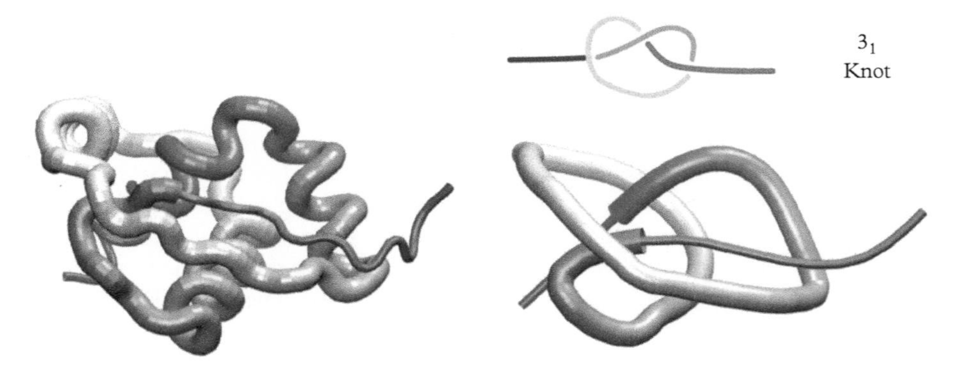

Figure 4.12 *Nowadays several protein structures have been found to contain knotted* α*-helices, like the trefoil example shown here.*[30]

(Courtesy Professor S. E. Jackson)

REFERENCES

1. W. L. Bragg, J. C. Kendrew, and M. F. Perutz, *Proc. Roy. Soc. A*, **1950**, *203*, 321.
2. L. Pauling, R. B. Corey, and H. R. Branson, *Proc. Natl. Acad. Sci. U. S. A.*, **1951**, *37*, 205.
3. L. Pauling and R. B. Corey, *Proc. Natl. Acad. Sci. U. S. A.*, **1951**, *37*, 235.
4. L. Pauling and R. B. Corey, *Proc. Natl. Acad. Sci. U. S. A.*, **1951**, *37*, 241.
5. L. Pauling and R. B. Corey, *Proc. Natl. Acad. Sci. U. S. A.*, **1951**, *37*, 251.
6. L. Pauling and R. B. Corey, *Proc. Natl. Acad. Sci. U. S. A.*, **1951**, 37, 261.
7. L. Pauling and R. B. Corey, *Proc. Natl. Acad. Sci. U. S. A.*, **1951**, 37, 272.
8. L. Pauling and R. B. Corey, *Proc. Natl. Acad. Sci. U. S. A.*, **1951**, 37, 282.
9. L. Pauling and R. B. Corey, *J. Am. Chem. Soc.*, **1952**, *74*, 3694.
10. J. C. Kendrew, G. Bodo, H. M. Dintzis, R. G. Parrish, H. Wyckoff, and D. C. Phillips, *Nature*, **1958**, *181*, 662.
11. D. C. Phillips, *Proc. Natl. Acad. Sci. U. S. A.*, **1967**, *57*, 484.
12. D. Sirohi, Z. Chen, L. Sun, T. Kloze, T. C. Pierson, M. G. Rossmann, and R. J. Kuhn, *Science*, **2016**, *352*, 467.
13. D. Eisenberg, *Proc. Natl. Acad. Sci. U. S. A.*, **2003**, *100*, 11207.
14. Protein Data Bank, http://www.pdb.org/pdb/home/home.do
15. L. Pauling, *Protein Sci.*, **1993**, *2*, 1060.
16. L. Pauling and R. B. Corey, *J. Am. Chem. Soc.*, **1950**, *72*, 5349.
17. F. H. C. Crick and J. C. Kendrew, *Adv. Protein Chem.*, **1957**, *12*, 133.
18. J. D. Dunitz, *Angew. Chemie. Int. Ed.*, **2001**, *40*, 4167.
19. W. T. Astbury, *Br. J. Radiol.*, **1949**, *22*, 259.
20. F. H. C. Crick, *Nature*, **1952**, *170*, 882.
21. F. H. C. Crick, *Acta Cryst.*, **1953**, *6*, 689.
22. L. Pauling and R. B. Corey, *Nature*, **1953**, *171*, 59.
23. M. F. Perutz, '*I Wish I'd Made You Angry Earlier*', Oxford University Press, Oxford, **2002**.
24. M. F. Perutz, *Nature*, **1951**, *167*, 929.
25. M. F. Perutz, *Nature Struct. Biol.*, **1994**, *1*, 667.
26. T. Hager, '*A Force of Nature: The Life of Linus Pauling*', Simon and Schuster, New York, NY, **1995**, p. 379.
27. R. B. Corey and L. Pauling, *Proc. Roy. Soc. B*, **1953**, *141*, 10.
28. D. N. Woolfson in '*Fibrous Proteins: Structures and Mechanisms*' (eds. D. A. D. Parry and J. M. Squire), Springer International Publishing AG, Cham, **2017**, p. 35.
29. S. E. Jackson, private communication, June **2018**.
30. S. E. Jackson, A. Suma, and C. Micheletti, *Curr. Opin. Struct. Biol.*, **2017**, *42*, 6.

5

Perutz and Kendrew: The Heroic Era of Structural Molecular Biology

5.1 Introduction

Nowadays, with the ready availability of ever more powerful synchrotron radiation sources covering wide ranges of X-ray wavelengths and with ultra-powerful computers and associated, well-proven rapid programs, based on anomalous absorption of X-rays for the determination of the phase of an X-ray diffracted beam, comparatively little difficulty is encountered experimentally—compared with the methods described in Chapter 2—in determining the detailed structure of protein crystals. Small wonder, therefore, that the international repository of experimentally obtained protein structures—the branch of the Protein Data Bank in Rutgers University in the United States—now contains some 150,000 detailed structures.

When, however, Max Perutz and John Kendrew set out on their mammoth and marathon intellectual journey to determine the three-dimensional folded (or coiled) structures of haemoglobin and myoglobin in the late 1940s, the situation was very different. Their work, culminating in success in the period 1957–1960, constitutes a major landmark in macromolecular structural chemistry, and it had repercussions throughout the corpus of molecular biology. Indeed, as late as 1965, when their mentor Sir Lawrence Bragg addressed an audience of physicists, he reflected[1] upon the fact that more than five years had elapsed since the epic achievements of Perutz and Kendrew and that no other protein had had its detailed structure solved—such was the measure of the technical virtuosity of Perutz and Kendrew.

In this chapter, we first recall why it was that haemoglobin and myoglobin were the targets at which these two pioneers aimed. It is a fact that when they embarked on their investigations, no other physico-chemical technique, other than X-ray diffraction, stood a chance of solving the structures of these giant molecules; haemoglobin contains some 10,000 and myoglobin some 2500 atoms of carbon, nitrogen, oxygen, sulfur, and, more important than all the others, iron. Chemical methods, like an extension of the work of Sanger (see Section 3.3.3) on insulin,[2] which were able to determine the sequence of the amino acid units that formed the polypeptide chains of these materials had slowly become available to them. But such sequences do not give clues to the three-dimensional structure, and hence the function, of a protein. Only the X-ray route involving diffraction offered a solution to their problems. (It is interesting to note that, in his Nobel Lecture of 1962,[3] John Kendrew expressed the view that, in due course, once the sequence of the amino acids is known, it would be possible to predict the three-dimensional structure of the protein. This is still not possible today, although much encouraging progress has recently been reported. At the University of Washington, Seattle, the group led by David Baker claims to be able to predict the precise folded structures of proteins containing up to sixty amino acid residues. Enormous computational efforts are currently being made to predict the stable three-dimensional structure of larger proteins.[4]) Using the recently available procedures of machine learning and artificial intelligence, such as the so-called *Deep Mind*, results of impressive quality have emerged.

Another noteworthy point is that, nowadays, chemists use not only X-ray diffraction, but also nuclear magnetic resonance, as well as sophisticated mass spectrometry, and increasingly cryo-electron microscopy (for this yields the spatial atomic arrangements, as explained in Chapter 10) to determine the structure of materials of interest to molecular biologists. In the heroic days of Perutz and Kendrew, however, these alternative techniques were simply not at their disposal.

In the ensuing sections of this chapter, we first recall the cardinal importance of haemoglobin and myoglobin; we then summarize the strategy and tactics deployed by Perutz and Kendrew. And, finally, we outline some of the molecular biological consequences of their work.

5.2 Prior Knowledge of Haemoglobin and Myoglobin

As mentioned in the Preface, each human being carries around her or his own supply of haemoglobin, amounting to approximately one per cent of the body weight. This supply is in the red corpuscles of blood, and haemoglobin is the pigment of blood. It has a beautiful red colour in arterial blood, and a purple colour in venous blood. The red corpuscles are flattened discs of about 70,000 Å in diameter and 10,000 Å in thickness. A single red blood cell contains about 250 million molecules of haemoglobin, and the cells are suspended in the plasma of blood and constitute about one-third of blood itself. The molecular weight of haemoglobin—determined early on by osmotic pressure measurements and then from density and unit cell measurements (by X-rays, as explained in Sections 2.3 and 3.5.1)—is 68,000, compared to forty-six for ethyl alcohol, seventy-eight for benzene, and 416 to vitamin D. Each of the iron atoms in both the metalloproteins, haemoglobin, and myoglobin lies at the centre of a group of atoms that form a pigment called haem, the structure of which was worked out by the German chemist Hans Fischer and is shown in Figure 5.1.

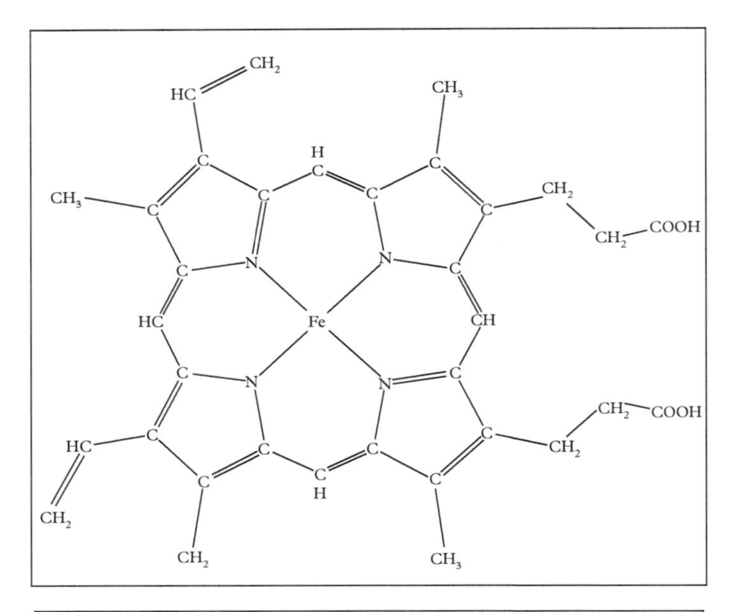

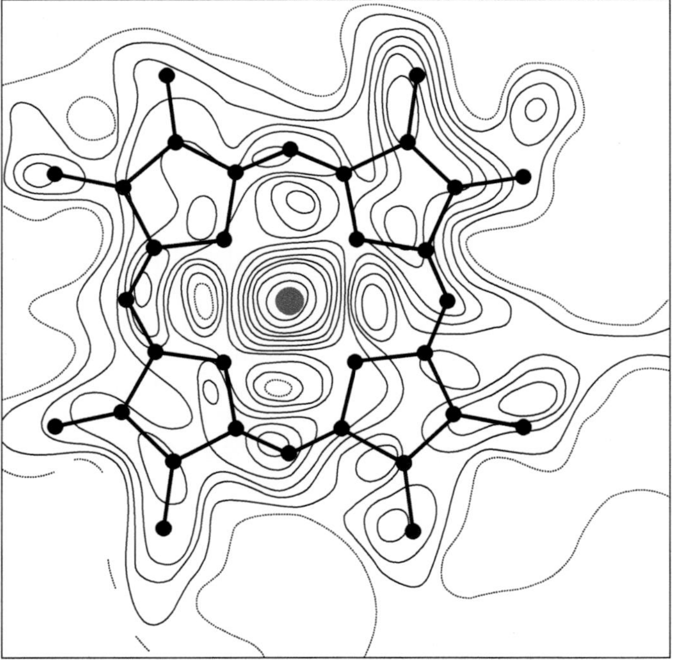

Figure 5.1 *(Top) Skeletal outline of the haem group and (bottom) the electron density distribution in the haem group of myoglobin. At the centre of the organic molecule is an iron atom. Four of these are present in haemoglobin, and one in myoglobin.*

(Courtesy J. C. Kendrew/MRC-LMB)

Haem is what confers a red colour to blood, as well as its ability to combine with oxygen.

The main work done by blood is to carry oxygen from the lungs to tissues, and carbon dioxide and other products of the breakdown of tissues and foods back to the lungs and excretory organs. A haemoglobin molecule can combine with four molecules of oxygen; the resultant oxyhaemoglobin is bright red. In tissues, where the partial pressure of oxygen is less than that in the lungs, it gives up part of its oxygen, which is then used in oxidation reaction of various kinds. The carbon dioxide produced by this oxidation is then carried by blood back to the lungs and released by exhalation.

The four molecules of oxygen that are taken up by haemoglobin attach themselves to the four iron atoms in haem. The nature of the bonds in the haemoglobin molecule was elucidated in 1936 by Pauling and Coryell[5] in the study of magnetic properties, following on from the pioneering work of Faraday (see Section 3.2). Magnetic studies showed that venous blood is paramagnetic—that is, it is attracted into a magnetic field—whereas arterial blood is diamagnetic.

There are many different kinds of haemoglobin. All vertebrate animals and many invertebrate ones use haemoglobin as an oxygen carrier, and the molecule is slightly different for every animal from that of others. As pointed out earlier, all healthy human beings contain precisely the same amino acid constituents and sequences in their polypeptide chains (the globins), and the haem is also the same for all haemoglobins. (Later in this chapter, we shall briefly discuss haemoglobin molecules of an abnormal kind that give rise to a so-called molecular disease.)

Apart from the many contributions that Perutz and Kendrew had made, using a multiplicity of physico-chemical techniques, concerning the nature of haemoglobin and myoglobin, many eminent early investigators had made several prior crucial observations. Thus, the Cambridge physiologist Barcroft, who suggested to Perutz that Kendrew should compare adult and fetal sheep blood, was the first to discover the striking sigmoid shape of the oxygen equilibrium curve of haemoglobin (see Figure 5.2).

Other eminent scientists, such as A. V. Hill, J. S. Haldane, and his son J. B. S. Haldane, were among the first to attempt to explain the nature of this curve (see Section 5.2.1). The Danish workers Bohr and Hasselbalch elucidated the importance of acid groups in haem to account for haemoglobin's ability to carry carbon dioxide (Christian Bohr was the father of the famous Niels Bohr)—and it was the Harvard scientist J. B. Conant who first established that the iron in functional haemoglobin is in the ferrous (Fe^{2+}) state. (To most chemists, it seems, at first, enigmatic that, upon exposure to oxygen, the iron of haem remains in the Fe^{2+} state. As we shall discover, it is the interaction between the haem and one of the amino acids of the polypeptide chain that is responsible for this behaviour.)

Myoglobin is also a protein like haemoglobin and is found in the muscles of animals, and it acts as an oxygen storage unit for the oxygen brought by haemoglobin and provides oxygen when required by the working muscles. It is, as Kendrew used to say, a junior relative of haemoglobin, being a quarter of its size and containing a single polypeptide chain of about 150 amino acid units, together

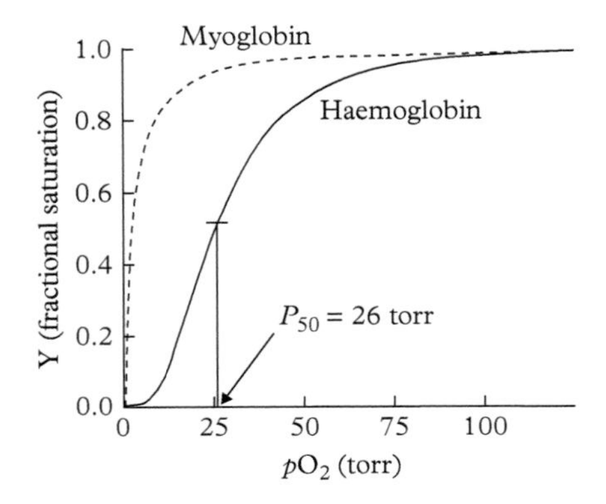

Figure 5.2 *The curve obtained for haemoglobin when it takes up oxygen as a function of oxygen pressure is 'S'-shaped and termed sigmoid. It was realized early on that a sigmoid shape indicates that distinct, but interacting, oxygen-binding sites are present in each haemoglobin molecule. Note that half saturation for haemoglobin occurs at 26 Torr pressure. For comparison, the binding curve for myoglobin is shown as a dashed curve.*

(By kind permission of J. M. Berg, et al., *in* 'Biochemistry', *(8th Edn), Freeman, Macmillan, New York, 2015, p. 196)*

with its single haem group. In haemoglobin itself, there are four chains of amino acid units that collectively constitute the protein part of the molecule. These four chains of the so-called globin consist of two identical pairs. The members of one pair are called α-chains and those of the other known as β-chains. Together, the four chains contain a total of 574 amino acid units.

Kendrew began his protein studies by investigating sheep haemoglobin; then he embarked on the structure of myoglobin. After a long struggle with myoglobin from horse heart, which refused to yield crystals large enough for X-ray analysis, Kendrew realized that diving mammals and birds offered a better prospect. A chance encounter by Perutz led to the arrival of a large chunk of sperm whale meat from Peru and, to their joint delight, its myoglobin yielded large sapphire-like crystals that gave beautiful X-ray diffraction patterns.

Admirable fuller accounts than the present have been given by Dickerson and Geis[6] on the structure, function, evolution, and pathology of haemoglobin, and Dickerson's monograph entitled '*Present at the Flood*' deals in a pedagogically instructive manner with original reprints of articles (and comments thereon) on topics such as the folding and coiling of polypeptide chains and how to solve protein structures.[7] Valuable information pertaining to haemoglobin and myoglobin is also given as an atlas by Fermi and Perutz, which appeared in 1981.[8]

5.2.1 Oxygen uptake and cooperativity

To understand the physiological properties of both haemoglobin and myoglobin, it is essential to enquire further into the nature of the dependence of the fractional saturation of these two globins as a function of the partial pressure of oxygen (see Figure 5.2). What follows below owes much to the description given in the Fermi–Perutz atlas.[8]

The most fundamental physiological property of myoglobin and haemoglobin is their ability to combine reversibly with oxygen. The basic measurement *in vitro* of this property is the *oxygen equilibrium curve*, wherein the fraction of haem groups bearing oxygen (the fractional saturation) is measured as a function of the partial pressure of oxygen. Typical oxygen equilibrium curves for haemoglobin and myoglobin are shown in Figure 5.2. The curve for myoglobin is hyperbolic, as expected for the combination of one molecule of myoglobin with one molecule of oxygen, but the curve for haemoglobin has a more complex sigmoid shape. The physiological advantage of the sigmoid curve is that it allows haemoglobin to deliver more oxygen to tissues. As oxygen partial pressure is reduced from arterial to venous levels, haemoglobin saturation is reduced from nearly full saturation to about three-quarters saturation, so that haemoglobin can deliver about a quarter of a mole of oxygen per mole of haem to tissues where it is taken up by myoglobin, which has a higher oxygen affinity. A further advantage of the sigmoid curve is that its steepness at venous pressure enables the amount of oxygen delivered to tissues to adjust sensitively to changes in oxygen partial pressure there.

The sigmoid shape of the oxygen equilibrium curve for haemoglobin indicates that the oxygen affinity of haemoglobin rises with uptake of oxygen. This phenomenon is known as cooperativity or haem–haem interaction. Since haemoglobin has four binding sites for oxygen, its equilibrium with oxygen is defined by four equations:

$$\mathrm{Hb}(\mathrm{O}_2)_{n-1} + \mathrm{O}_2 \rightleftharpoons \mathrm{Hb}(\mathrm{O}_2)_n$$

where $n = 1, 2, 3, 4$.

The four corresponding equilibrium constants K_n are known as the Adair constants. The sigmoid oxygen equilibrium curve indicates that K_1, the equilibrium constant for the uptake of the first oxygen molecule, is less than K_4, the constant for uptake of the last oxygen molecule. This means that the ligation state of one haem in a molecule affects the affinity of the other haems in the molecule.

As stated earlier, the shape of the haemoglobin uptake (of oxygen) curve suggests that binding of oxygen at one site within the haemoglobin tetramer increases the likelihood that oxygen will bind at the remaining unoccupied sites. Conversely, the process of liberation of oxygen from one haem facilitates the liberation of oxygen at the other sites. In essence, this phenomenon (of cooperativity) indicates that the binding reactions at individual sites in each haemoglobin molecule are not independent of one another.

It is useful to recall the manner in which pioneers like Perutz and Monod, especially the latter, have dealt with features pertaining to the mechanism of cooperativity between the two haem groups. The arrangement of subunits in fully liganded haemoglobin is known as the *R*, or relaxed, quaternary structure; that of the unliganded (deoxy) haemoglobin is known as the *T*, or tense, quaternary structure. The reason for this terminology is that the subunits are joined more loosely in *R* than in *T*. There is evidence that the structure of individual subunits (the α- or β-chains) in the *T*-state is distorted, relative to that of the *R*-state subunits. All this is related to so-called allosteric changes—an allosteric protein is one that changes from one conformation to another when it binds another molecule (such as oxygen).

An important contribution to the understanding of cooperativity came from the landmark paper of Monod, Wyman, and Changeux in 1965.[9–11] Their model (abbreviated MWC) describes allosteric transitions of proteins made up of identical subunits. It assumes, for simplicity, that in haemoglobin the four subunits are functionally identical. The reasons for the success of the MWC model in explaining the cooperativity that occurs in haemoglobin have been extensively discussed—see Perutz,[12,13] Schulman,[14] and others—but lie outside the scope of this book. The key point to note is that the MWC model explains the sigmoidal binding property shown in Figure 5.2.

What follows below is a summarized account of the achievements of Perutz and Kendrew in arriving at the detailed three-dimensional structures of myoglobin and haemoglobin. This work, far ahead of that of their contemporaries, initiated a new era in protein crystallography and structural biology generally.

5.3 A Summarized Account of the Successes Registered by Perutz and Kendrew

5.3.1 The importance of the heavy atom method

According to Dorothy Hodgkin,[15] in the history of the study of the structure of proteins by crystallography, Max Perutz's X-ray photograph of a mercury derivative of haemoglobin taken in 1953 stands out as of special importance. The idea that dawned on Perutz, on receiving an unsolicited letter from Austin Riggs of Harvard, is what led to this advance. Riggs had discovered that, on introducing mercury atoms within molecules of haemoglobin, there was essentially no difference in the uptake characteristics towards oxygen of haemoglobin; in other words, the structure of the giant polypeptide was left intact. Perutz established that, even though a mercury atom (with its eighty electrons) used as a label is very small, compared with that of a complete haemoglobin molecule, the X-ray diffraction pattern differed significantly from that of the unlabelled crystals. The excitement associated with this observation is captured in Perutz's own words: '*As I developed the first X-ray photograph of mercury haemoglobin my mood altered between sanguine hopes of immediate success and desperate forebodings of all possible causes of failure. I was jubilant when the diffracted spots appeared in exactly the same positions as in the mercury-free protein, but with slightly altered intensity, exactly as I had hoped.*'

What Perutz had discovered was that the method of isomorphous replacement, until then applied rather rarely in crystallography generally and never in the field of proteins, was ideally suited to solve, ultimately, their structures, as this procedure yielded the all-important phases of the X-ray diffraction spots (see Chapter 2, especially Section 2.4.3).

5.3.2 The three-dimensional structure of myoglobin as revealed by X-ray crystallography

Detailed popular[16] and technical[3] accounts have been given by Kendrew of how he and his team solved the structure of myoglobin, first at low resolution[17] and later at high resolution.[18] We now summarize these papers. First, in Figure 5.3, typical crystals of myoglobin, together with the X-ray diffraction patterns of the unlabelled (heavy atoms) and labelled myoglobin are shown.

The lower-resolution (6 Å Fourier synthesis) study[17] yielded the picture of myoglobin, with its prominent haem group shown in Figure 5.4.

During Kendrew's progress towards achieving the 2 Å resolution structure of myoglobin, knowing that such an image would reveal clearly the nature of the individual amino acids constituting the single polypeptide chain of the molecule, he initiated a chemical study by Edmundson, first at the Rockefeller Institute in New York and later at the Laboratory of Molecular Biology (LMB) in Cambridge, by tryptic digestion (see Section 3.3.3). A comparison between the chemical and

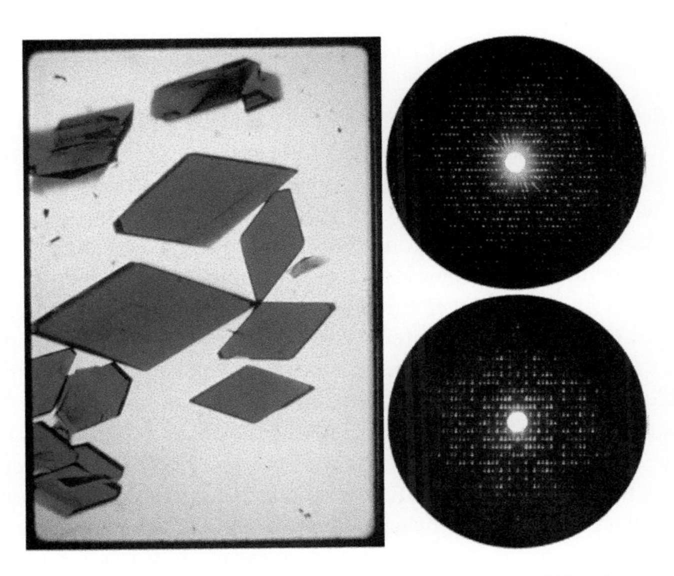

Figure 5.3 *A typical crystal of myoglobin, together with its X-ray diffraction pattern without (top) and with (bottom) a heavy atom label.*

(After J. C. Kendrew, Copyright MRC-LMB)

Figure 5.4 *Low-resolution picture of myoglobin, with its prominent haem group shown as a dark grey disc.*

(After J. C. Kendrew[3]. Copyright MRC-LMB)

X-ray evidence for part of the amino acid sequence of myoglobin is shown in Figure 5.5, where the agreement is remarkable.

In Figure 5.6 is shown part of the 1.4 Å Fourier synthesis of the myoglobin structure, where the haem group (end on) is clearly visible and so is the so-called distal histidine attached to the iron atom of the haem (see later). Also seen both along and perpendicular to their local axes are α-helices. The Kendrew team were able to discover, in the first ever evidence of the existence of the α-helix predicted by Pauling and Corey (see Chapter 4), that 118 out of a total of 157 amino acids that make up the polypeptide chain in myoglobin make up eight segments of the right-handed α-helix, seven to twenty-four amino acids long.

The details shown in this remarkable image have very recently (2017) taken on a fresh significance, since, in the hands of Professor Wolfgang Baumeister and his team in Martinsried, Germany, these workers, using the powerful new technique of cryo-electron microscopy, have confirmed the spatial proximity of histidine and water molecule to the haem group, as we elaborate further in Chapter 10.

So far as the side chains in myoglobin are concerned, Kendrew *et al.*[3,17,18] found that almost all those containing polar groups are on the surface of the molecules. Thus, with very few exceptions, all of lysine, arginine, glutamic acid, aspartic acid,

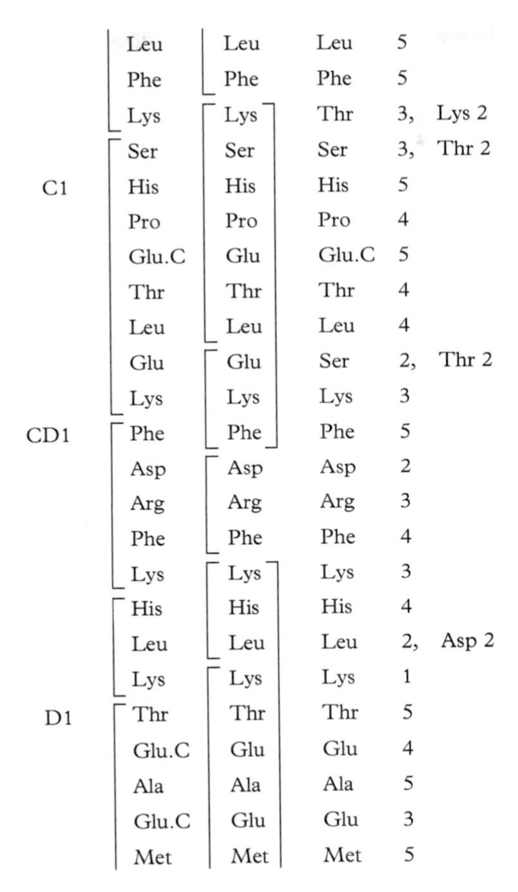

	Leu	Leu	Leu	5	
	Phe	Phe	Phe	5	
	Lys	Lys	Thr	3,	Lys 2
	Ser	Ser	Ser	3,	Thr 2
C1	His	His	His	5	
	Pro	Pro	Pro	4	
	Glu.C	Glu	Glu.C	5	
	Thr	Thr	Thr	4	
	Leu	Leu	Leu	4	
	Glu	Glu	Ser	2,	Thr 2
	Lys	Lys	Lys	3	
CD1	Phe	Phe	Phe	5	
	Asp	Asp	Asp	2	
	Arg	Arg	Arg	3	
	Phe	Phe	Phe	4	
	Lys	Lys	Lys	3	
	His	His	His	4	
	Leu	Leu	Leu	2,	Asp 2
	Lys	Lys	Lys	1	
D1	Thr	Thr	Thr	5	
	Glu.C	Glu	Glu	4	
	Ala	Ala	Ala	5	
	Glu.C	Glu	Glu	3	
	Met	Met	Met	5	

Figure 5.5 *A comparison between chemical and X-ray evidence for part of the amino acid sequence of myoglobin. First column, tryptic peptides; second column, chymotryptic peptides (see Section 3.3.3); third column, X-ray evidence. Peptides are enclosed in brackets. The figures give the degree of confidence in the identification (5, complete confidence; 1, a guess). This comparison was shown in Kendrew's Nobel Lecture in December 1962.*

(Courtesy The Nobel Foundation)

histidine, serine, threonine, tyrosine, and tryptophan have their polar groups on the outside. The interior of the molecule, on the other hand, is almost entirely made up of non-polar residues, generally close-packed and in van der Waals contact with their neighbours.

The interaction of the haem group with histidine (see Figures 5.6 and 5.7) is responsible for the characteristic function of myoglobin, since an isolated haem group does not exhibit the phenomenon of reversible oxygenation, which is displayed both by myoglobin and haemoglobin.

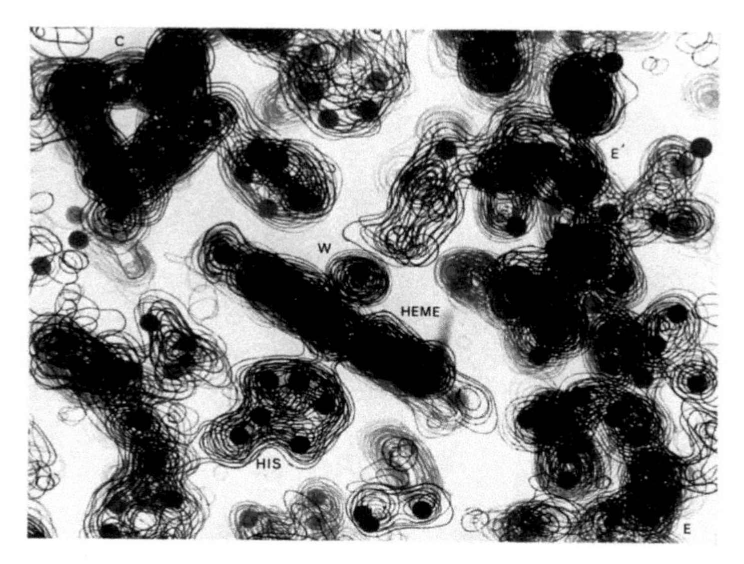

Figure 5.6 *Part of the high resolution (1.4 Å Fourier version) of myoglobin. In the centre, the haem group is seen end on, and the links between it and adjacent histidines are clearly visible, as is a water molecule attached (marked 'W') to the iron atom. Top left shows a helix end on; on the right, a helix is seen longitudinally, together with several side chains. This image was also shown in Kendrew's Nobel Lecture in 1962.*

(Courtesy The Nobel Foundation)

5.3.3 The three-dimensional structure of haemoglobin

As mentioned earlier, a complete molecule of haemoglobin[19–21] is made up of four subunits (polypeptide chains), each of which also contains one haem. The two kinds of subunits are designated α and β (see Figures 5.8 and 5.9), which have different sequences of amino acid residues, but similar three-dimensional structures. The β-chain also has one short extra helix. Each haem lies in a separate pocket at the surface of the molecule.

This structure was mapped out by Perutz and his colleagues in 1960 at a Fourier synthesis of 5.5 Å resolution. This was later extended to 2.8 Å resolution.[22] It was possible for Perutz *et al.* to achieve this major goal in protein science and structural biology generally—as it was for Kendrew *et al.* to achieve their goal with myoglobin—by involving several experimental procedures that, together, enabled them to fulfil their ambitions. Below are given the salient ones.

First, in order to record their X-ray diffraction data reliably, they used a high-intensity rotating anode source of X-rays, which was so well designed that it could operate without interruption for very long times. (Perutz's colleagues U. W. Arndt and J. F. W. Mallet were able to keep the rotating anode source in running order, day and night, during the fifteen months which it took to measure the intensities

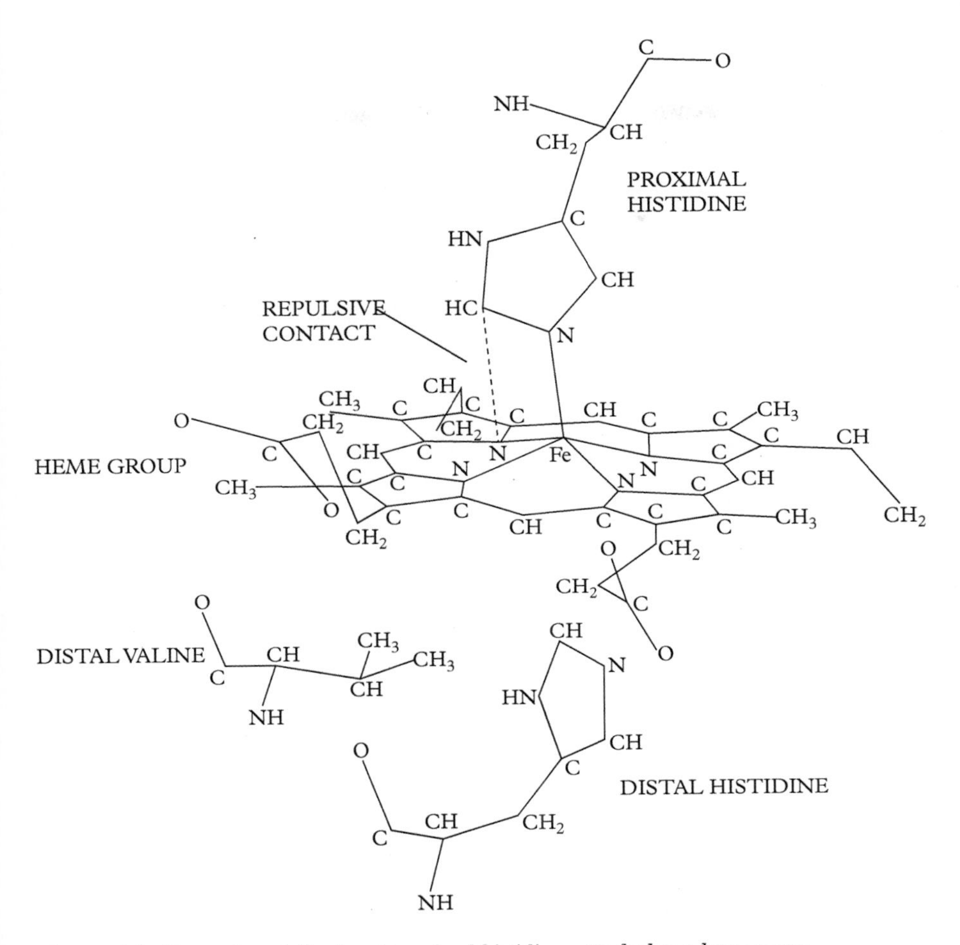

Figure 5.7 *Illustration of distal and proximal histidines attached to a haem group. (After M. F. Perutz, copyright Joan Starwood)*

of some 100,000 reflections.) Second, they assembled a large team of women co-workers to measure, by densitometry, the thousands of intensities of diffraction spots—sometimes these women were required to work 'night shifts' to complete their tasks. Third, both Kendrew and Perutz, thanks to the advances made by Arndt and Phillips at the Davy-Faraday Research Laboratory (DFRL), were able to accelerate their rate of collection of data by using a diffractometer, rather than relying on photographic films or plates (see Figure 5.10). Often Perutz himself utilized the diffractometer.

Fourth, because of the massive amount of computation (required in their Fourier analysis), it was necessary for Kendrew and Perutz to have access to arguably

Figure 5.8 *The complete molecule of haemoglobin, as deduced from X-ray diffraction studies, is shown from above (left) and the side (right). The drawings follow the representational scheme used in three-dimensional models built by Perutz and his co-workers. The irregular blocks represent electron density patterns at various levels in the haemoglobin molecule. The molecule is built up from four subunits: two identical α-chains (light blocks) and two identical β-chains (dark blocks). Haem groups are coloured red, the iron-containing structure that binds oxygen to the molecule.*

(Courtesy of the MRC-LMB, Cambridge)

Figure 5.9 *Another view of a model of the complete haemoglobin held by Max Perutz.*
(Copyright MRC-LMB)

Figure 5.10 *Photograph of Max Perutz using a four-circle X-ray diffractometer to record X-ray intensities. This photograph appeared on a British postal stamp to celebrate its fiftieth anniversary in 2014, the centenary of Perutz's birth.*
(Courtesy of Cambridge Evening News*)*

the fastest electronic computers in the world, which, to their good fortune, were being developed in Cambridge at the time in the Mathematics Laboratory—EDSAC-I and then EDSAC-II. (EDSAC stands for *Electronic Delay Storage Automatic Calculator* and was built by Maurice Wilkes in Cambridge.) Fortunately, as early as 1952, Kendrew had realized from conversations with his PhD student Hugh Huxley that a young Australian computer scientist, named J. M. Bennett (who shared the same lodgings as Huxley), possessed the expertise to help him formulate a program that could rapidly compute Fourier syntheses.[23] Lastly, both Kendrew and Perutz, the former especially, had expert co-workers at postdoctoral level working alongside them, notably Rossmann, Dickerson, Strandberg, and Dintzis, as well as D. C. Phillips and A. C. T. North from the DFRL.

5.4 Chemical and Medical Consequences Deduced from the Three-Dimensional Structure of Haemoglobin

In a popular article published[20] in 1978, Perutz has given a detailed account of what could be concluded from the structure of the haemoglobin molecule. At one stage, he expressed the thought that '*our much-admired model did not reveal its inner workings – it provided no hint about the molecular mechanism of respiratory transport. Why not? Well-intentioned colleagues were quick to suggest that our hard-won structure was merely an artefact of crystallisation and might be quite different from the structure of haemoglobin in its living environment, which is the red blood cell.*'

The question posed by Perutz was: '*How was it that, in effect, haemoglobin functions as a molecular lung?*' It was not until early in the 1970s that a satisfying answer to this question came and the fundamentals of the mechanism were elucidated.[24] The firm experimental evidence on which the now accepted mechanism of so-called allosteric change[19] came not only from crystallographic work, but also from Perutz's imaginative use of other techniques, such as spectroscopic and magnetic measurements, as well as extended X-ray absorption fine structure (EXAFS).[25] He found that the structural changes in haemoglobin accompanying oxygenation were large. In the 'deoxy' form, the iron atom of haem is displaced a little from the plane of the haem group, whereas in the 'oxy' structure, it lies almost in the plane (see Figure 5.11).

Perutz recognized that this is because of a change in the electronic spin state of the iron atom—from so-called 'high-spin' in the deoxy to 'low-spin' in the oxy state—and hence to a diminution in the radius of the iron. When the iron centre moves closer to the plane of the haem in the oxy state, it drags with it the α-helix of the protein to which it is connected.[25–27] This is the trigger that initiates a sequence of 'molecular levers' that loosen and rearrange the subunits in the (tense) deoxy structure into a new (relaxed) oxy structure. This is the basis of the molecular mechanism, which Perutz described as '*infinitely rewarding in its simple beauty*'. It is what governs the oxygen affinity of haemoglobin in response to

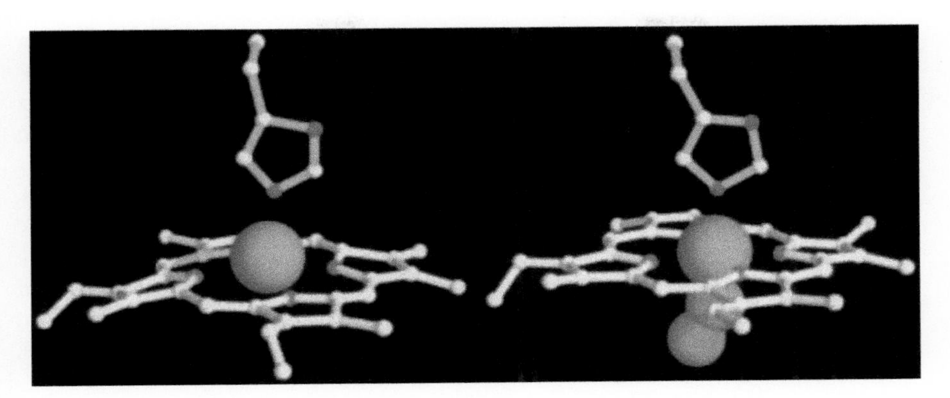

Figure 5.11 *(Left) In human deoxyhaemoglobin, the iron ion attached to histidine (above) lies slightly above the plane of the haem group. (Right) When the iron binds oxygen, as in the oxyform of haemoglobin, it moves into the plane of the haem. This process, according to Perutz, is what initiates the sequence of resulting changes—see text for further information.*

(Copyright Wiley-VCH)

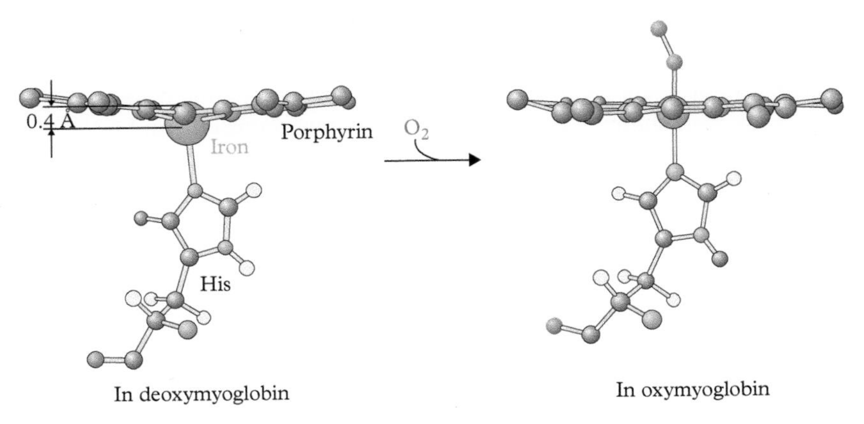

Figure 5.12 *Illustration of how the position of the iron ion changes from being slightly outside to be inside the plane of the haem when oxygen is bound to either myoglobin or haemoglobin.*

(Courtesy, J. M. Berg, et al., *reference as in Figure 5.2. Copyright Freeman Macmillan, New York)*

physiological needs and the release of blood oxygen when it is needed under conditions of oxygen scarcity. Figure 5.12 shows how binding of oxygen changes the position of the iron ion. In the deoxy state, the iron lies slightly outside the plane of the porphyria (left). In the oxy state, the iron moves into the plane of the haem.

This mechanism was to lead, in due course, through work done on haemoglobin mutants, to a fuller understanding of several inherited diseases, and it opened up a new field of *molecular pathology*, a subject which relates a structural abnormality to a particular disease. Perutz and colleagues gained new insights into *molecular evolution* and into the delicate (sometimes major) differences exhibited by haemoglobin in a wide range of living species. With his colleague G. Fermi,[8] Perutz chronicled the amino acid sequences in the α and β subunits of haemoglobin in the following creatures: man, rhesus monkey, orang-utan, slow loris, tupai, savannah monkey, capuchin monkey, spider monkey, rabbit, dog, horse, cat, pig, camel, llama, Indian elephant, opossum, rat, chicken, goose, carp, goldfish, caiman, Nile and Mississippi crocodile, tadpole, frog, and shark.

Because of his long friendship with Herman Lehmann of the Department of Clinical Biochemistry in Addenbrooke's Hospital, Perutz was totally conversant with an enormous range of properties exhibited by all manner of haemoglobins. Lehmann was a world authority on abnormal haemoglobins.

5.4.1 Sickle-cell anaemia and other molecular diseases

In 1949, Linus Pauling *et al.*[28] described the discovery of sickle-cell anaemia as the first molecular disease; and, in due course, it was established by Perutz's colleague Vernon Ingram[29] that this disease arose from a single amino acid residue in one of the polypeptide chains of haemoglobin—the replacement of glutamic acid by valine. (The disease itself was discovered as early as 1910 by Herrick.[30]) The red cells of patients with this disease, when partially deoxygenated (as in venous blood), undergo a change in shape, as shown in Figure 5.13.

This change in shape is called sickling (deformation resembling a sickle). The sequence of pathological events in sickle-cell disease is: (i) polymerization of abnormal haemoglobin to form a viscous, fibrous gel upon deoxygenation of red cells in the tissue; (ii) a large decrease in the deformability of the cells; followed by (iii) occlusion of the vessels of microcirculation by the stiffened cells, and finally, (iv) tissue damage.[31]

Max Perutz and Herman Lehmann pursued the question of abnormal haemoglobins still further. In 1968, they published a landmark paper,[32] about which eminent haematologists subsequently remarked[33] that '*very few concepts have been established which are not to be found in this paper*'. As had been commented upon earlier, this paper[32] was a curtain-raiser for the era of molecular medicine, which nowadays represents a substantial proportion of medical research. The abstract of the Perutz–Lehmann paper[32] merits repetition: '*The haemoglobin molecule is insensitive to replacements of most amino acid residues on its surface, but extremely sensitive to even quite small alterations of internal non-polar contacts, especially those near the haems. Replacements at the contacts between α and β subunits affect respiratory function.*'

In the first table contained in this paper, they give the amino acid substitutions of thirty-eight variants of haemoglobin, explaining each in terms of the structure alterations that the 'new' amino acid would cause. All this became possible as a result of Perutz's triumph in obtaining a three-dimensional picture of the haemoglobin molecule and Lehmann's encyclopaedic knowledge of abnormal haemoglobins.

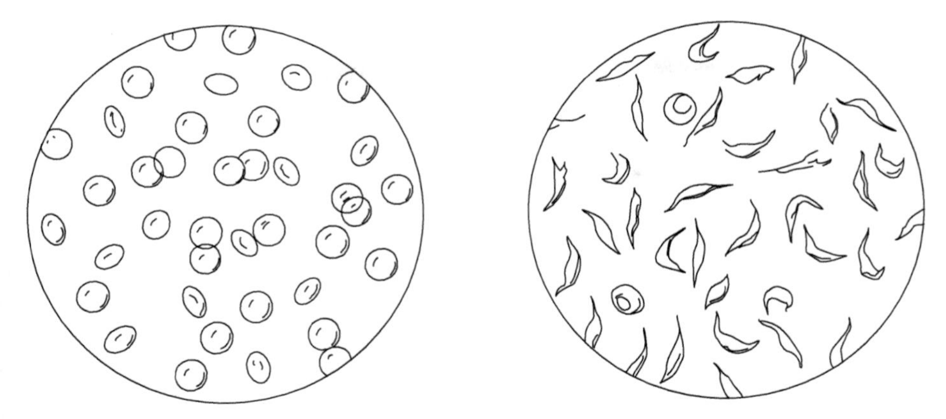

Figure 5.13 *Illustration of the appearance of normal (left) and sickled (right) red blood cells. (Copyright American Philosophical Society)*

This major contribution by Perutz and Lehmann has led, in association with the work of others, to the era of molecular pathology, which focuses on the study and diagnosis of disease through the examination of mutated molecules within organs, tissues, or body fluids. An increasing number of human diseases is now thought to be caused by abnormal peptide or protein aggregation (as discussed in Chapters 9 and 10), most notably cystic fibrosis, Parkinson's disease, and prion disease.[33,34]

5.4.2 Molecular evolution

Perutz gained new insights into molecular evolution[2] and into the delicate (sometimes major) differences exhibited by haemoglobin in a wide range of living species. For example, the frogs of Lake Titicaca, high in the mountains (above 13,000 feet) of Bolivia, have evolved a form of haemoglobin that can absorb oxygen better than that in the frogs of Lake Michigan at lower (approximately 700 feet) altitudes. The enhanced absorbability of oxygen is also a feature of the haemoglobin of migrating geese that fly at high altitudes. Crocodiles are able to remain under water for more than an hour without surfacing to breathe.[35] How do crocodiles stay under water for so long? It was a German zoologist—Christian Bauer—and his colleagues who discovered that bicarbonate ions drastically reduce the oxygen affinity of the haemoglobin of crocodiles.[36] Kiyoshi Nagai and Noboru Komiyama and others[37] showed which residues in the haemoglobin of crocodiles are responsible for the binding of bicarbonate ions. The oxygen affinity of crocodile haemoglobin is not regulated by diphosphoglycerate (DPG), as in human haemoglobin which binds the oxygen to its central cavity. Nagai and Komiyama found[37] that bicarbonate ions bind at the interface between the $\alpha1$ and $\beta2$ subunits, which slide over one another upon the binding of the oxygen.

REFERENCES

1. W. L. Bragg, *Rep. Progr. Phys.*, **1965**, *28*, 1.
2. F. Sanger, *Biochem. J.*, **1945**, *39*, 507.
3. J. C. Kendrew, *Science*, **1963**, *139*, 1259.
4. For a typical protein of 200 amino acid residues in length, there are a staggering 10[200] possible structures to explore!
5. L. Pauling and C. R. Coryell, *Proc. Natl. Acad. Sci. U. S. A.*, **1936**, *22*, 210.
6. R. E. Dickerson and I. Geis, '*Haemoglobin: Structure, Function, Evolution and Pathology*', The Benjamin/Cummins Publishing Co. Inc., Menlo Park, CA, **1983**.
7. R. E. Dickerson, '*Present at the Flood: How Structural Molecular Biology Came About*', Sinauer Associates, Sunderland, MA, **2005**.
8. G. Fermi and M. F. Perutz, in '*Atlas of Molecular Structures in Biology: 2. Haemoglobin and Myoglobin*' (eds. D. C. Phillips and F. M. Richards), Clarendon Press, Oxford, **1981**.
9. J. Monod, J. Wyman, and J. P. Changeux, *J. Mol. Biol.*, **1965**, *12*, 80.
10. J. Wyman, *J. Mol. Biol.*, **1969**, *14*, 523.
11. J. P. Changeux, *Neuron*, **1998**, *21*, 959.
12. G. Fermi, M. F. Perutz, and R. G. Schulman, *Proc. Natl. Acad. Sci. U. S. A.*, **1987**, *84*, 6167.
13. M. F. Perutz, A. J. Wilkinson, M. Paoli, and G. G. Dodson, *Am. Rev. Biophys. Biomol. Struct.*, **1998**, *27*, 1.
14. R. G. Schulman, in '*Memories and Consequences*' (ed. H. E. Huxley), Laboratory of Molecular Biology, Medical Research Council, Cambridge, **2013**, p. 41.
15. D. H. Hodgkin, *Ann. N. Y. Acad. Sci.*, **1979**, p. 66.
16. J. C. Kendrew, *Scientific American*, **1961**, *205*, 96.
17. J. C. Kendrew, G. Bodo, H. M. Dintzis, R. G. Parrish, H. W. Wyckoff, and D. C. Phillips, *Nature*, **1958**, *181*, 662
18. J. C. Kendrew, R. E. Dickerson, B. E. Standberg, R. G. Hart, D. R. Davies, D. C. Phillips, and V. C. Shore, *Nature*, **1960**, *185*, 422.
19. M. F. Perutz, *Scientific American*, **1964**, *208*, 64.
20. M. F. Perutz, *Scientific American*, **1978**, *222*, 92.
21. M. F. Perutz, M. G. Rossmann, A. F. Cullis, H. Muirhead, G. Will, and A. C. T. North, *Nature*, **1960**, *185*, 4711.
22. M. F. Perutz, H. Muirhead, J. M. Cox, L. C. G. Goaman, F. S. Mathews, E. L. McGandy, and L. E. Webb, *Nature*, **1968**, *219*, 29.
23. J. M. Bennett and J. C. Kendrew, *Acta Cryst.*, **1952**, *5*, 109.
24. M. F. Perutz, *New Scientist*, **1971**, *49*, 676.
25. M. F. Perutz, S. S. Hasnain, P. J. Duke, J. L. Sessler, and J. E. Hahn, *Nature*, **1982**, *295*, 535.
26. In some cases, the binding of a molecule to the protein produces little conformational change. Sperm whale myoglobin[26] is a good example; the oxy and deoxy structures may be superimposed almost exactly.
27. M. Lesk, '*Protein Architecture: A Practical Approach*', Oxford University Press, Oxford, **1991**, p. 121.
28. L. Pauling, H. Itano, J. Singer, and I. Wells, *Science*, **1949**, *110*, 548.
29. V. M. Ingram, *Nature*, **1957**, *180*, 326.

30. J. B. Hernick, *Arch. Int. Med.*, **1910**, *6*, 517.
31. W. A. Eaton and J. Hofrichter, *Science*, **1995**, *268*, 1142.
32. M. F. Perutz and H. Lehmann, *Nature*, **1968**, *219*, 902.
33. G. Ferry, '*Max Perutz and the Secret of Life*', Chatto and Windus, London, **2007**.
34. C. M. Dobson, *Nature*, **2003**, 426, 884.
35. G. W. Christoph, J. Hofrichter, and W. A. Eaton, *Biophys. J.*, **2005**, *88*, 1371.
36. C. Bauer, M. Forster, G. Gros, *et al.*, *Biol. Chem.*, **1981**, *256*, 8429.
37. N. H. Komiyama, G. Miyazaki, J. Tame, and K. Nagai, *Nature*, **1995**, *373*, 244.

6

Sir Lawrence Bragg at the RI (1953–1966) and the Determination of the First Three-Dimensional Structure of an Enzyme at the DFRL (1965)

6.1 Introduction

In the early 1950s, the Royal Institution (RI) experienced a catastrophic turbulence unknown in its previous sesquicentennial existence. It all arose because its then Director, the famous physicist E. N. da Costa Andrade made life and work very unhappy for those who were employed both at the RI and the Davy-Faraday Research Laboratory (DFRL), of which he was also Director. Andrade was described once by W. L. Bragg (WLB) as '*a stormy petrel*'. He was a rather aggressive individual, who, inter alia, treated the technical and other staff in dismissive ways. His volatility made it difficult for others to work with him. He ran foul of the managers of the RI also. W. H. Bragg used to say that the constitution of the RI ran only on goodwill. There was a great deal of that when he and his successors were around, but little under the reign of Andrade.

In 1952, Andrade lost a vote of confidence from the Members and resigned. In April 1953, after strong persuasion from Lord Adrian, the then President of the Royal Society Lawrence Bragg resigned from one of the most prestigious chairs of physics in the world—the Cavendish Professorship at Cambridge—and at the age of 63, he was appointed to the Fullerian Professorship of Chemistry and the Directorship of the DFRL, just as his father had been thirty years earlier. Shortly thereafter, a new sunlit era dawned in Albemarle Street, London.

6.2 W. L. Bragg at the RI

After Andrade's departure, Sir Lawrence set about assembling a new research team at the DFRL devoted to protein crystallography. He wanted both Max

Perutz and John Kendrew to move with him from Cambridge to Albemarle Street in Mayfair, London. They refused to do so but agreed to come to the DFRL on an approximately one day-per-week basis as Honorary Readers, posts that they each held for some thirteen years, until Sir Lawrence was succeeded by Sir (later Lord) George Porter. This was an admirable compromise, as it harmonized with Sir Lawrence's other action, namely the appointments of Drs David C. (later Lord) Phillips, A. C. T. (Tony) North, and Roberto Poljak, whom he attracted to the DFRL.

David Phillips had graduated in physics, mathematics, and electrical communications at the University College of South Wales, Cardiff, and had completed his PhD there under the expert supervision of the Canadian-born crystallographer A. J. C. Wilson. After postgraduate work at the National Research Council in Ottawa, he was attracted by Sir Lawrence to the DFRL. Tony North, who had completed his PhD in King's College, London, had worked there with Sir John Randall and alongside the future Nobel Laureate M. H. F. Wilkins. There were two other key individuals in Bragg's team of crystallographers. One was the Argentinian R. J. Poljak, who, while working at the Massachussetts Institute of Technology (MIT) in the United States, in 1953, heard lectures by both John Kendrew and Max Perutz when they visited MIT and Harvard. These talks convinced him that the three-dimensional structures of proteins were the most challenging and fruitful research that a crystallographer could undertake. He soon focused his attention on egg white lysozyme, which Sir Alexander Fleming had discovered[1] at St Mary's Hospital in London in 1922 (see Section 6.4) six years before his epochal discovery of penicillin.[2] Poljak quickly learned how to grow good crystals of lysozyme, and so he wrote to Perutz with the aim of joining the group there on molecular biology. Max Perutz had no room for him, but he advised Poljak to approach Sir Lawrence. In 1960, he arrived at the DFRL. Uli Arndt, German-born, but educated in physics in Cambridge, had joined the DFRL in 1950 with Andrade; they formed a wonderful partnership with David Phillips and Tony North. Also appointed by Bragg at that time were Jack Dunitz—who was not at all interested in protein crystallography, especially as he was doing fruitful work at the time with Leslie Orgel on crystal field theory—Colin Blake and several others who are mentioned later in this chapter (see Figure 6.1).

The impressive work accomplished at the DFRL by W. L. Bragg's team is described in Section 6.3 below. Here we focus first on the beautifully composed and balanced way in which he maintained the unique traditions of the RI as a showcase of science.

It has often been stated that for all the nineteenth and most of the twentieth centuries, the greatest scientists on earth, as well as many other individuals of distinction, at one time or another, spoke about their science and culture to lay audiences in the famous lecture theatre built by its founder Count Rumford in 1800 (see Figures 6.2 and 6.3).

A reflection of the intellectual life of the UK may be gleaned from some of the names of the persons who were invited to present Friday Evening Discourses at

Figure 6.1 *Group photograph of the team at the DFRL that solved the three-dimensional structure of lysozyme. From left: Gareth Mair, Colin Blake, Louise Johnson, Tony North, David Phillips, and Raghupathy Sarma.*

(By kind permission of Prof A. C. T. North)

the RI in WLB's day as Director. In chronological order, the names and topics of their Discourses in the period 1953–1966 are tabulated below (see Table 6.1).

In addition to the impressive speakers that he invited to give Friday Evening Discourses, he also arranged a spectacular list of lecturers to present the famous Christmas Lectures '*for a juvenile auditory*'—the term used by Michael Faraday, who introduced the annual series in 1826. Among the Christmas Lecturer chosen by W. L. Bragg was the inventor of the jet engine Sir Frank Whittle (see Figure 6.4), whose subject was '*The Story of Petroleum*'. Two Nobel Prize winners—Sir Martin Ryle and Anthony Hewish—were joined, in1957, by two other (Mancunian) radio-astronomers—Sir Bernard Lovell and Sir Francis Graham-Smith—who, between them, gave their series on '*Exploration of the Universe*'. In 1957, Sir Julian Huxley and James Fisher dealt with '*Birds*' as the topic of their Christmas Lectures, and W. L. Bragg himself chose '*Electricity*' as the subject of his in 1961.

In the mid 1960s, W. L. Bragg, strongly supported by his Deputy Professor Ronald King (sitting to his left in the front row in Figure 6.5), initiated a new series of lectures, aimed at schoolchildren of London and its environs. These turned out to be extraordinarily successful, with audiences numbering some 500 on most occasions. Bragg became extremely adept at speaking to children. His

Figure 6.2 *A view of the entrance to the lecture theatre of the Royal Institution taken on the occasion of the eightieth birthday celebration of Max Perutz. To his immediate right is Sir Aaron Klug and next to him is Michael Rossmann, and on his right are, first, Wim Hol and then Jim Wells. On his left is David Blow, followed by Peter Colman, Don Wiley, and Wayne Hendrickson. Behind the group is a painting of a typical Friday Evening Discourse audience in the early 1900s, at which Sir James Dewar demonstrated the properties of liquid hydrogen. In Dewar's audience are Lords Rayleigh and Kelvin, A. J. Balfour (the Prime Minister), and G. Marconi.*

(By kind permission of Dr Richard Henderson)

numerous elegant demonstrations—like the illustration of how one could sail a boat on a fluidized bed of sand (see Figure 6.6)—are still remembered by all those fortunate enough to have heard and seen him and who are now in prominent academic and public life.[3]

A favourite mantra recalled by WLB when talking about lecturing was also regularly used by the Italian–American physicist Enrico Fermi: '*Never underestimate the pleasure that people derive in hearing the things that they already know!*' This is but one reason why his lectures were so memorable. One of the gems in his (and Faraday's) advice to lecturers starts with the following words: '*A lecture is made or marred in the first ten minutes. This is the time to establish the foundations to remind the audience of things they half knew already and to define terms that will be used.*'

As Director of the RI, he and Lady Bragg were frequently invited to banquets and private dinners in London. He preferred the latter to the former. On one occasion, in 1954, Sir Alexander Fleming, the discoverer of penicillin, invited him to a private dinner where the principal guest was Marlene Dietrich, the German actress, singer, and film star. At dinner, seated between two Nobel Prize winners,

Figure 6.3 *Dmitri I. Mendeleev, who gave a Friday Evening Discourse at the Royal Institution in 1889 and again a few years later. The stamp reproduction here, created by Evgeni Egorov, was printed in Russia to commemorate Mendeleev's 175 years after his death and 140 years after the publication of his periodic table of the elements.*

(The author thanks Erling Norby and Ulf Lagerkvist for drawing this stamp to his attention.)

Table 6.1 *A selection of Friday Evening Discourses arranged by Sir Lawrence Bragg*

Date	Speaker	Title
1953	Edwin Hubble	*The observational evidence for an expanding universe*
	Alexander Fleming*	*Antibodies*
	Nevill Mott*	*Dislocations, fatigue, and fracture*
1954	A. H. Compton*	*Science and Man's unfolding view of himself*
	W. L. Bragg*	*Models of metal structure*
1955	Dorothy L. Sayers	*Oedipus simplex: freedom and fate in folklore and fiction*
1956	P. M. S. Blackett*	*Rock magnetism and the movement of continents*
	Harold C. Urey*	*Diamonds, meteorites, and the origin of the solar system*
1957	Kathleen Kenyon	*Excavations at Jericho*
1958	J. M. Allegro	*The Dead Sea Scrolls*
	J. C. Kendrew*	*The architecture of a protein molecule*
1959	Yehudi Menuhin	*Art and science as related concepts*

Date	Speaker	Title
1960	L. Pearce Williams	*Michael Faraday and the evolution of the concept of electric and magnetic fields*
1961	Lord Adrian*	*Dreaming*
	Kenneth Clarke	*Rembrandt's self-portraits*
	Sydney Brenner*	*The handing on of genetic information in living matter*
1962	Konrad Lorenz*	*The function of colour in coral and reef fishes*
	Audrey Richards	*The pragmatic value of magic in primitive societies*
1963	C. P. Snow	*The two cultures and the sorcerer's apprentice*
1964	R. D. Keynes	*The electric eel*
1965	A. F. Huxley*	*Muscle fibres under the microscope*
1966	B. F. Skinner	*Conditioning responses by reward and punishment*
	Fred Hoyle	*The quasi-stellar objects*
	Murray Gell-Mann*	*Elementary particles*

* Those marked with an asterisk were Nobel Prize winners.

Figure 6.4 *Sir Frank Whittle standing at the Lecturer's desk before giving his RI Christmas Lecture in 1954 on the story of petroleum. His jet engines and associated apparatus are in the foreground. All the clothes and the shoes worn by the lady standing to his left were made from petroleum.*

(By kind permission of Shell International)

Figure 6.5 *Group photograph of the scientists, technicians, and administrative staff of the Royal Institution and the Davy-Faraday Research Laboratory shortly after Sir Lawrence Bragg took over his responsibilities there (David Phillips is encircled left in the middle row, and Bill Coates, right, and Uli Arndt is encircled, back row). Ronald King is on WLB's left. Almost half the team were women.*

(By kind permission of Professor A. C. T. North)

Figure 6.6 *Sir Lawrence Bragg illustrating how a fluidized bed of sand behaves so much like a liquid that a toy boat can be made to sail on it.*

(By kind permission of the BBC Photo Library)

she discussed educational matters. On another occasion, W. L. Bragg was delighted to meet at a private dinner Agatha Christie whose crime novels he had always admired.

6.3 Research at the DFRL in WLB's Day

Even though Max Perutz and John Kendrew could not be persuaded by Bragg to join him full-time at the DFRL, other crystallographers, notably David Phillips, Tony North, and the others mentioned earlier, along with expert technicians and instrument makers, were there to assist him.

With an excellent team of technical staff in the workshop at the DFRL, and many visiting scientists, some semi-retired, but still active in research, a happy family atmosphere pervaded both at the RI and the DFRL (see Figure 6.5).

It was my good fortune, as successor-but-one of Sir Lawrence, as Director of the RI and of the DFRL, to get to know many of the individuals pictured in Figure 6.5. One exceptional member of the support staff of the RI and DFRL was William (Bill) Coates—seen in the second row in Figure 6.5—who became a nationally recognized figure, once Sir (later Lord) George Porter (Sir Lawrence's successor) arranged for the RI Christmas Lectures to be televised by the BBC (see Section 8.4.7). Bill Coates seemed to know, and was able to demonstrate, all of Michael Faraday's prodigious range of lecture demonstrations and experiments.

Uli Arndt, the expert creator of X-ray generators and detectors (seen in Figure 6.5, centre top row), and David Phillips built at the DFRL their linear automatic X-ray diffractometer, the first in the world.[4,5] This instrument, adapted to make multiple simultaneous measurements of the intensities of diffraction peaks, was to have profound consequences. It played a role both in Kendrew *et al.*'s.[6] landmark advance in the determination of the three-dimensional structure of sperm whale myoglobin at 2 Å resolution in 1960, and in Perutz's parallel determination of horse haemoglobin at 5.5 Å resolution.[7] Moreover, Phillips and Blake[8] at the DFRL used the linear diffractometer to extend the myoglobin measurements to 1.4 Å resolution.[9]

6.4 Lysozyme: Its Discovery

When, in 1922, while suffering from a cold, Alexander Fleming allowed a few drops of his nasal mucus to fall on a culture of bacteria with which he was then working, he decided to put the plate to one side to see what would happen. He became excited when he discovered, some time later, that the bacteria near the mucus had dissolved away. Shortly thereafter, he found out that the antibacterial action of the mucus was due to the presence in it of an enzyme. He named it lysozyme,[1] because of its ability to 'lyse', or dissolve, the bacterial cells. Lysozyme was soon discovered in many tissues and secretions of the human body. As well as

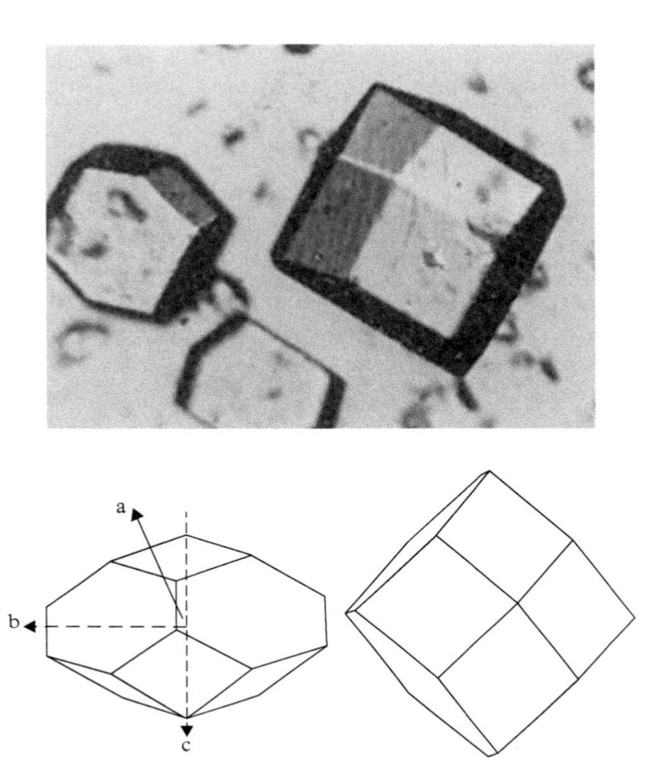

Figure 6.7 *Tetragonal lysozyme crystals with well-developed faces.*
(By kind permission of the late Professor Michael Rossmann)

nasal mucus, lysozyme has been found in tears, saliva, and the exudate from infections; in extracts from the spleen, kidney, liver, lungs, and lymph; and in especially high concentrations in cartilage. (Plants such as turnip, cabbage, and cauliflower contain it.) However, egg white is the best source of lysozyme; good yields are obtained by adding a small quantity of salt to egg white and homogenizing and acidifying it, whereupon the enzyme crystallizes out in pure form (see Figure 6.7).

6.5 Lysozyme: Its Three-Dimensional Structure and Mode of Action as an Enzyme

Important as the determination of the structure of the folded lysozyme molecule was when it was announced in the 1960s it must not be thought that this was the birth of the mechanistic understanding of enzymatic action. (The sequence of amino acid residues had been determined, as described in Chapter 3.[10]) A full account of the various studies of enzyme action has been given in the monographs by Fersht and Petsko.[11] But the first enzymatic reaction mechanism appeared as long ago as 1961, when Mathias and Robin of University College, London, published their paper on '*The Active Site and Mechanism of Action of Bovine Pancreatic Ribonuclease*'.[12] This was the first paper dealing with the roles assigned to functional groups of a protein. Shortly thereafter, mechanistic work on lysozyme (from the DFRL) and chymotrypsin (from the LMB) was soon to follow.

We first begin with the conditions that prevailed in W. L. Bragg's DFRL in the early 1960s. Apart from the linear automatic diffractometer, the DFRL, in 1961, was also well equipped with several X-ray sources (including high-powered rotating-anode ones) and a Joyce–Loebl scanning densitometer (like the one used by Kendrew and Perutz in Cambridge), and it had access (at night) to the University of London Ferranti MERCURY computer.

When Phillips, Poljak, North, and their team set out to determine the structure of lysozyme, they were aware that at least two other groups were already working on lysozyme—Dickerson and Steinrauf at the University of Illinois, Urbana, and Pauling and Corey at the California Institute of Technology. The DFRL team were also aware that a common feature of the career of their Director W. L. Bragg was competition with Linus Pauling (see Chapter 7).

The chemical make-up of lysozyme had been established (using the techniques described in Section 3.3.3) by Pierre Jollès and his colleagues of the University of Paris and by Robert Canfield[10] of the Columbia University College of Physicians and Surgeons. The molecule of lysozyme obtained from egg white consists of a single polypeptide chain of 129 amino acid subunits of twenty different kinds, as depicted in Figure 6.8. The 129-residue lysozyme molecule is cross-linked in four places by disulphide bridges formed by the sulfur-containing side chains in different parts of the molecule.

The DFRL team succeeded in preparing stoichiometric amounts of heavy atom derivatives of lysozyme, the elements in question being: platinum (introduced via K_2PtCl_4), palladium (via K_2PdCl_4), uranium [introduced via $UO_2(NO_3)_2$], and mercury (via p-chloromercuribenzene sulphonate and also via K_2HgI_4). Using the methodology described earlier (see Chapters 2 and 5) to obtain the intensities and phases of the diffraction peaks, Blake *et al.*[8] were able to derive a Fourier map of the electron density at 6 Å resolution of the structure of lysozyme in 1962. And in 1965, this DFRL team, led by Phillips and North,[9] published a three-dimensional electron density structure at 2 Å, in what became the first enzyme the three-dimensional architecture of which had been determined.

Figure 6.9 shows, in a striking manner, the way in which the folding of the sequence of amino acid forms its compact structure. The function of lysozyme is to split a particular long-chain molecule, a complex carbohydrate (sugar) found in the outer membrane of many living cells (see later). The sugar (i.e. the substrate) fits into a cleft, or pocket, formed by the three-dimensional structure of the

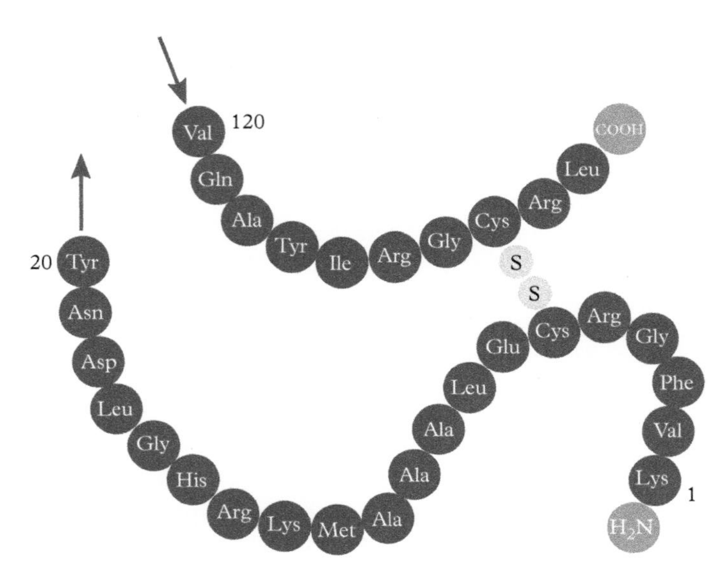

Figure 6.8 *Part of the amino acid sequence, 129 in total (see Figure 6.9), determined by chemical methods by Canfield and Liu,*[10] *in hen egg white lysozyme.*

(Courtesy, Dr D. N. Johnstone)

lysozyme molecule. In Figure 6.9, the amino acid residues that line the pocket are shown in dark green.

Louise Johnson, who was a PhD student supervised by David Phillips at the DFRL, has given a graphic account[13] of the Friday Evening Discourse at the RI on 5 November 1965 when the first public presentation of the structure was made (see Figure 6.10).

This photograph shows a model of lysozyme hanging from the ceiling, with its 129 amino acids in an extended conformation. Constructed from components on a scale of 2 cm to the angstrom, the chain was just over 10 m long. To the speaker's right (David Phillips) is a model (see Figure 6.10) of the folded protein structure illustrating the remarkable property of proteins to fold into compact structures. In Louise Johnson's words:[13] '*During the lecture the model of the protein structure was lowered from the ceiling, the lecture lights dimmed and a light path traced the chain, composed of light bulbs sequentially lit for every amino acid, with blue and red for the ionizable groups and yellow and green for the polar and non-polar groups respectively. This was a magical moment.*'

The lysozyme molecule (molecular weight 14,300), composed of 1000 or so non-hydrogen atoms, contrasted with the model of sodium chloride, shown (enlarged) in Figure 6.10, to the lecturer's left. It was W. L. Bragg, as we saw in Chapter 2, who had solved the structure of sodium chloride in 1913.

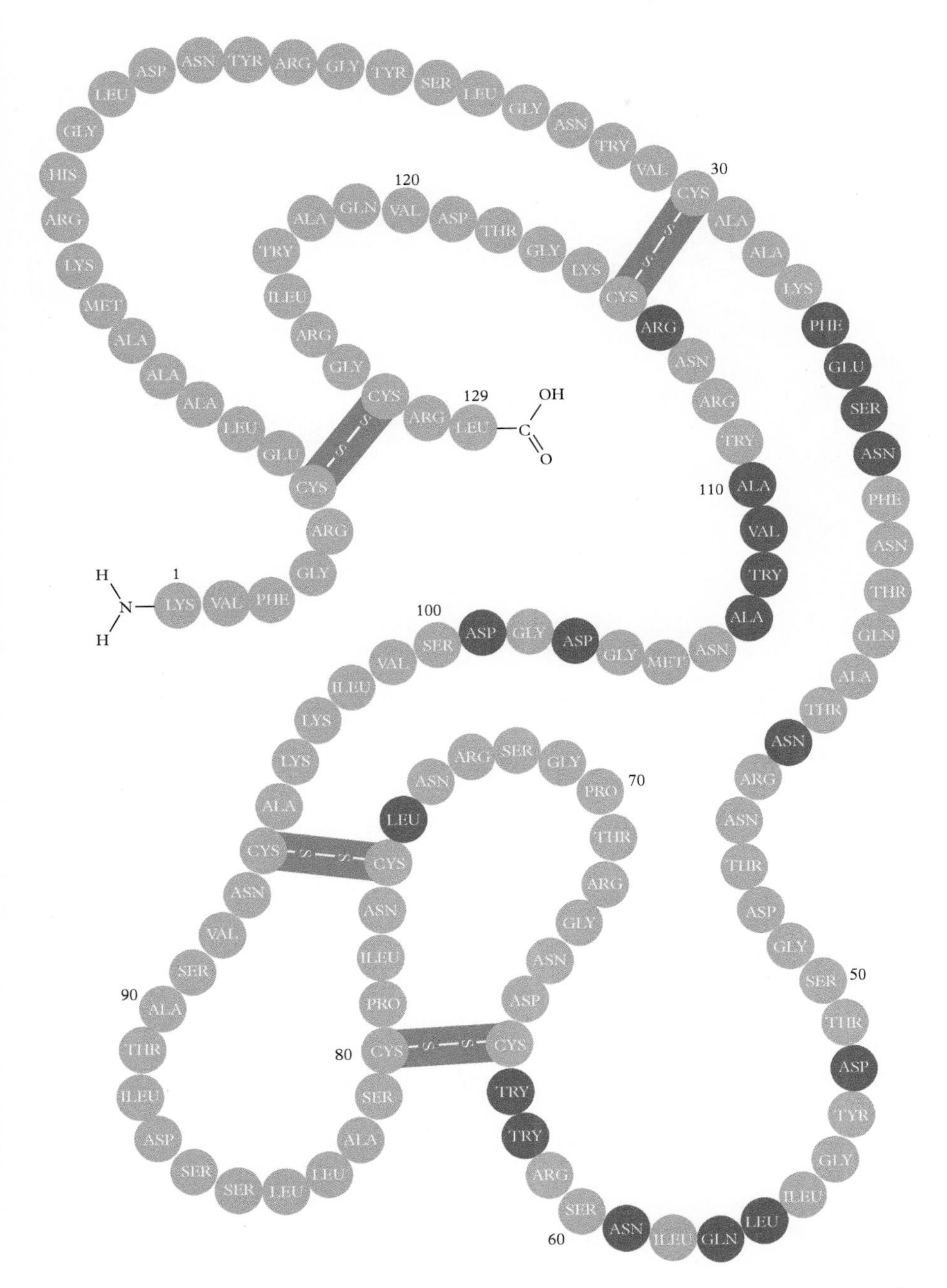

Figure 6.9 *Illustration (by D. C. Phillips) of the manner in which the 129 constituent amino acid residues are folded in the three-dimensional structure. The dark green residues line the cleft (that houses the active site) that is formed in the folded three-dimensional structure.*

(Copyright Joan Starwood)

Figure 6.10 *David Phillips presenting his famous Friday Evening Discourse at the Royal Institution in November 1965 (see text).*
(By kind permission of Prof L. N. Johnson)

6.5.1 A description of the lysozyme molecule

The main polypeptide chain appears as a continuous ribbon of electron density running through the image, with regularly spaced promontories on it that are characteristic of the carbonyl groups (CO) that mark each peptide bond. In some regions, the chain is folded in ways that are familiar from theoretical studies of polypeptide configurations and from the structural analysis of myoglobin and fibrous proteins such as keratin of hair. The amino acid residues in lysozyme can be designated by number; the residues numbered 5 through to 15, 24 through to 34, and 88 through to 96 form three lengths of 'α-helix', a confirmation that was proposed by Pauling *et al.* (as described in Chapter 4) and that was found by Kendrew *et al.* to be the most common arrangement of the chain in myoglobin. Because of some other structural irregularities that lysozyme exhibits, the proportion of its polypeptide chain in the α-helix conformation is difficult to calculate in a meaningful way, but it is clearly less than half the proportion observed in myoglobin, in which helical regions make up about seventy-five per cent of the chain (see Chapters 4 and 5). The lysozyme molecule does include, however, an example of another regular conformation predicted by Pauling and Corey in 1951, and described in Chapter 4. This is the 'anti-parallel pleated sheet' (see Figure 4.9 of Chapter 4). In it (also called 'β-sheet'), two lengths of the polypeptide chain run parallel to each other in opposite directions. This structure, like that of the α-helix, is stabilized by hydrogen bonds between the NH and CO groups of the main chain.

Louise Johnson (see Figure 6.1), who was D. C. Phillips' PhD student at the DFRL, made a detailed study[14] of the way in which lysozyme operates as an enzyme upon the components of the cell wall of the bacterium that it destroys—*Micrococcus lysodeikticus* (see Figure 6.11). This bacterium contains equi-molecular amounts of N-acetylglucosamine (abbreviated NAG) and N-acetyl-muramic acid (NAM)[14] (see Figures 6.11 and 6.12). A tetrasaccharide of the two components (as shown in Figure 6.12), on exposure to lysozyme, is converted into two dimers, as a result of the enzyme's action in severing the β-(1-4) linkage (see arrow in Figure 6.12). In effect, lysozyme hydrolyses the acetal link between the monomers of the carbohydrate. The substrate shown in Figure 6.12 also shows, like in Figure 6.11, the joined NAM and NAG components of the bacterial cell wall. The active site of the lysozyme is at the cleft that contains two key amino acids—aspartate and glutamate, labelled Asp 52 and Glu 35, respectively (in Figure 6.12, the numbers referring to the positions of these amino acids in the polypeptide chain).[15]

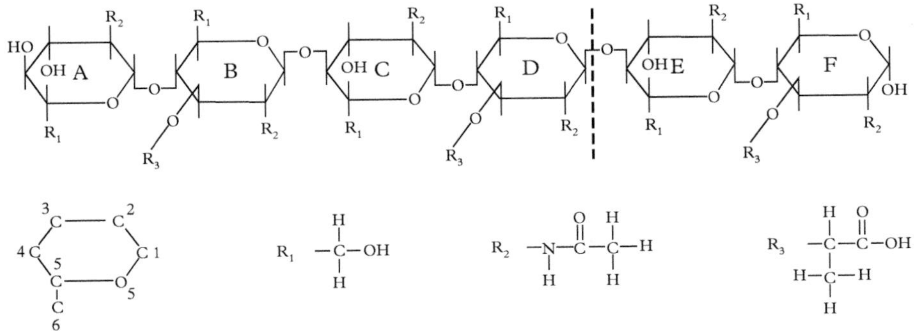

Figure 6.11 *The substrate that lysozyme breaks and destroys is a polysaccharide molecule, found in the walls of certain bacterial cells, shown here as connected rings A, B, C, D, E, and F. The polysaccharide consists of residues of two kinds of amino sugar: N-acetyl glucosamine (abbreviated NAG) and N-acetyl muramic acid (NAM). In the six constituents of the chain shown here, the first, third, and fifth (i.e. A, C, and E) residues are NAG, whereas the second, fourth, and sixth are NAM residues. The inset at the bottom left depicts the numbering scheme for identifying the principal atoms in each sugar ring. Six rings of the polysaccharide fit into the cleft of the lysozyme, which effects a catalytic cleavage between the fourth and fifth rings (i.e. D and E).*

(Courtesy, D. C. Phillips, copyright Joan Starwood)

6.5.2 Author's note: fuller details of the mode of action of lysozyme

The following section is primarily directed at those readers who are acquainted with modern ways of describing detailed mechanisms of chemical transformation. It supplements what has already been adequately described in Figure 6.11, which give the quintessential facts pertaining to the mode of action of lysozyme as an enzyme that destroys the bacteria it attacks. This section can be omitted by non-technical readers.

In its protonated form, Glu 35 acts as a weak acid and donates a proton to the carbohydrate C–O–R groups (where R is the next sugar in the chain), breaking the C–O bond. Aspartic acid 52, in its ionized form, functions as a means of stabilizing the positive charge in the transition state that builds up on the sugar (forming an oxo-carbonium ion) during catalysis. This mechanism, illustrated in Figure 6.12, was accepted as valid for many decades after it was proposed in

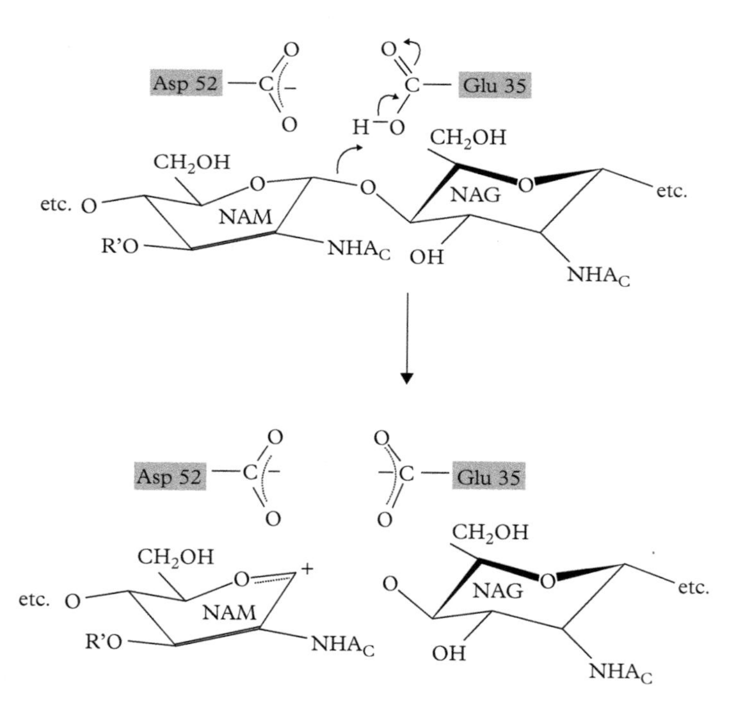

Figure 6.12 *Lysozyme hydrolyses the acetal links between NAG and NAM. Two carboxylic acid side chains (Glu 35 and Asp 52) line the active site of lysozyme. The microenvironment of the cleft that circumscribes the active site affects the pK_a of these acids. Fuller details of this mechanism are given in G. A. Petsko and D. Ringe's* Protein Structure and Function, *New Science Press, Cambridge, MA, 2004 (ref [11(b)]).*

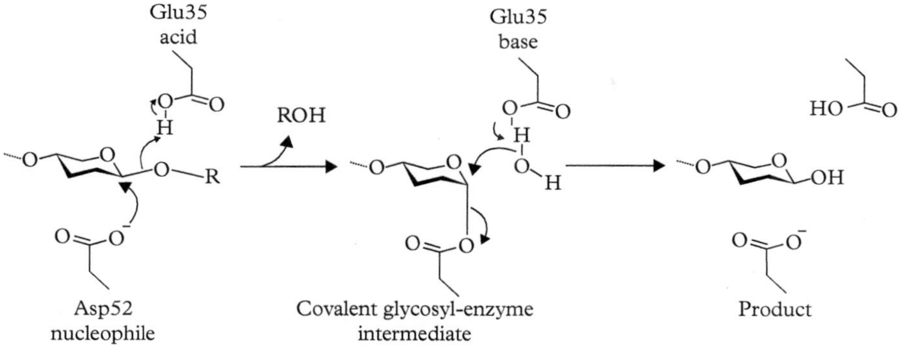

Figure 6.13 *Simplified reaction mechanism of lysozyme in which Asp 52 acts as a nucleophile to form a covalent glycosyl enzyme intermediate. Glu 35 acts as both an acid and subsequently a base in the reaction.*

(By kind permission of Professor Gideon Davies)

1965. Recently, it has come to light[16] that the mechanism may involve a transient covalent intermediate between aspartic acid 52 and the substrate. What the recent work has shown,[16] thanks to some elegant, very powerful mass spectrometry and X-ray diffraction, is that the purported lifetime of the ion pair intermediate in that proposed by the DFRL group mechanism is much less than the time required for diffusion events that are necessary for the catalytic turnover (see Figure 6.13).

Detailed insights into the precise nature of the amino acids that constitute the active site in lysozyme, as well as the locations of the substrate within the cleft that circumscribes the active site, were gained by Phillips and Johnson in a series of experiments where they added, individually, a range of inhibitor carbohydrate molecules (like NAG, NAM, and related species) to the crystalline enzyme and then determined the enzyme inhibitor structures by X-ray diffraction.[17] This was invaluable in formulating the mechanism[15] for the lysozyme's mode of action depicted in Figure 6.13. But many enzymologists now feel it has been refined by the work of Vocadio and Davies *et al.*,[16] as illustrated in Figure 6.13. One of the key general points to emerge from the DFRL work is that knowledge of structure alone is not enough to reveal the reaction mechanism.

6.6 Lysozyme: Its Role in the Evolution of Protein Crystallography and Enzymology

The achievement of the DFRL team in determining the structure of lysozyme gave particular pleasure to W. L. Bragg, not only because it coincided, in 1965, with the fiftieth anniversary of his receipt of the Nobel Prize (see Figure 6.14).

Figure 6.14 *Photograph taken in 1965 at the fiftieth anniversary of W. L. Bragg's Nobel Prize. Sitting in the row behind Sir Lawrence and Lady Bragg are (from left to right): Sir George Thomson, Dorothy Hodgkin, Sir Howard Florey, Sir Peter Medawar, Sir John Kendrew, and Sir Alan Hodgkin, all Nobel Prize winners. Sitting behind them are the following Nobel Laureates: Fred Sanger, Sir Hans Krebs, Max Perutz, E. T. R. Walton, Sir Ernst Chain, F. H. C. Crick, M. F. H. Wilkins, R. L. M. Synge, A. J. P. Martin, Sir A. V. Hill, Sir Andrew Huxley, Sir John Cockcroft, Lord P. M. S. Blackett, and C. F. Powell, as well as the Swedish Ambassador to the UK and Sir Gerald Gardiner, the Lord Chancellor of the UK. The technical and scientific staff of the DFRL were also in the RI theatre (and in this photograph) that night.*

(By kind permission of the BBC Photo Library)

W. L. Bragg was especially pleased that it was in his own laboratory, not at Caltech or the University of Illinois, that the major breakthrough in determining an enzyme's three-dimensional structure was made. It also was of significance that it was at the RI, in his Friday Evening Discourse in 1948, that Linus Pauling drew attention to the fact that scientists were then ignorant of the structure of any enzyme. It gave further pleasure to both Bragg and Perutz[18] that the lysozyme story was prominently discussed at an important Royal Society Discussion Meeting on 3 February 1966. Indeed, Max Perutz's concluding remarks at that meeting made specific reference to the fact that enzymologists and molecular biologists now had a mechanism for a particular enzymatic process.[18,19]

In Perutz's words:[18] '*What these three proteins*'—he was referring to lysozyme and to both haemoglobin and myoglobin (which he liked to call '*honorary enzymes*')—'*have in common is the distribution of polar and non-polar side chains; the interior of the lysozyme as well as that of the globin chains is made up largely of hydrocarbons, providing a medium of low dielectric constant. In each case the active centre lies in a pocket formed by the non-polar medium, and the pocket contains some polar residues which are essential for activity.*

We may now ask ourselves why chemical reactions, which normally require powerful organic solvents or strong acids and bases, can be made to proceed in aqueous solution near neutral pH in the presence of enzyme catalysts. Organic solvents have the advantage over water of providing a medium of low dielectric constant, in which strong electrical interactions between the reactants can take place. The non-polar interior of enzymes provide the living cell with the equivalent of the organic solvents used by the chemists. The substrate may be drawn into a medium of low dielectric constant in which strong electrical interactions between it and specific polar groups of the enzymes can occur.'

Perutz goes on to note that hydrolysis of polysaccharides does take place in aqueous solution, but that it would normally require much stronger acids than the β-carboxyl groups of aspartic acid, or the γ-carboxyl group of glutamic, can provide. Fundamentally, the key message conveyed by the Phillips mechanism[18,19] of lysozyme action was that the non-polar medium in the cleft of lysozyme increases the strength of the interaction between the carbohydrate and the carboxylic groups of the enzyme.

Reflecting[13] upon that period, in 1998, Louise Johnson found it appropriate to recall one of Lawrence Bragg's beautiful passages in an article that he wrote for the popular satirical British magazine *Punch*:[20]

> '*And when after the long search, some new fragment of truth has been captured, it comes not as something discovered but as something revealed. The answer is so unexpected, and yet so simple and aesthetically satisfying, that it carries instant convictions, and that wonderful never-to-be forgotten moment comes when one says to oneself "Of course, that's it".*'

6.6.1 Subsequent work on enzyme structures

Whereas only three protein structures had been solved by 1965, and only eleven by 1970, since then—thanks to the ready availability of synchrotron radiation sources and easier methods for producing crystalline proteins (by biotechnology), as well as more powerful computers—there has been an exponential rise in the annual number of protein structures solved (see Figure 6.15).

In 1990 alone, over a hundred new protein structures came to light and by mid 1991, approximately 300 protein structures had been solved, many of them of

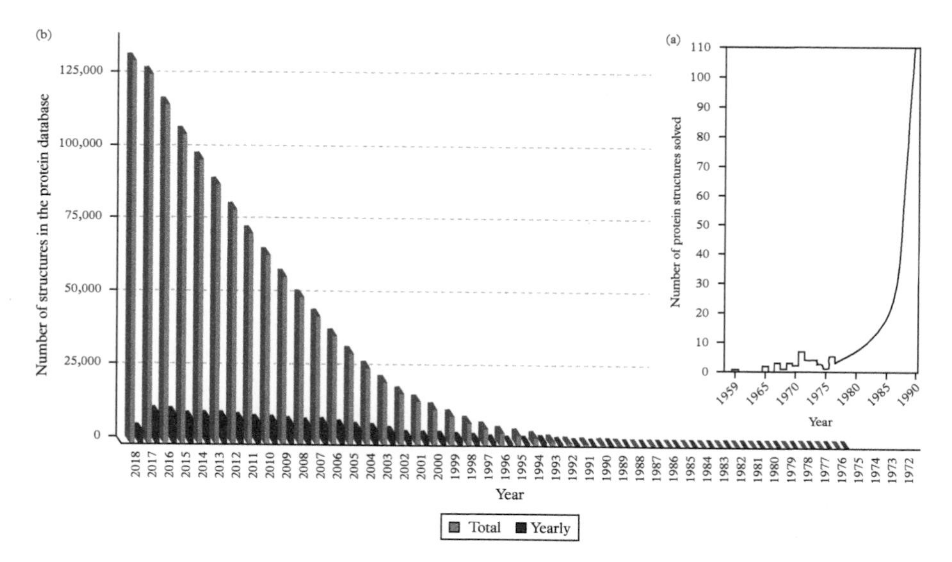

Figure 6.15 *(a) Graphical indication of the rapid growth of enzyme structural determination used by Perutz in his Faraday Discussion opening address in 1992.*[18] *Alongside it (b) is the present status of the protein structure in the Protein Data Bank. There has been an exponential growth in determined structures since the 110 or so in 1990.*

(By kind permission of the Royal Society of Chemistry) (See the Appendix to this chapter.)

practical interest to medicine. Very many other enzyme structures and the mechanisms of their mode of operation followed quite swiftly after the lysozyme breakthrough had been made.[21] Thus, the structure of crystalline complexes of α-chymotrypsin with several substrates or inhibitors was published by David Blow's group in 1970.[22–24] In the current, urgent, and worldwide quest to identify the causes and nature of misfolding of proteins—which is the root cause of neurodegenerative diseases like those associated with Alzheimer's disease, Parkinson's disease, and Pick's disease (see Section 10.6.1)—the Protein Data Bank (PDB) proves invaluable, as it is the widely consulted repository of the known, regular, folded nature of both giant and smaller protein molecules, the behaviour of which govern neural processes. Amyloids are aggregates of proteins that become folded into a shape that allows many copies of that protein to stick together, forming harmful fibrils. Pathogenic amyloids form when previously healthy proteins lose their normally physiological functions and form fibrous deposits in so-called plaques around cells which tend to disrupt the normal function of tissues and organs. Amyloids are known to arise from many different proteins, and these generally form β-sheet structures (see Section 4.3.2) that aggregate into long (harmful) fibres.

6.7 Concluding Remarks and Reflections on Current Work at the DFRL

When the date of W. L. Bragg's retirement[25] from the RI approached, David Phillips was persuaded by Dorothy Hodgkin and others, including Rex (later Sir Rex) Richards, Sir Hans Krebs, and Sir Lawrence, to take up the post of Professor of Molecular Biophysics at the University of Oxford. This he did in 1966, and from there, he accomplished further admirable work. Louise Johnson retired in 2009 after a distinguished Professorship of Molecular Biophysics at Oxford. She later took up new responsibilities at the DIAMOND Synchrotron Facility.

David Phillips was not only an outstanding research scientist and an exceptionally good scientific advisor—he shouldered, for many years, the role of a key post in British science known as Director General of the Research Councils—but he was also a fine lecturer. Always proceeding without notes, he moved smoothly from the familiar to the unfamiliar, from the old to the new, from the known to the unknown. His practice of not writing out a script for his talks caused some distress to the BBC cameramen and producers when he gave the 1980–1981 RI televised Christmas Lectures. How can such broadcasters place their cameras correctly if they do not know precisely what to expect next from the lecturer? These lectures were, however, a great success. David loved telling the following tale involving Max Perutz, who gave one of the six lectures in David's series. Max had always suffered from acute back problems (see Chapter 8), and his way of coping with it involved lying down on the floor for substantial periods. David said that when the BBC team entered the lecturer's room before Max was due to perform, they nearly had apoplectic fits when they saw him flat on the floor. They thought he had fainted just before going on the air. But Max was simply relieving his back pain before his performance in the lecture theatre!

When David Phillips took over his responsibilities at Oxford, he was a great success. Not only did he lead a team of gifted co-workers, he also played an important role as chairman of the so-called Oxford Enzyme Group that regularly brought together chemists, biochemists, and other natural scientists (see Section 8.3.3). He also found the time (for five years) to serve as the Biological Secretary of the Royal Society, and he was a founding member of the Cabinet Office Advisory Council for Applied Research and Development (ACARD) on which I also served (1982–1986). He also wrote one of the very best Biographical Memoirs of Fellows of the Royal Society on Sir Lawrence Bragg.[26]

6.7.1 What has happened subsequently to the centre of research excellence at the RI?

The viability of the DFRL, from its outset, following the bequest of Ludwig Mond, was always a little precarious. Generous as the action of Mond was in founding it at the RI, it still required great ingenuity on the part of those elected as Directors of the DFRL to ensure that it was well equipped and also peopled by the right innovative technical staff who assisted the regular research teams.

In this regard, W. H. Bragg was eminently successful, as may be gleaned from Chapter 3. He ensured that his researchers were backed up by a team of expert technicians. For example, he recruited to the DFRL a technical supremo—Jenkinson,[27] who had constructed for him his renowned, transformative X-ray spectrometer (see Section 2.2) at Leeds University, and then moved with him to the Physics Department at University College London. In addition to making the crucial appointment of Jenkinson, W. H. Bragg also had several other technicians who could build equipment that was not available commercially but was seen to be essential for use by his team of gifted co-workers.

On reading the recollections of Bernal, Astbury, Robertson, Cox, and Lonsdale (see Section 3.3 and references therein—see also ref [27])- it becomes apparent that the major advances these protégés of W. H. Bragg were able to achieve depended crucially upon the fact that the DFRL was an exceptionally well-equipped laboratory. When Sir Eric Rideal[28] relinquished his Professorship of Colloid Science at the University of Cambridge to become Fullerian Professor of Chemistry at the RI, he, like W. H. Bragg before him (and Lawrence Bragg after him), also established a very well-equipped laboratory there. Outstanding research work in heterogeneous catalysts emanated from the DFRL in the short period that he served at the helm.

One of the most judicious of Rideal's appointments to the technical staff was Bill Coates (see Section 8.4.7), who was adept not only in the construction of new instruments (like the linear diffractometer conceived by David Phillips and Uli Arndt—see Section 6.3), but also in imparting his own knowledge to younger associate technicians.

Research scientists who were in Rideal's team[29] told me that a noteworthy feature of the DFRL at that time was Rideal's habit of visiting, at least once every working day, each member of his research team, at the bench (so to speak), to thrash out difficulties and to plan future experiments.

As mentioned earlier (see Section 6.1), Rideal's successor as Fullerian Professor and Director of the DFRL was E. N. da C. Andrade whose tenure was a troubled one, during which the morale of the technical and other staff was at a low ebb. When Lawrence Bragg took over the responsibilities at the RI and DFRL, a centre of research excellence was in due course re-established. A happy atmosphere (see Figure 6.5 above and Figure 2.19), which Max Perutz regarded as a *sine qua non* for a successful research centre, pervaded the DFRL and this nurtured top-quality research.

The DFRL continued to be a world-class centre under Lawrence Bragg's successor Sir George (later Baron) Porter. Porter's work on flash photolysis, started in Cambridge in his days as a graduate student of R. G. W. Norrish and later developed further by him as Head of the Department of Chemistry at the University of Sheffield, was universally regarded as seminal in tracking the progress of fast chemical reactions.

For twenty years during Porter's tenure, the DFRL, though small by most university standards, continued to execute research work of the highest quality, largely in fields allied to photosynthesis, where the techniques developed by Porter and his colleagues in tracing the lifetime of short-lived intermediates were used to great effect.[30] Two of Porter's former colleagues at the DFRL became Heads of some of the most prestigious research centres in the world: David Phillips (no relation to Lord Phillips) at Imperial College London; and Graham Fleming at the University of California, Berkeley, where he also served as Vice Chancellor.

In 1986, I took over as Fullerian Professor of Chemistry and Director of the RI and the DFRL. Shortly thereafter, I was joined by Professor C. R. A. Catlow[31] and his team of researchers at University College London. In my day, a thriving team of researchers contributed significantly to the solution of many physico-chemical problems, especially in the design and application of new heterogeneous catalysts required for green chemistry, clean technology, and sustainability.[32,33] Single-site heterogeneous catalysts (SSHCs) are the inorganic analogues of enzymes. They are synthesized in such a manner that there is only one crystallographically distinct kind of active site present, just as in any enzyme. The great advantage of SSHCs is that they give molecularly pure products, can be studied readily under *in situ* conditions, and are of great value in enabling selective oxidations and reductions under environmentally responsible conditions, like the synthesis of a vitamin B_3 (e.g. niacin) using benign reagents.[34]

Like Sir Eric Rideal, I was obliged to resign as Director because of family illness, but I continued as a researcher at the DFRL until 2006.

For seven years, from 1991, my successor was Professor Peter Day, formerly of the University of Oxford. He continued his pioneering work to good effect as an inorganic chemist at the DFRL. His successor as Director of the RI was a pharmacologist—Susan Greenfield, later Baroness Greenfield, at the University of Oxford; she did not pursue research work at the RI. The DFRL was led by Professor Catlow until he migrated to University College London. The decision was made by the Director of the RI, with the support of its Council in this period, to cut down drastically the number of technical staff working at the RI. The workshop was closed, and part of its premises was used to house a restaurant. With the passage of time, comparatively little original research has been prosecuted at the DFRL. Whereas typically several hundred original papers were published per annum in premier journals in 1998, by 2008, only a handful were published. At present, the DFRL is in abeyance, owing to a major financial crisis that the RI faced for several years after the departure of the Director of the RI in 2010.

In conclusion, one recalls that, notwithstanding the outstanding successes in research carried out at the RI—by Davy, Thomas Young, Faraday, and Tyndall, all in the period prior to the establishment of the DFRL—and subsequently by Dewar, Rayleigh, the two Braggs, and their successors, this small research centre has essentially come to what is, effectively, a halt. It may again be resurrected.[35] This is the hope of all those who have followed its fortunes.[36] The art of planning and managing a research laboratory is a subject that is fully discussed in Section 8.2.2, which deals, in particular, with Max Perutz's policies in this regard.

APPENDIX

Origin of the Protein Data Bank

As J. D. Bernal became increasingly aware that X-ray diffraction could yield the detailed, atomically resolved structure of both small and large molecules, he urged several colleagues to consider establishing central databases. Early on, Dr Olga Kennard, a crystallographer who had solved the structure of adenosine triphosphate (ATP),[37] was inspired by Bernal to establish the Cambridge Structural Database (CSD), of which she became the first Director in 1965.[38] At the CSD, which was based at the University Chemical Laboratory, Kennard was also involved in the founding of the PDB in 1969, when workers at Brookhaven National Laboratory and Texas A and M University collaborated on the so-called Brookhaven Raster Display to visualize protein structures in three dimensions. The inauguration of the PDB, announced in the journal *Nature New Biology* in 1971, involved a joint venture between the Cambridge Crystallographic Data Centre (CCDC—its now name), UK, and the Brookhaven National Laboratory, USA.

Olga Kennard was also involved in the founding of the EMBL nucleotide sequence data library, later called the European Nucleotide Archive.

The CCDC is the repository of the structural details of 'small' molecules—in comparison to those of proteins—but it does contain the structure of many natural products. At present (March 2019), the details of nearly one million structures are housed in the CCDC.

REFERENCES

1. A. Fleming, *Proc. Roy. Soc. B*, **1922**, *93*, 306.
2. A. Fleming, *Br. J. Exp. Path.*, **1929**, *10*, 226.
3. The brilliant lectures, given by Sir Lawrence, delivered in his characteristic avuncular style, may be viewed on *YouTube* where his insights and lucidity can be savoured anew. See also '*Crystal Clear: The Autobiography of Sir Lawrence and Lady Bragg*' (eds. A. M. Glazer and Patience Thomson), Oxford University Press, Oxford, **2015**.
4. U. W. Arndt and D. C. Phillips, *Acta Cryst.*, **1961**, *14*, 807.
5. U. W. Arndt, A. C. T. North, and D. C. Phillips, *J. Sci. Instrum.*, **1964**, *41*, 421.
6. J. C. Kendrew, R. E. Dickerson, B. E. Strandberg, *et al.*, *Nature*, **1960**, *185*, 422.
7. M. F. Perutz, M. G. Rossmann, A. F. Cullis, H. Muirhead, G. Will, and A. C. T. North, *Nature*, **1960**, *185*, 416.
8. C. C. F. Blake, R. H. Fenny, A. C. T. North, D. C. Phillips, and R. J. Paljak, *Nature*, **1962**, *196*, 1173. See also H. C. Watson, J. C. Kendrew, C. L. Caulter, and C. Branden, *Acta Cryst.*, **1963**, *16*, A81.
9. C. F. Blake, D. F. Koenig, G. A. Mair, A. C. T. North, D. C. Phillips, and V. R. Sarma, *Nature*, **1965**, *206*, 757.
10. R. E. Canfield and A. K. Liu, *J. Biol. Chem.*, **1965**, *240*, 1997.

11. (a) A. R. Fersht, '*Enzyme Structure and Mechanism*', third edition, W. H. Freeman, New York, NY, **2018**.
    (b) G. A. Petsko and D. Ringe, '*Protein Structure and Function*', New Science Press, Cambridge, MA, **2004**.
12. D. Findley, D. G. Hermes, A. P. Mathias, B. R. Robin, and C. A. Ross, *Nature*, **1961**, *190*, 781.
13. L. N. Johnson, *Nature Struct. Biol.*, **1998**, *5*, 842.
14. L. N. Johnson and D. C. Phillips, *Nature*, **1964**, *202*, 588.
15. L. N. Johnson, *Acta Cryst.*, **1966**, *21*, 885.
16. D. J. Vocadio, G. J. Davies, R. Laine, and S. G. Withers, *Nature*, **2001**, *412*, 835.
17. L. N. Johnson and D. C. Phillips, *Nature*, **1965**, *206*, 761.
18. M. F. Perutz, *Faraday Disc.*, **1992**, *93*, 1.
19. M. F. Perutz, *Proc. Roy. Soc. B*, **1967**, *167*, 448.
20. W. L. Bragg, *Punch*, **1968**, *235*, 352.
21. C. C. F. Blake, D. E. P. Grace, L. N. Johnson, *et al.*, in '*Molecular Interactions and Activity of Proteins*' (eds. R. Porter and D. W. Fitzsimons). CIBA Foundation, **1979**, p. 137.
22. R. Henderson, *J. Mol. Biol.*, **1970**, *54*, 341.
23. T. A. Steitz, R. Henderson, and D. M. Blow, *J. Mol. Biol.*, **1969**, *46*, 337.
24. R. Henderson and J. H. Wang, *Ann. Rev. Biophys. Bioeng.*, **1972**, *1*, 1.
25. Sir Lawrence remained as a researcher in the DFRL during (Lord) George Porter's term as Director until the time of his death in 1971.
26. D. C. Phillips, *Biographical Memoirs of Fellows of the Royal Society, Sir Lawrence Bragg*, **1979**, *25*, 1.
27. See J. M. Thomas, *Notes and Records of the Royal Society*, **2011**, *65*, 163.
28. Sir Eric Rideal (1890–1974) was an exceptionally versatile and gifted physical chemist, who made significant contributions to a wide range of science, including electrochemistry, chemical kinetics, catalysis, electrophoresis, colloids, and surface chemistry. He became Professor of Colloid Science at Cambridge in 1930, and his laboratory there was acknowledged as a world centre for surface science. He left Cambridge at the cessation of World War II and took up the post of Fullerian Professor of Chemistry at the RI and the Directorship of the DFRL. He set up an excellent research centre; he also gave the RI Christmas Lectures on '*Chemical Reactions: How They Work*' and initiated the still continuing (roughly annual) series of '*Advances in Catalysis*'. For personal reasons, he resigned from his posts at the RI. He was followed by E. N. da C. Andrade (see Sections 3.6 and 6.1).
29. Notably the late B. M. W. Trapnell, who later wrote the first definitive university text om the topic of chemisorption, the formation of sub-monolayer amounts of species on solid surfaces.
30. G. R. Fleming and D. Phillips, *Biographical Memoirs of Fellows of the Royal Society*, **2004**, *50*, 257.
31. C. R. A. Catlow, apart from being a solid-state chemist and an expert on the use of synchrotron radiation for the elucidation of numerous chemical problems, was also one of the earliest computational chemists in the UK. This work has earned him many honours. He now holds the post of Foreign Secretary of the Royal Society, and he carries out his research work at Cardiff University and University College London.
32. J. M. Thomas, '*Design and Applications of Single-Site Heterogeneous Catalysis: Contributions to Green Chemistry, Clean Technology and Sustainability*', World Scientific Publishing, Singapore, **2012**.

33. K. D. M. Harris and J. M. Thomas, '*Selected Scientific Papers of Sir John Meurig Thomas*', World Scientific Publishing, Singapore, **2015**.
34. R. Raja, J. M. Thomas, M. Greenhill-Hooper, S. V. Ley, and F. A. Almeida Paz, '*Single-site catalytic green production of niacin*', *Chemistry: A European Journal*, **2008**, *14*, 2340.
35. This is the aim of the newly elected (in 2018) Director Professor Shaun Fitzgerald.
36. Of the RI, a distinguished former President of the US National Academy of Sciences—Joseph Henry—once said: 'More light has issued from that establishment in proportion to its means than perhaps from any other on the face of the earth'.
37. O. Kennard and N.-W. Isaac, *Nature New Biology*, **1970**, *225*, 333.
38. O. Kennard, '*Bernal's vision: From Data to Insight*', J. D. Bernal Lecture, Birkbeck College, London, **1995**.

7

Lawrence Bragg and Linus Pauling: Comparisons and Rivalries

7.1 Introduction

Among their contemporaries, and also among successive generations of scientists, these two giants elicited admiration that sometimes included reverential awe and outright adulation. Bragg and Pauling each contributed enormously to the growth and power of crystallography and its influence on our understanding of the nature of both the inorganic and biological world. They each exhibited great respect for one another, but there were subterranean rivalries—even envy—between them.

As very young individuals, they both displayed a precocity that was unusual; in later life, they each won Nobel Prizes—Bragg at the age of twenty-five!—and were accorded numerous awards throughout the world. As a teenager, while pursuing his interests in seashells on the shores of his native South Australia, Bragg discovered a new species of cuttlefish, now named *Sepia Braggi*. And, in 1910, Pauling's father wrote a letter to *The Oregonian*, a local newspaper, asking for advice about suitable reading matter for his nine-year-old son, who had already read the Bible and Darwin's '*Origin of Species*'.[1]

As has been abundantly apparent from preceding chapters, both Lawrence Bragg and Linus Pauling possessed exceptional skills as scientists; their penetrating intellects and imaginative insights, no less than their leadership qualities in two of the foremost university laboratories in the world, were, and continue to be, universally acclaimed. But in personal habit, mode of working, and style of teaching, as well as in their range of interests and the public images that they projected, they were very different.

Whereas Bragg focused very largely on X-rays and the information that could be gleaned from their diffraction by numerous different kinds of solids—more than eighty per cent of the 260 or so articles and all but three of the thirteen books that he wrote (or co-edited) were devoted to X-ray-related topics; Pauling, on the other hand, exhibited an exceptionally wide range of scientific interests: quantum mechanics, mineralogy, structural chemistry, medicine, immunology, anaesthesia, and evolution. He wrote over 1200 articles and numerous books dealing with quantum mechanics, as well as social, political, and medical topics, including his monumental '*The Nature of the Chemical Bond and the Structure of Molecules and Crystals*',[2] which almost all chemists and many others regard as one of the most influential texts in chemistry published in the twentieth century. Max Perutz, in particular, is on record as saying that this text by Pauling was one of the most transformative books he ever read.

Bragg and Pauling overlapped in their interests in silicate minerals and also in the structure of proteins, two areas of study where their rivalry was most apparent, as is described in Section 7.2.

Bragg was a devoted family man, very much in his element when surrounded in his garden by his children and grandchildren. He gave superb public lectures, and his enviable gift of metaphor and analogy won the hearts of his listeners, especially young children. All this pedagogic skill was combined with other commendable qualities, as Max Perutz explained in a 1965 BBC broadcast to celebrate Lawrence Bragg's fiftieth anniversary of his winning the Nobel Prize: '*I have always been tremendously impressed by the speed and clarity of Bragg's mind and his power of scientific judgement. While others doubt and hesitate, Bragg would see the importance of a scientific discovery in a flash. If he conceives a scientific idea of his own, he goes home in the evening to write it up and comes back in the morning with a finished paper ready for the typist – rather like Mozart writing the overture of the Marriage of Figaro in a single night. Not a word needs changing. His mind leaps like a prima ballerina with perfect ease. What is so unique about it, and this is what made his lectures so marvellous, is the combination of penetrating logic and visual imagery. Many of his successes in crystal structure analysis were due to this power of visualising the aesthetically and visually most satisfying way of arranging a complicated set of atoms in space and then, having found them, with a triumphant smile, he would prove the beauty and essential simplicity of the final solution.*'

As described in Chapter 6, Lawrence Bragg was more comfortable presenting science to young people in the illuminating manner that Michael Faraday pioneered at the Royal Institution (RI), where he produced exciting lecture-demonstrations '*to a juvenile auditory*', than he was in addressing large audiences at international symposia, although he rose to those occasions with relative ease.

Linus Pauling, on the other hand, being blessed with highly developed histrionic skills, tinged with a touch of vanity (even conceit[11]), relished giving spellbinding performances to all the audiences that he was called upon to address. And he was certainly fully aware of his own intellectual and communicative prowess. His former postdoctoral research associate and author of his biographical memoir for the Royal Society Professor Jack Dunitz of the ETH, Zurich, remembered Pauling's lectures as Eastman Visiting Professor of Chemistry in Oxford in 1948: ' *...I had never heard anyone quite like him, with his jokes, his relaxed manner, his seraphic smile, his slide-rule calculations and his spontaneous flow of ideas. (Only much later did I realise that much of that apparent spontaneity was carefully studied.)*' (see the Appendix to this chapter).

7.2 Rivalry Concerning the Structures of Silicates and Other Minerals

Shortly after Lawrence Bragg took over from Rutherford as Head of Physics at the University of Manchester in 1919, he undertook, with a group of gifted co-workers, a series of beautiful crystallographic studies of minerals, especially silicates and alumino-silicates that, together, constitute a large fraction of the crust of the Earth. Many gemstones and other important materials, like beryl, chrysoberyl, olivines, and feldspars, and chain structures such as amphiboles and pryoxenes like diopside ($CaMgSi_2O_6$), and layered materials like mica—of which there are more than thirty different varieties—were studied by Bragg and his school. One of the micas—muscovite [idealized formula (K,Al_2) ($AlSi_3O_{10}$) ($F,OH)_2$]—the most common mica, found in granite schists and as a metamorphic rock, was investigated in Bragg's school, as were many other important oxidic minerals such as rutile, corundum, and topaz (see Section 2.4). All these minerals were also of great interest to Linus Pauling, who had determined their atomic structure, along with the structures of many other naturally occurring inorganic solids, as a young professor at the California Institute of Technology, Pasadena (see Section 2.4.4).

By the early 1930s, both Bragg and Pauling had published numerous elegant papers on the structures of silicates and the other materials cited above. Whereas Bragg was the first to appreciate that the oxygen ion O^{2-} occupied a relatively large volume in inorganic solids consisting of it, it was Pauling, through his so-called *Pauling Rules* (see below)[3–5] who brought order and coherence into the seemingly incomprehensible body of factual information pertaining to the full range of silicate and alumino-silicate structures that both the Bragg and Pauling schools, between them, had assembled.

Whereas simple ionic substances, such as alkali halides, are limited in the types of crystal structure they can adopt, the possibilities open to more complex materials such as silicate minerals may appear to be immense. *Pauling's Rules* themselves are concerned with radius ratios, electrostatics, and the sharing of polyhedra (e.g. whether a hexagonal array of oxide ions are corner-sharing, edge-sharing, or face-sharing with other like polyhedra). These *Rules*, as Pauling emphasized when he enunciated them, are neither rigorous in their derivation nor universal in their application—they were obtained, in part, by induction from known structures and, in part, from theoretical considerations.

Pauling's second *Rule* states essentially that electrostatic lines of force stretch only between nearest neighbours. Lawrence Bragg, who felt ambivalent about *Pauling's Rules* because the body of experimental material from which they were deduced had been generated in considerable measure by Bragg's group in Manchester, was generous enough to state later:[5] '*The second rule appears simple, but it is surprising what rigorous conditions it imposes upon the geometrical configuration of a silicate... To sum up, these rules are the basis of the stereochemistry of minerals.*'

Both Bragg and Pauling, as well as others, had noted early on that the observed interatomic distances in simple inorganic crystals were consistent with the approximate additivity of characteristic radii associated with the various cations and anions. Among the many sets that were proposed, Pauling's are not merely designed to reproduce the experimental observations but are derived from a mixture of approximate quantum mechanic calculations and experimental data. Pauling's values, derived nearly ninety years ago, are still in common use. Moreover, Pauling's realization that isomorphous substitution, in silicate and alumino-silicate minerals especially, could be readily understood in terms of equivalence, or near-equivalence, of ionic radii made it possible to understand the wide varieties of stoichiometry exhibited by these materials. Thus, Li^+ and Mg^{2+} ions have similar ionic radii, as do Al^{3+} and Si^{4+}. Such simple facts, highlighted by Pauling, brought a greater understanding to the huge variety of compositions that silicates and alumino-silicates display.

7.3 Pauling's Indebtedness to, and Contacts with, Lawrence Bragg

In correspondence from Linus Pauling that Sir David Phillips and I received at the RI in the early summer of 1990,[6] we learnt that, shortly after he had been accepted as a graduate student at Caltech in 1922, he acquired a copy of the book that Bragg had written with his father entitled '*X-Rays and Crystal Structures*'[7] and that he benefitted greatly from that monograph in his studies for his doctorate. He also revealed that, in 1922–1923, he read the literature, including Bragg's papers and efforts to assign radii to atoms.

He then disclosed that, in the spring of 1930, Pauling and his wife and eldest son spent a month in Manchester where Lawrence Bragg was particularly helpful in making the Paulings feel at home. But despite the warm welcome, Pauling felt bitterly disappointed that Bragg entered into no scientific discussions with him, and he was distressed that Bragg never invited him to present a seminar at Manchester. Pauling's own words were: '*It astonished me to discover that the scientific atmosphere in Manchester was much different from that in Pasadena. In Pasadena, I was accustomed to attending about three seminars per week, in chemistry, physics and astronomy, whereas there were no seminars in Manchester during the month that I spent there. Cambridge was different. Science was livelier there, and when we stopped in Cambridge for a few days, Bernal asked me to give a seminar on the rotational motion of molecules in crystals,*[8] *which provoked a lively discussion.*'

In spring of 1948, during his tenure of the Eastman Professorship at Oxford, Pauling also visited Lawrence Bragg at Cambridge. Pauling said: '*At that time, too, Bragg and I did not have any serious discussions about science.*' When he visited Bernal that year, there were animated scientific discussions. Then Pauling revealed: '*Some years later I was told that Bragg resented my having intruded into the fields of crystallography and mineralogy in which he was working, and that he considered me to be a competitor. This information came as a shock to me. I had thought of Lawrence Bragg as a member of an older generation of great scientists, who had made very important discoveries, and in a sense I had thought of him as one of my scientific heroes. In retrospect, I am sure that in 1930 I would have rejected the idea that I was a competitor of Bragg. I did not think of my own scientific work as being competitive; I found it engrossingly, interesting for its own sake. In 1930 I was still thinking of myself as a student, with Bragg the professor, the member of the older generation who had made the great discovery of determining the structure of crystals by X-ray diffraction. Bragg was always courteous and gentlemanly. I now regret that I did not have enough insight to have enabled me to take such action as would have permitted a more intimate friendship to have been developed between us.*

Without having much supporting evidence, I surmise that circumstances during the second half of the 1920s may have influenced Bragg's life. At that time, although still quite young, he held an important position in science, involving administrative and teaching duties, as well as the direction of research. As a result, when quantum mechanics was discovered he was not in a position to devote enough time to this rather complicated and somewhat abstruse subject to master it. I suggest that he may have felt handicapped by this lack, and that it may have kept him from entering into lively scientific arguments and discussions.

I am glad to remember my association with a fine man, Lawrence Bragg. I wish that it had been closer.'

Acquaintances of mine, who knew Lawrence Bragg very well, especially Jack Dunitz [who had the privilege of being a professional colleague of both Pauling at Caltech and Lawrence Bragg at the Davy-Faraday Research Laboratory (DFRL)] have disclosed that, indeed, Bragg was disappointed that the rational and illuminating rules of Pauling had not occurred to him first, especially since many of the substances to which they applied were studied by him. Others have pointed out to me that, had Bragg entered into active scientific discussions with Pauling in 1930, it would have become clearer to him that Pauling had used many of the data acquired by Bragg and his team.

7.3.1 A genius who knew nothing

A book review of the biography by G. K. Hunter of Lawrence Bragg,[9] by Sir Brian Pippard (1920–2008), who was Lawrence Bragg's successor-but-one as Cavendish Professor of Physics at Cambridge (1971–1982), appeared in the journal *Physics World* bearing the above title. Pippard was also an undergraduate at Cambridge when Lawrence Bragg was Head of the Cavendish Laboratory and, later, they collaborated on a problem in crystal physics. Still later, he became Bragg's friend and knew a great deal about Bragg's academic distinction, his inner doubts and occasional sense of inadequacy and '*the rare, but explosive angers of the suave and kindly Edwardian gentleman who had charge of our destinies*'.[10]

Pippard, in this article, also quotes the distinguished geophysicist Sir Edward Bullard (1907–1980), who once remarked: '*Bragg can't stand having anyone cleverer around, and it's lucky for the Cavendish he's so clever himself*'. In his article,[10] Pippard also echoes the surmise of Linus Pauling concerning Bragg's unfamiliarity with quantum mechanics when he recited the story of the German scientists in the 1930s wondering: '*How does Bragg discover things? He doesn't know anything!*' (see Section 9.4 for specific details).

This story was also related to me by Sir Aaron Klug, who had heard it from one of his mentors Professor R. W. James (a former associate of Bragg's at Manchester). James had heard these remarks in the 1950s when he visited the University of Leipzig, after Bragg had lectured there earlier.

7.4 Polypeptide Structures and the α-Helix

In Chapter 4, attention was drawn to the two cardinal errors that Bragg, Kendrew, and Perutz made in their 1950 article on '*Polypeptide Chain Configurations in Crystalline Proteins*':[11] (i) they were oblivious to the planarity of the peptide link; and (ii) they discounted the possibility that non-integral numbers of turns of amino acids could yield satisfactory helical structures.

Bragg himself was particularly irked to learn, not only from Pauling's rebuttal,[12,13] but also from his Cambridge colleague Lord Todd (see Figure 7.1), that he and his co-workers had not incorporated the planarity of the peptide link into their elegantly presented paper. It is known that all three authors regarded this paper in retrospect as their weakest, notwithstanding the compelling arguments and candid confessions that they had articulated about the need for more work that was expressed in its beautiful narrative. It added salt to their wounds when Pauling, in a later article, exulted in the fact that he, unlike the Bragg team, had been trained as a chemist, who had acquired crystallography as a necessary technique for his chemical adventures. He also asserted that the Cavendish authors were physicists, who would have been blind to the fact that non-integral numbers of residues were perfectly acceptable. As it happened, both Perutz and Kendrew *were* trained as chemists, which added further embarrassment to the Cavendish trio.

When, in 1950, Perutz pointed out[14] that Pauling's proposed structure for the α-helix was further substantiated by the observation (made by Perutz, but not by Pauling) of a 1.5 Å repeat distance, observed by X-ray diffraction, Pauling himself became furious (see Section 4.5), and Bragg was gratified—see '*I wish I'd made you angry earlier*'.[15]

We also saw in Section 4.5 that Pauling, in 1948, when he was already gestating his picture of the α-helical structure of polypeptides, visited the Cavendish Laboratory and was shown (by Perutz) the models that were to appear in their controversial 1950 paper,[11] and Pauling made no comment. Pauling did not disclose his picture of the α-helix until later.

Figure 7.1 *This photograph taken in 1948 shows two of the greatest chemists of the twentieth century: Alexander (later Lord) Todd and Linus Pauling. Todd was awarded the Nobel Prize in 1957 for his work on nucleic acid, and Pauling in 1954 for his on the chemical bond. These two giants (Todd was forty-one at the time, and Pauling forty-seven) are seen in action on a punt on the Cam River when Pauling visited Cambridge.*

(Courtesy Oregon State University Special Collection and Archive)

Lawrence Bragg, as well as all of the Kendrew–Perutz team in Cambridge, was jubilant when Kendrew's prolonged studies of myoglobin revealed unmistakable evidence[16]—the first ever experimental one—of the existence of the α-helix as a structural motif.[16] (See Figure 4.2 of Chapter 4.) Dickerson[17] has described how, at the drinks celebratory party held on the lawns of Kendrew's College, Peterhouse, shortly after the myoglobin structure was solved, Lawrence Bragg, with delight, ushered many of those present to examine the α-helix that Kendrew and his team had uncovered.

7.4.1 DNA and lysozyme

The model-building approach pioneered by Pauling, which led him to the discovery of both the α-helix and the β-pleated sheet motifs of polypeptides was also the approach that led to the discovery of the Watson–Crick structure of DNA. But Pauling himself did not succeed in arriving at the correct structure of DNA. Why was this so?

Dunitz[1] has given plausible reasons why Pauling's model-building approach was so unsuccessful (in his hands) with DNA. First, the time factor. Whereas Pauling had thought about polypeptide structures for more than a decade before he published his conclusions, for DNA, he had thought about its structure for only a few months. Second, massive amounts of available quantitative information pertaining to simple and progressively more complicated peptides—metrical and stereochemical—were at hand (garnered largely in Pauling's laboratory). '*There is no doubt in my mind*', said Dunitz,[1] '*if Pauling had had access to Rosalind Franklin's X-ray photographs he would immediately have drawn the same conclusion as Crick did, namely, that the molecule possesses a two-fold axis of symmetry, thus pointing to two chains running in opposite directions and definitely excluding a three-chain structure.*'

Some molecular biologists of my acquaintance have told me that if the US State Department had not prevented Linus Pauling from attending the Royal Society Discussion Meeting on Proteins in 1952, Pauling might well have got the right structure for DNA before Crick and Watson, as he would have seen, on his visit to London, the two-fold axis of symmetry alluded to above. Interestingly enough, according to his son Peter's account,[18] some twenty years later, he never felt he was in any sense '*in a race*'. There is also another factor—Pauling, at that time, was under severe harassment from the Federal Bureau of Investigation (FBI) and other agencies for his political views and activities, which must have taken up much emotional and mental energies in those months (see also Section 4.5).

Turning to lysozyme, it was a source of great delight to Lawrence Bragg that it was the team that he had assembled at the DFRL, led by David C. Phillips,[19] that succeeded in solving the first ever three-dimensional structure of an enzyme by X-ray diffraction. There were many reasons for Lawrence Bragg's jubilation when lysozyme's structure was solved at the DFRL. First, the detailed structure revealed the existence of both the β-pleated sheets (not discovered hitherto) and the α-helices proposed by Pauling, but not observed by him. Second, Bragg was aware of the extremely impressive Friday Evening Discourse that Pauling gave at the RI on 27 February 1948[20] on '*Nature of the Forces Between Large Molecules of Biological Interest*'. In that Discourse, Pauling bemoaned the fact that scientists were ignorant of the structure and mode of action of any enzyme. Bragg's favourite scientific son Sir David Phillips and his team had now provided the answer[21] (see Chapter 6, Sections 6.5 and 6.6).

Third, the general notion of complementariness in molecular structure—an idea proposed in the famous Pauling–Delbrück paper[22] of 1940—was beautifully vindicated by the lysozyme work done at the DFRL. It is appropriate to recall Pauling's precise words in his 1948 Discourse: '*I think that enzymes are molecules that are complementary in structure to the activated complexes of the reaction that they catalyse, that is, to the molecular configuration that is intermediate between the reacting substances and the products of reaction for these catalysed processes. The attraction of the enzyme molecule for the activated complex would thus lead to a decrease in its energy, and hence to a decrease in the energy of activation of the reaction, and to an increase in the rate of the reaction*' (see Figure 7.2).

Figure 7.2 *Pauling, Delbrück, and Perutz at the California Institute of Technology on the occasion of Linus Pauling's seventy-fifth birthday celebrations.*

(Courtesy Oregon State University Special Collection and Archive)

This powerful interpretive message—not generally applicable in inorganic catalysis, it is now realized—had been largely forgotten for a quarter of a century. (It is not even mentioned in one of the classic texts on enzymatic catalysis by Jencks[23] in 1969.)

7.5 Summarizing Verdicts

7.5.1 Linus Pauling

First, we deal with Pauling's legacy in structural molecular biology, and a convenient place to begin is to recall some other of his remarks in his famous 1948 Friday Evening Discourse. Whereas, in Pauling's earlier '*Nature of the Chemical Bond*',[2] there were intimations in his chapter on hydrogen bonds, it was later, especially in the RI Discourse,[20] that he expanded on the likely supreme role of hydrogen bonding in molecular biological phenomena. To cite Pauling himself on this topic: '*Because of its small bond energy and the small activation energy involved in its formation and rupture, the hydrogen bond is especially suited to play a part in reactions occurring at normal temperatures. It has been recognised that hydrogen bonds restrain protein molecules to their native configuration, and I believe that as the methods of structural chemistry are further applied to physiological problems it will be found that the significance of the hydrogen bond for physiology is greater than that of any other single structural feature*.'[20]

The notions encapsulated in this passage had, in a manner of speaking, been articulated by Pauling and Mirsky even earlier in 1936,[24] but they lay dormant for many years. The brilliant investigator J. D. Bernal, for example, in his critique in 1939 on the structure of proteins,[25] made no mention of hydrogen bonds in the entire function of polypeptides and other macromolecular participants in biomolecular transformation. (Bernal was, however, fully familiar with the importance of the hydrogen bond, as is apparent from his classic paper on the structure of water, with R. H. Fowler, in 1933.)

In yet another prophetic passage presented in his 1948 RI Discourse, Pauling expresses thoughts that are highly relevant to what came later, in 1953, with the structural elucidation and subsequent genetic consequences of the Watson–Crick DNA double helix. As the following excerpt shows, Pauling surmises what had not yet been discovered: '*The detailed mechanism by means of which a gene or virus molecule produces replicas of itself is not yet known. In general, the use of a gene or a virus as a template would lead to the formation of a molecule not with identical structure, but with complimentary structure...If the structure that serves as a template (the gene or virus molecule) consists of, say, two parts, which are themselves complementary in structure, then each of these parts can serve as the mould for the production of a replica of the other part, and the complex of two complementary parts thus can serve as the mould for the production of duplicates of itself.*'

In the realm of antibodies, Pauling proposed an early theory[26] which, although wrong in parts, contained one vital claim: '*atoms and groups which form the surface of the antigen attract certain complementary points of the globulin chain and repel other parts*'.[20]

Pauling's standing as one of the principal founders of structural molecular biology rests partly on his identification of sickle-cell anaemia (a hereditary disease) as a molecular disease[27] (see Chapter 5). In their compilation of '*The Selected Papers of Linus Pauling*', edited by Barclay Kamb, Linda Pauling Kamb, Peter J. Pauling, Alexander Kamb, and Linus Pauling Jr (World Scientific Publishing, Singapore, 2001), it is stated that, without doubt, this was one of Linus Pauling's greatest contributions to medical science—the concept of molecular disease. The discovery was made in human haemoglobin, but the concept of molecular disease, as described in Chapter 5, extends far beyond it and has become one of the foundations of modern medicine in general and in molecular pathology in particular.

A decade or so after his introduction of the notion of a molecular disease arising from a single mutation, Pauling, with Zuckerandt,[28] made yet another fundamental contribution to molecular biology, namely the concept of a '*molecular clock in evolution*'. Round about that time (1962), the sequencing of the amino acid constituents of proteins had become standard. And so, Pauling and Zuckerandt[28] analysed the different haemoglobins obtained from humans, horses, gorillas, and several other animals. Based on palaeontological evidence, it is known that the common ancestor of man and horse lived some time around 130 million years ago. The α-chains of horse and human haemoglobin contain about 150 amino acids and differ by some eighteen amino acid substitutions, that is, about nine evolutionary effective mutations for each of the chains, or, in other words, about one per fourteen million years. Based on these observations, the differences between gorilla and human haemoglobin (two substitutions in the α-chain and one in the β-chain) suggest a relatively recent divergence between the species, of the order of only ten million years. On the other hand, differences between the haemoglobin α- and β-chains of several animals suggest divergence from a common chain ancestor of about 600 million years ago—in other words, during the Precambrian era. (Nowadays, Pauling's molecular clock has been supplanted by the technique of DNA sequencing, which is a more convenient and reliable method of dating.)

There are many other major contributions that Pauling made to structural chemistry. These include the concept of hybridization. Indeed, Pauling himself felt that his introduction of hybrid orbitals and the concept of hybridization were his most important contributions to chemistry. It is this idea of hybridization that has added immeasurably to the everyday language of the chemist.

7.5.2 Lawrence Bragg

One of the most handsome tributes paid to Lawrence Bragg is the letter that his contemporary J. D. Bernal wrote to him after the Copley medal—the highest award made by the Royal Society—was awarded to Bragg in 1946: '*This is only to congratulate the Royal Society for giving you at last the Copley that you have deserved many times over...crystal structures may seem now an old story, and it is; but you its only begetter are still with us. Three subjects, mineralogy, metallurgy and molecular biology, all first sprang from your head.*'

There is no doubt that, following the discovery of the phenomenon of X-ray diffraction by von Laue *et al.* in 1912, it was principally Lawrence Bragg that transformed it into the new, powerful technique of exploring the atomic composition of matter. He and, to a lesser degree, his father changed the face of many branches of twentieth-century and current science.

It is interesting to recall that, at about the time when Lawrence Bragg became aware of von Laue *et al.*'s discovery of X-ray diffraction, Japanese workers in Tokyo, notably T. Takeda,[29] had already (quite independently of the work of the Braggs) concluded that the results of their investigations of crystals of borax, rock salt, quartz, alums, mica, fluorspar, and cane sugar indicated that reflection of X-rays from crystal planes was involved in the process of diffraction. It is a measure of Lawrence Bragg's speed and clarity of mind (alluded to above by Perutz) that the work carried out in Leeds and Cambridge far outstripped what was done in Japan.

So far as his scientific endeavours were concerned, as distinct from his pedagogic ones, his achievements came, unlike those of Linus Pauling, by concentrating on one major area—that of X-ray crystallography. Unlike his predecessor-but-two as Cavendish Professor Lord Rayleigh, he did not contribute extensively to other branches of science. Many eponymous topics and equations signifying Rayleigh's phenomenal contributions abound—as may be seen from the following (short) list of topics and phenomena in which he achieved profound advances: Rayleigh scattering; Rayleigh criterion; Rayleigh disc; Rayleigh waves; Rayleigh law; and Rayleigh–Jeans Law.

While Bragg's versatility, as a physical scientist, does not match the extraordinary intellectual sweep of Rayleigh, or in chemical science that of Pauling, it can nevertheless be indisputably argued that his skills in fostering fundamentally new areas of scientific enquiry are well-nigh matchless. In responding to the invitation that Sir David Phillips and I sent to Sir Brian Pippard (Cavendish Professor at Cambridge from 1971 to 1982), we were told:[30] '*The choice of a crystallographer,*[30,31] *however distinguished, was a blow to many hopes. This was not simply a reaction of disappointed personal ambition, for the shock was felt afar – as late as 1955 I was told by a leading American physicist that since Rutherford's death, the Cavendish had produced nothing of significance. Yet, when one looks back on Bragg's tenure of the Chair one sees that the seeds sown then had already, by the time he left, shown great potential, and shortly afterwards came to fruition in advances and discoveries that rivalled, perhaps even eclipsed, any from Rutherford's Cavendish. The radioastronomy of Ryle's group, and the partnerships in molecular biology of Perutz and Kendrew, Crick and Watson, have changed our view of the world more radically than the discovery of the neutron and of artificial disintegration; and these are only the most conspicuous successes from the post-war era of intellectual ferment. Bragg deserves great credit for creating an environment in which a multitude of ideas could prosper, and for his enthusiastic support (or at least, in the case of Crick and Watson, long-suffering tolerance) of every promising venture, whether or not it was directed at obviously fundamental problems.*'[30]

He was not simply a passive observer in the pioneering work of Perutz and Kendrew—it is to be noted that he was a co-author or author of sixteen original and popular articles on haemoglobin and other protein molecules—but he was used as a consultant, by Perutz especially, whenever he ran into difficulties. Moreover, both Kendrew and Crick[32] were extremely impressed by Bragg's intellectual acumen. The following remarks, broadcast by Kendrew on the BBC in 1965, reflect his admiration: '*Bragg is a man with a rather simple mind, but at the same time it is one of the most powerful minds I have ever come across. As a very young man, 50 years ago, he invented a complete new branch of science and he has lived to see the day when that branch of science has worked revolutions in many other branches, including one I am most interested in, which is biology and medicine.*'

Much has been written and said about the strained relationship (at one time) between Bragg and Crick. At that time, Bragg forbade Crick and Watson from working on the structure of DNA. This was tinged by the fact that Bragg's acute sense of fair play made him uncomfortable that two of his staff were engrossed with DNA, which he, Bragg, felt was Wilkins' problem.

The reply that David Phillips and I received from Francis Crick reinforced the admiration that Bragg received from his junior, brilliant co-workers: '*I have learned a lot from Bragg. He was a major influence on my scientific career. Perhaps the most useful and lasting lesson was how to approach a scientific problem... The other thing I learnt from Bragg was how to give a lecture, or rather, how not to give a lecture. Bragg had been asked to speak to a small meeting at the Cavendish on a topic in which I also was closely interested. Without realising it, I had outlined in my head how I would have given the lecture. When Bragg delivered his short talk it was not like my version at all. He made a few simple points that set out the main aspects of the work. Each was explained very clearly in a straightforward way, without too many technical details. I realised that had I given the talk all the emphasis would have been on the details and the broad conclusions would have been submerged beneath a mass of technicalities.*'[22]

Notwithstanding all the public and private praise with which Lawrence Bragg was festooned during most of his scientific career, he still found it necessary (at the age of seventy-five) in a lecture that he gave in Stockholm in December 1965 to begin his talk with the following statement: '*It is sometimes said that my father and I started X-ray analysis together, but actually this was not the case.*' He then proceeded to tell his audience that it was he alone who had first analysed the von Laue photographs and that he had discovered *Bragg's Law* and used it to determine the first structures using X-ray diffraction photographs. Precisely why, in the autumn of his life, when the whole world revered him, he felt it necessary to emphasize this fact remains an enigma.

Many commentators have said that the great discoveries of W. H. Bragg and W. L. Bragg strained relationships between them for the rest of their lives. As Perutz wrote in 1990: '*W. L. Bragg was fond of defining the exact roles played by himself and his father, but he never hinted to those strains until a few days before his death when he wrote to me: "I hope that there are many things your son is tremendously good at which you can't do at all because that is the best foundation for a father–son relationship*"'.[33]

Recently, a distinguished US structural biologist, who worked at the Laboratory of Molecular Biology (LMB) in Cambridge in 1962–1963 and then spent most of his scientific career in California, told me about his views on Pauling and Bragg. In regard to Pauling, he said that he had a rapier-sharp mind and an exceptionally retentive memory and was always on top of his work. But he lacked human warmth. Bragg, as others (e.g. Jack Dunitz, who was a postdoctoral research associate of both Pauling and Bragg) have said, did not have a particularly retentive memory: '*When he spoke to you about your work, he had forgotten what you had told him earlier about it – in contrast to Pauling.*' But my US friend said that a scientific discussion with Bragg was full of human warmth. Max Perutz conveyed the same message whenever he recalled the avuncular influence that Bragg exerted upon him.

APPENDIX

On the topic of Pauling's apparent spontaneity, I was recently told by the Nobel Prize winner Dudley Herschbach, the Harvard University chemist, the following story. Herschbach was one of E. Bright Wilson's graduate students at Harvard, and, in his early years, Bright Wilson (1908–1992) served as a Teaching Fellow at Caltech at the time when Pauling was lecturing in quantum mechanics to graduate students. (Later, Pauling and Wilson wrote one of the first books on *Introduction to Quantum Mechanics*—it was a brilliant success and is still in widespread use some 80 years after it was produced.)

> '*Bright told me that Linus, in response to a question, liked to pretend he had not considered it before. He would work out the mathematical solution on the blackboard, step-by-step, with commentary suggesting it was a fresh excursion. Often, however, Bright knew that Linus had carefully prepared the derivation before class.*'

As many other friends of mine who knew him well have told me, Pauling always loved the opportunity of injecting drama and excitement into a particular situation. This is but one of the reasons why his lectures were so memorable.

REFERENCES

1. J. D. Dunitz, *Biographical Memoirs of Fellows of the Royal Society, Linus C. Pauling (1901–1994)*, **1996**, *42*, 317.
2. L. Pauling, '*The Nature of the Chemical Bond and the Structure of Molecules and Crystals*' (second edition, **1940**; third edition, **1960**), Cornell University Press, Ithaca, NY.
3. L. Pauling, *J. Am. Chem. Soc.*, **1929**, *51*, 1010.
4. *Pauling's Rules*, https://en.wikipedia.org/wiki/Pauling%27s_rules
5. W. L. Bragg, '*Atomic Structure of Minerals*', Cornell University Press, Ithaca, NY, **1937**.
6. J. M. Thomas and D. C. Phillips, '*The Legacy of Sir Lawrence Bragg: Selections and Reflections*', Science Reviews Ltd, London, **1990**.
7. W. H. Bragg and W. L. Bragg, '*X-rays and Crystal Structures*', G. Bell and Sons Ltd, London, **1915**.
8. This is the phenomenon known as 'rotator phase solids'.
9. G. K. Hunter, '*Light is a Messenger: The Life and Science of William Lawrence Bragg*', Oxford University Press, Oxford, **2004**.
10. Brian Pippard, *Physics World*, **2005**, p. 41.
11. W. L. Bragg, J. C. Kendrew, and M. F. Perutz, *Proc. Roy. Soc. A*, **1950**, *203*, 321.
12. L. Pauling and R. B. Corey, *J. Am. Chem. Soc.*, **1950**, *72*, 5349.
13. L. Pauling and R. B. Corey, *Proc. Natl. Acad. Sci. U. S. A.*, **1951**, *37*, 235.
14. M. F. Perutz, *Nature*, **1951**, *168*, 653.
15. M. F. Perutz, '*I Wish I'd Made You Angry Earlier*' (expanded edition), Cold Spring Harbor Press, Cold Spring Harbor, NY, **2003**, p. 189.
16. J. C. Kendrew, R. E. Dickerson, B. E. Strandberg, R, *et al.*, *Nature*, **1960**, *185*, 422.
17. R. E. Dickerson, *Protein Sci.*, **1992**, *1*, 182.

18. P. Pauling, *New Scientist*, **1973**, *58*, 558.
19. C. C. F. Blake, A. C. J. North, D. C. Phillips, R. J. Poljak, G. A. Mair, and V. R. Sharma, *Nature*, **1965**, *206*, 757.
20. L. Pauling, *Nature*, **1948**, *161*, 707.
21. D. C. Phillips, *Scientific American*, **1966**, *215*, 5, 78.
22. L. Pauling and M. Delbrück, *Science*, **1940**, *92*, 77.
23. W. P. Jencks, '*Catalysis in Chemistry and Enzymology*', McGraw-Hill, New York, NY, **1969**.
24. L. Pauling and A. E. Minsky, *Proc. Natl. Acad. Sci. U. S. A.*, **1936**, *22*, 439.
25. J. D. Bernal, *Nature*, **1939**, *143*, 663.
26. L. Pauling, *J. Am. Chem. Soc.*, **1940**, *62*, 2643.
27. L. Pauling, H. A. Itano, S. J. Singer, and I. C. Wells, *Science*, **1949**, *110*, 543.
28. L. Pauling and E. Zuckerandt in '*Horizons in Biochemistry*' (eds. M. Kasha and D. Bullman), New York Academic Press, New York, NY, **1962**, p. 189.
29. T. Takeda, *Nature*, **1913**, *91*, 135—see also the work of Nishikana as given in '*Fifty Years of X-Ray Diffraction*' (ed. P. P. Ewald).
30. B. Pippard, '*Bragg: the Cavendish Professor*', ref [6], p. 78.
31. W. L. Bragg as successor to Lord Rutherford as Cavendish Professor in Cambridge, 1938.
32. F. H. C. Crick, '*W. L. Bragg: A Few Personal Recollections*', ref [6], p. 110. See also M. S. Bretscher and G. Michison, *Biographical Memoirs of Fellows of the Royal Society*, **2017**, *63*, 159.
33. M. F. Perutz, '*How Lawrence Bragg Invented X-ray Analysis*', ref [6], p. 75.

8

Biographical Sketches

8.1 Introduction

We begin by comparing the attributes of Perutz and Kendrew and identifying some of their differences. Then, based on my personal interaction (and those of others) with them, I describe the qualities of each separately, emphasizing their unusual characteristics and their major impacts on the national and international scenes. In so doing, I cover ground that has received little attention hitherto in the published literature. Dorothy Hodgkin's work and personality, along with noteworthy attributes of a number of influential figures, are described in Part II.

PART I

8.2 Perutz and Kendrew

A good deal has been written already about these two pioneers, covering not only their landmark advances in molecular biology, but also their different styles of working. A comprehensive account of Max Perutz's life is contained in the monograph by Georgina Ferry,[1] and numerous detailed obituaries appeared in 2002 when Perutz passed away and in 1997 after the passing of Kendrew. Two definitive *Biographical Memoirs* have been published by the Royal Society—one by Blow[2] on Perutz, and the other by Holmes[3] on Kendrew. In addition, brief reminiscences by Dickerson[4] and by Rossmann,[5] both of whom worked in the Perutz–Kendrew team in the 1950s, have also appeared. In de Chadarevian's book,[6] as well as in Olby's definitive study[7] of Francis Crick, there is a good deal of information about both the content of their work as well as their mode of promulgating it. The American science writer the late Horace Judson[8] wrote about Perutz, Kendrew, and their contemporaries, and Wassarmann's[9] long-awaited study of Kendrew is to appear soon.

Much has been said in previous chapters of this book about the nature and significance of the revolution wrought by Perutz and Kendrew. But the opinion of one of the recent occupants of the Cavendish Chair of Physics in Cambridge—Sir Brian Pippard[10]—bears repetition: '*The revolution initiated in W. L. Bragg's day by*

Perutz and Kendrew and their colleagues eclipsed in its significance even the pioneering work of an earlier Cavendish Professor, Lord Rutherford.'

Apart from their intrinsic intellectual excellence as scientists, what is exceptional about both Perutz and Kendrew is the astonishing degree of world-leading advances that they accomplished. They each exhibited unusual skills in planning, collaborating with one another, and negotiating with several organizations and other scientists, and they each possessed brilliant powers of expression, both in the written and in the spoken word. They were each blessed with an unusual degree of brain power—they were exceptionally gifted individuals who also had the uncanny knack of choosing important topics of research and had the attendant abilities to solve them. The words used by Perutz in his obituary[11] of Kendrew in *Protein Science* apply equally to him: '*I found Kendrew an outstandingly able, resourceful, meticulous, brilliantly organised, knowledgeable, and hard worker, and a stimulating companion with wide interests in science, literature, music and the arts.*'

Whereas Max Perutz loved listening to classical music and had a liking for Haydn, Mozart, Schubert, Bach, and Beethoven, John Kendrew was extremely knowledgeable about music; he was an avid and discriminating listener and possessed an enormous stock of recordings. He was one of the first persons in Britain to equip himself with Hi-Fi reproduction; and in the early 1950s, he helped his colleagues and students at Peterhouse to set up their own musical systems. (While he was at St John's College, Oxford, the students there sometimes complained about the volume of music that emanated from the President's Lodge!)

Perutz and Kendrew—each blessed with questing minds and perceptive genius—also possessed an extraordinary degree of single-minded determination to stick to the problems that they considered important. They exhibited, to an exceptional degree, an amalgam of pertinacity and perspicacity, and these qualities shone through to, and profitably influenced, their collaborators and contemporaries.

They were also loyal to each other, whenever they spoke or wrote about the other's work. But there were some differences of opinion which caused the occasional minor disagreement. Perutz thought, in 1959, that Kendrew should not have frittered away his creative energies by inaugurating a new scientific journal (*Journal of Molecular Biology*). But, as shown in Chapter 1, this quickly became a triumphant success[12] and highly influential (as it still remains) as one of the principal repositories of advances in molecular biology. Kendrew often felt that he could have organized the LMB more efficiently than the gentle, avuncular, and unconventional way it was done by Perutz (see Section 8.2.2). Perutz was also unhappy about the fact that the structural co-ordinates of Kendrew's high-resolution structure determination of myoglobin (to 1.4 Å—far ahead of anyone else in the world) achieved post-1962 (the date of their Nobel Prize) had not been deposited by Kendrew in the Protein Data Bank, which Perutz had played a major role in initiating.[13] Perutz was also unhappy that, from *c.*1963 onwards, Kendrew, who was Deputy Director, was away from the laboratory a great deal, largely because he was involved in Government advisory work.

Perutz and Kendrew cooperated brilliantly in setting up the *European Molecular Biology Organization (EMBO)*, partly because they were each ardent Europeans. Each felt, from their respective experiences during World War II, that supporting trans-European activities and organizations were conducive to a safer, more peaceful, and saner world.

Perutz lived the full, committed life of a family man, whose devotion to his German–Jewish wife Gisela extended to his cooperation between them in some of his scientific adventures (see Figure 8.1). Thus, for over twenty years, Gisela Perutz ran the canteen at the LMB, which, by bringing together all the scientists in the laboratory three times a day, facilitated discourse among the investigators there. Numerous eminent visitors to the LMB over the years have testified[14] that the canteen at the LMB was the focus and seedbed and the breeding ground of a multiplicity of new ideas—see the remarks of the Yale Nobel Laureate Tom Steitz[15] and the testimony of many other distinguished alumni of the LMB in Chapter 10.

After a short-lived marriage that ended in divorce, Kendrew lived the life of the urbane bachelor, who enjoyed participating in public affairs in his own country

Figure 8.1 *Max Perutz, Gisela, his wife, and John Kendrew on the day the Nobel Prize was announced in October 1962.*

(By kind permission of the MCR-LMB, Cambridge)

and on the world scene. At one time or another, he served as President of the British Association for the Advancement of Science; as a Trustee of the Science Museum, London; as Scientific Advisor to the BBC; and, successively, as General Secretary, Vice-President, and President of the *International Council of Scientific Unions*, based in Paris, where his extraordinary skills as a committee member became legendary. More information about Kendrew's extraordinary busy life is given in Section 8.2.3 and in Appendix 1 to this chapter.

Perutz was disappointed not to have been made a Fellow of his College (Peterhouse) prior to his election as an Honorary Fellow there (without teaching duties) when he was awarded the Nobel Prize in 1962. Thereafter, Perutz participated fully in collegiate life and was especially prominent in attending the Kelvin Club—the scientific society of the students, to whom he frequently delivered popular lectures.

8.2.1 Max Perutz: a summary of his scientific and humane legacy

In numerous respects, Perutz's character and his humanity, kindness, scientific acumen, writings, and lectures, as well as his many other qualities, remind me of one of his (and my) great idols—Michael Faraday. I lived and worked in the home that was Faraday's during the time I was Director of the Royal Institution and of the Davy-Faraday Research Laboratory in London, and welcomed Max Perutz there often; and in Cambridge, I lived a few doors away from the Perutz home for the last quarter-century or so of his life. I also had leadership responsibilities for nine years in his Cambridge College, Peterhouse. Our families talked frequently and shared noteworthy experiences in our respective homes; and I went on many long walks with him in and around Cambridge. We got to know one another intimately, and I gained insights into why he was a great man, a great scientist, and a committed champion of social justice and intellectual honesty. He also shared with me some of his disappointments, as well as his successes; and all these experiences taught me why he was universally admired, respected, and loved.

I vividly remember the first day my wife and I entered the Perutz home. Prominent on Max Perutz's desk was the photograph of W. L. Bragg, whom Max admired and venerated (see Figure 8.2), and also of his Viennese–Jewish father.

In the Perutz living room, there was a special item of furniture, reminiscent of a music stand, which enabled Max to do his reading while standing up (which he found more convenient owing to his back problems). From my very first conversation with him in 1978, his extraordinary breadth of interests became apparent. Leaving aside his fascinating observations about science and scientists and other eminent public figures, it was soon apparent that he had read extensively the works of George Eliot, Tolstoy, Chekhov, Dickens, Shakespeare, Bacon, and Manzoni. One of his favourite novels, he revealed, was Anna Karenina. He was the first to draw my attention to the extraordinary (and solitary) novel of Giuseppe

Figure 8.2 *The photograph of Sir Lawrence Bragg that graced the Perutz home. Perutz liked placing this photograph of his scientific father, alongside his own biological father.*
(Courtesy The Royal Society)

Tomasi di Lampedusa '*Il Gattopardo*'. And it was in the Perutz household that I first heard about Iris Origo and her riveting book '*The Merchant of Prato*'.[16] It was there also that I became captivated by the American monologist Ruth Draper.[17]

Although in those first few meetings that we had as families, it was apparent that he was proud of his Austrian heritage, and its contributions to world culture, he was also exceptionally happy being a citizen of Cambridge—see below the excerpt of his letter to Queen Elizabeth II. I frequently heard him extol the beauty, charm, and historicity of Cambridge as a centre of learning. He sometimes talked about the things associated with the University and its environs that have changed very little over the years—the College and University buildings, the bridges, the churches, the lawns, the river, the gardens, and the trees. Isaac Newton possibly, and certainly Charles Darwin, Clerk Maxwell, and Gowland Hopkins, would have heard the same bells, looked up at the same spires, perceived the same patterns of light and shade in the beautiful East Anglian skyscapes, and heard the rustle of the wind through the same weeping willows.

Max Perutz delighted in the beauty of the natural world, especially the mountainous landscapes. He was the kind of man who, before starting his laboratory work at the

LMB on a spring morning, would occasionally take a walk on the Gog-Magog hills[18] outside Cambridge, filling his heart and soul, in so doing, with pantheistic pleasure. (This is the impression he gave me. But his daughter assures me that his habit was animated more by his desire to have plenty of fresh air and to see green trees.)

It must not be thought that Max Perutz was primarily a cerebral individual. As a teenager, he was the Austrian downhill skiing champion. And, all his life, he loved mountain walking, especially in the Italian Dolomites.

8.2.1.1 Letters and the Commonplace Book

As mentioned earlier, Perutz was similar to his hero Michael Faraday in many respects. Not only did each possess prodigious intellectual energy, contagious enthusiasm, and youthful acuity and thirst (throughout their lives) for new knowledge, but each was also able to live with what seemed to them a daunting task. In Faraday's case, it was his conviction, encapsulated in a compelling opening paragraph in one of his famous papers, that '*I have long held an opinion amounting almost to conviction that the various forms under which the forces of nature are made manifest have one common origin*...'[19,20] It was Faraday's belief in the unity of the forces of nature—still a topic of perennial debate—that fuelled much of his curiosity.

In Max Perutz's case, it was his conviction, declared as a fresh graduate, that the almost impossible task of determining the structure of haemoglobin—a molecule consisting of about 10,000 non-hydrogen atoms—could be solved by X-ray crystallography, when, at the time of his embarking on this arduous task, the largest molecule that had yielded its structure to X-ray crystallography contained only a few dozen atoms. As that proved successful, he moved on to investigate its mode of action. Just as Faraday's indomitable quest for evidence for unification of natural forces remained undiminished,[21] so Perutz explored every conceivable new technique in his lifelong quest to establish the nature, mode of action, and function of haemoglobin molecules in humans and other creatures.

Faraday and Perutz were also compulsive letter writers. The endearing compositions that Faraday sent to his wife when he was away from her are mirrored by the expressive missives that Perutz sent to his wife on numerous occasions when he was away from her (and their two children)—see the examples cited below, where, for example, he describes listening to a memorable lecture by Hans Bethe.

Whereas very many of Faraday's letters were penned out of a sense of duty and were, as a consequence, sometimes aridly factual and platitudinous, Perutz's were redolent of human warmth and of social, political, historical, and literary awareness. A measure of the skill and charm that Perutz displayed as a letter writer is conveyed in the fascinating collection that his daughter Vivien has produced in her evocative selected letters of her father.[14]

Perutz wrote to a wide range of public figures, as well as to members of his family, including his scientist son, art historian daughter, and grandchildren.[22] An appropriate initial example is his letter to Hugh Trevor-Roper (Lord Dacre[23]) with whom, during Trevor-Roper's tenure as Master of Peterhouse, Perutz, as a Fellow of that College, became close friends. Trevor-Roper, according to his contemporaries (and even his enemies[24]), was the greatest stylist in the English language. Late in 1989, Perutz learnt that Trevor-Roper, in an article in the *Daily Telegraph* newspaper, had selected his recent book as the *Book of the Year*. Later Trevor-Roper reviewed it for *The Spectator*. This elicited the following letter from Perutz: '*I was touched by your choosing "Is Science Necessary?" as the Book of the Year…I was particularly pleased by your last sentence alluding to the human values that I try to express in my writing about science and scientists…I was disappointed that none of the reviewers of "Is Science Necessary?" commented on my review of Weizmann and my quotations from his speeches. "One law and one manner shall be for you and for the stranger that sojourneth among you". "I am certain that the world will judge the Jewish state by what it does to the Arabs." My piece originally appeared in Nature and was met by stony silence on the part of my colleagues in Israel.*'

Two of the letters that Perutz wrote within a few weeks of his death were to Margaret Thatcher and to the Queen. In his message to the former, he begins: '*It seems that my days are numbered, and I thought that I should let you know how much I enjoyed our meetings. I enormously enjoyed the lunch party you gave for the Nobel Laureates and the privilege of sitting next to you at lunch. You had just come back from a trip to Africa. I remember you telling me that the one African leader who impressed you was Mugabe, because he was honest! That must have been the worst misjudgement of your life! I can match that. You asked me for my views about cold fusion…I replied that I believed that there must be something in it…soon afterwards I learnt that it was all nonsense and felt very ashamed of my reply.*'

Privately, during one of our walks, Perutz told me that he admired Margaret Thatcher for her ability and determination, but that he disagreed with her politics, which, he felt, led to the dismantling of the manufacturing base of the nation and dire social consequences.

In a letter to his wife Gisela in June 1980, he describes his reaction to a lecture by Hans Bethe (1906–2005), the distinguished German–Jewish US Nobel Prize-winning physicist, who was then aged 74 and who lectured on '*Cessation of Nuclear Tests*'. '*Bethe impressed me immensely by his combination of fervent convictions for what is right and calm reasoning, which seems inexhaustible. He talked for two hours without either exhausting himself or the audience and never said a word too much!*'

In Perutz's letter to the Queen dated 14 January 2002, just a few weeks before he died, he again begins by referring to his '*numbered days*' and then proceeded to recall the numerous times that he spent with her, from her visit to the LMB in 1962 (see Figure 8.3) to his installation as a Companion of Honour (CH) in 1975, to the award of the Order of Merit (OM) in 1988 when he recalled: '*…while I was waiting in your anteroom I met Ernst Gombrich for the first time, to whom you were giving the OM on the same day. That same afternoon we discovered that we came from the same family background in Vienna with a similar education; that we both emigrated (as Austrian Jews) to England before Hitler overran Austria and that we were both anti-Zionists and had the same outlook on many things. Then I discovered that he knew everything that I didn't know. He had an encyclopaedic knowledge of classical literature…He was a delightful man who never opened his mouth without saying something either original or witty.*'

Figure 8.3 *John Kendrew showing the atomic model of myoglobin to the Queen, with onlookers (from the left): Mavis Blow, Gisela Perutz, Charlotte and Sir Harold Himsworth (Secretary of the MRC), and Lord Shawcross.*

(Courtesy of MRC-LMB, Cambridge)

Towards the end of his letter, he says: '*When I was 21 years old, I decided to leave my native town, Vienna, and came to Cambridge for my PhD. It was the best decision I have ever made in my life. I have had 65 years of productive research in Cambridge, the most wonderful centre of learning in the world, and I could never have accomplished what I did anywhere else.*'

The letters of Perutz eloquently reflect his multi-faceted character. Apart from his scientific distinction, he was also a passionate human rights activist,[25] a deeply cultured man, exceptionally wise and generous in his dealings with everyone with whom he interacted—from prime ministers, princes, princesses, and popes to young co-workers, undergraduate and graduate students, college employees, and laboratory technicians, and to the vast array of contact that he had worldwide. He could also make unpopular decisions. One or two of his LMB colleagues said he could be ruthless, but this is a quality of his I never observed.

Like Faraday, Perutz kept a *Commonplace Book*, a fascinating collection of the aphorisms, stories, stanzas and quotations, and philosophical verities that he had garnered over the years. (This is reproduced at the back of his book entitled '*I Wish I'd Made You Angry Earlier*'[26]). On more than one occasion, I heard him repeat the following statement (contained in the *Commonplace Book*) from the *Memories* of the ballerina Dame Margot Fonteyn: '*I cannot imagine feeling lackadaisical about a performance. I treat each encounter as a matter of life and death. The important thing I have learned over the years is the difference between taking one's work seriously and taking oneself seriously. The first is imperative and the second disastrous.*'

Perutz attended a lecture that I gave to the Cambridge Philosophical Society in the late 1990s on '*The Genius of Michael Faraday*'. To end that talk, I showed a slide of a verse in the '*Book of Job*'. It was Faraday's favourite verse, heavily marked by him in his copy of the Bible. Four days later, on a Saturday, Perutz came to lunch in Peterhouse—as he had done regularly on Saturdays for thirty years. He relished the conversations he had there with young scholars. Usually he made a point of sitting next to a young, previously unknown (to him), Fellow, just to be sociable and to enter into a new scholarly area. But, on this Saturday, he came straight to me and handed me pen and paper with a request that I write out Faraday's favourite biblical verse from the '*Book of Job*': '*If I justify myself, my own mouth shall condemn me: if I say I am perfect, it shall also make me perverse.*' This verse appealed greatly to Max Perutz. It is an injunction not to boast. Like Faraday, Perutz never boasted, and he disliked the company of those who did.

8.2.1.2 Perutz's table talk

The experience of sitting next to Max Perutz at table was greatly appreciated by numerous individuals and especially by the Fellows of Peterhouse where he came for lunch for most Saturdays of the year. In autumn of 2002, at the Perutz Memorial Meeting, a young German–Irish historian Fellow of Peterhouse—Brendan Simms[27]—gave an illuminating account of this experience. Simms was well aware that conversations with Max Perutz were redolent of the world of a Central European *Bildungsbürger*, in which Perutz had grown up, before its descent into barbarism and obscurantism. Here are some of Simms' observations: '*It should not be imagined that conversations with Max were just politely inconsequential. On the contrary, they could be nerve-wracking, because you knew that an insatiable intellectual curiosity lay behind the unfailing courtesy. He would never allow a loose assertion or sloppy thinking go unchallenged, not because he wished to embarrass his interlocutor, but because he genuinely wanted to know. Like many truly great men, he would never pretend to possess a knowledge he didn't have, and he was therefore never too proud to ask an obvious question. Knowledge for him was to be tested, extended and shared, not hoarded.*'

On one occasion, listening to him expound firmly to a visiting historian, I heard him reject in stentorian terms the '*nonsense*', as he put it, popular among modern sociologically oriented philosophers of science, that scientific truth is relative and shaped by a scientist's personal concerns, including his or her political, philosophical, even religious instincts. When he attacked such opinions, he sometimes quoted Max Planck's memorable assertion: '*There is a real world independent of our senses: the laws of nature were not invented by man, but forced upon him by the natural world. They are the expression of a rational order.*'[28]

Generous and magnanimous to friends and strangers, he was implacably opposed to any unfairness. His uncompromising defence of Pasteur in response to an American historian of science's insinuation that the great Frenchman 'cheated' still rings in one's ears. He was also forthright in his criticism of T.-S. Kuhn's argument[29] that science advances by a succession of paradigms. As he said:[29] '*The perusal of old textbooks of chemistry and mineralogy have convinced me that there was no paradigm for the atomic structure of solid matter before 1912. The results of X-ray analyses opened a new world that had not even been imagined before.*'

In another article in the *New York Review of Books*, he dealt at length with Heisenberg's attitude towards creating a nuclear bomb for the Nazis. Perutz had been greatly impressed, as a young man, on hearing the thirty-one-year-old founder of quantum mechanics lecturing in Vienna. Perutz agreed with others, notably the eminent Dutch physicist Hendrik Casimir (1909–2000), that Heisenberg was indeed a genius, and in his *London Review of Books* article, he quoted Casimir: '*A genius is someone who can create things that are initially beyond his own understanding. In that sense, Heisenberg was certainly a genius, and this goes rarely with a special gift for understanding the feelings and the ways of thinking of others. Heisenberg did not have that gift. Perhaps his greatest shortcoming was that he was unable to grasp the full measure of the depravity of what was then the ruling group in Germany.*'

Perutz, who entertained Heisenberg in his home in Cambridge after World War II, was unconvinced by Heisenberg's claim that he never did work on the design of an atom bomb, that he was only concerned with trying to create a nuclear reactor.

8.2.1.3 Perutz's human kindness

Numerous examples abound of the manner in which Perutz behaved towards others and, in particular, his unusual acts of kindness. Three examples will suffice.

In the early 1980s, one of the friends of Gisela and Max Perutz who lived close to them in Cambridge had to undergo quite a serious surgical operation. This lady had two musically oriented daughters, aged eleven and thirteen—she was my late wife. On the day of her operation, Max returned early from the LMB to join his wife in entertaining our two young girls and me, while my wife underwent surgical treatment. To entertain them—they were each competent violinists—he sat them opposite each other at Perutz's dining table and placed between them the musical score of Mozart's palindromic duet for two violins '*Der Spiegel*' (see Figure 8.4). The event was a great success—and the young girls and their parents never forgot the incident. Here was one of the busiest and most able scientists in the world setting time aside—mid afternoon—to demonstrate, in Wordsworth's words: '*little unremembered acts of kindness and of love.*'

In the early 1980s also, Max Perutz began to explore the use of synchrotron radiation to investigate (by fine-structure X-ray absorption spectroscopy) the nature of the bonding in oxyhaemoglobin. This brought him in contact with Dr Samar Hasnain.[30,31] He made several visits to the Daresbury synchrotron and became a strong friend and collaborator with Hasnain, and stayed in his host's home. There he encountered the two sons of Hasnain, aged eight and eleven, and he discovered that they were interested in collecting foreign stamps. For several

Figure 8.4 *Part of the score of Mozart's palindromic duet for two violins (see text).*

years thereafter, Perutz sent letters to each of them containing foreign stamps that had come his way.

The third example involves Dr Kiyoshi Nagai, now a distinguished senior member of the LMB in Cambridge. He was born in Osaka, Japan, but he came to join Perutz in Cambridge on two occasions and carried out innovative research with him over many years. Nagai took advantage of Michael Smith's Nobel Prize-winning invention of a method using recombinant DNA—genetic engineering—in putting amino acids into particular positions at will in a polypeptide chain. Nagai's advance was his capacity to express the polypeptides of human haemoglobin in *Escherichia coli*, so that it would then make numerous molecules of the protein which could then be altered at will using Smith's method, thus allowing its new properties to be investigated.

Perutz was greatly aroused by this new advance because it enabled Nagai to distinguish between-species differences (in haemoglobin) that had arisen simply by chance and had no effect on haemoglobin's function and those that offered some selective advantages. It was Kiyoshi Nagai who succeeded in making both of the α- and β-globin chains of haemoglobin from colonies of bacteria. Later work by a US company, using Nagai's method, succeeded in expressing both chains together, and the resulting bacterial culture actually turned red on reaction with oxygen, just like red blood cells. Kiyoshi Nagai made several other outstanding advances both shortly thereafter and up to the present. So outstanding was his work that he was elected a Fellow of the Royal Society, a rare honour for a Japanese émigré scientist in Britain. Max was jubilant when this election occurred. And typical of his humanity, Perutz felt impelled to send a congratulatory-cum-explanatory letter to Nagai's parents in Japan (see Figure 8.5).

8.2.1.4 Perutz's idiosyncrasies

Very many scientists, as well as non-scientists, who knew him, recall with affection certain memorable attributes that Max Perutz possessed, and that added to the admiration they exhibited towards him.

Owing to his many health, especially back and digestive, problems, Max Perutz found it more comfortable to stand, rather than sit, during the course of personal discussions or during public lectures. (On one occasion at the LMB, when Uli Arndt chaired a lecture by a visiting student, when Max listened to the talk while lying on the floor, the first statement made by the chairman at the end of the lecture was: '*Are there any questions from the floor?*').

Max Perutz also insisted on eating the food that best suited him: over-ripe bananas, properly cooked potatoes, and Swiss (not any other) Emmentaler cheese, among a number of edibles that he savoured. A Swiss scientist friend of mine remembers Perutz requesting in a Philadelphia restaurant that he be served with Swiss—not American—Emmentaler cheese. In another incident, he took some fifteen minutes to describe to a waitress how his potatoes ought to be cooked, only to leave the restaurant before the meal arrived, as he had remembered that he was due to attend an important meeting!

Cambridge 18 · 4 · 2000

Dear Professor & Mrs Nagai,

I feel that I must tell you how absolutely delighted I am that Kiyoshi has been elected a Fellow of the Royal Society. He has done brilliant work ever since he came to rejoin our laboratory many years ago, and I am nearly as pleased as you must be that his good work has been so recognised. You must be very proud of your son.

With kind regards,

Yours, Max Perutz

Figure 8.5 *Message of congratulation sent by Max Perutz to the parents of Kiyoshi Nagai, after the latter had been elected a Fellow of the Royal Society.*

(Courtesy Dr K. Nagai)

All these idiosyncrasies endeared him even more to his enormous circle of friends. He was a very human and humane scientist. He was patient with others; and they, in turn, were patient with, and admiring of, him. Max was also extraordinarily good in recalling a memorable saying or incident. I once asked him, over a cup of tea, '*How did you become such a skilled negotiator?*' In replying, he quoted what a former Fellow of Trinity College, Cambridge, had once said: '*In Cambridge, to reach your goal, you must learn to combine the linear persistence of the tortoise with the circuitous locomotion of the hare.*'

8.2.1.5 Why is the LMB not an integral part of the University of Cambridge?

Max Perutz often used to tell the story of how his small unit at the Cavendish Laboratory—created for him and John Kendrew by Lawrence Bragg to pursue research on biological materials—became the subject of lively (often bitter) debate among the administrators of the University when, in 1954, a new Cavendish Professor—Nevill Mott (1905–1995)—took over the headship of the Laboratory after Sir Lawrence had migrated to the Royal Institution.

Even as a fresh Cambridge graduate student, Mott had dazzled his superiors by his theoretical insights; Lord Rutherford, an arch-sceptic concerning theoreticians, had been greatly impressed by Mott's insights in deriving a new way of explaining Rutherford's scattering law. Mott's multi-faceted skills as a physicist gave fresh impetus, as soon as he took up his post, to a range of new studies being pursued at the Cavendish Laboratory. As a consequence, a large number of co-workers wished to join Mott's department; and, inevitably, space for newcomers soon became a serious practical problem. Max Perutz was advised to seek a new home. He duly approached, cap-in-hand, so to speak, all the Heads of Department in the Cambridge science community. He was turned down by the Head of Organic and Inorganic Chemistry—the Nobel Prize winner Alexander R. Todd (1907–1997); by the Head of the Department of Physical Chemistry R. G. W. Norrish (who became a Nobel Laureate in 1956); and by the Heads of the Departments of Metallurgy, Biochemistry, Earth Sciences, and several others.

However, Max Perutz, John Kendrew, and their growing team, which, by that time, had welcomed Watson, Crick, Hugh Huxley, and several scientists from European and US laboratories, could be housed in small temporary accommodation in huts placed in the courtyard outside the so-called Austin Wing of the Old Cavendish Laboratory on Free School Lane (see Figure 8.6).

Full details of the remarkable scientific advances made by members of the Perutz–Kendrew team in that 'hut' have been given by Judson,[9] Holby,[8] Dickerson,[5] and Ferry.[1]

In view of the acclaim, worldwide, that greeted the work of the Perutz–Kendrew team, especially after the detailed folded structure of myoglobin and haemoglobin had been elucidated (see Chapter 5), the Medical Research Council (MRC) offered a substantial sum of money to the University of Cambridge, with a view to establishing a new Department of Molecular Biology.

A high-powered committee was set up to consider this gesture, and three highly influential Heads of Department dominated the discussion within this body: Sir Alexander (later Baron) Todd, Sir Nevill Mott, and Sir F. G. Young (1908–1988), the Head of Biochemistry. (All three became Masters of Cambridge Colleges: Todd of Christ College; Mott of Gonville and Caius College; and Young of Darwin College.)

According to the scientific gossip that is still heard in Cambridge, none of these three were at all enthusiastic about the formation of a separate LMB. Each voiced the same kind of reservation, as expressed by Conrad Waddington (see Section 1.1). And there are still some members of the University who recall that each of these three 'wise men' spent as much time in attacking the views of the other two than on addressing the helpful offer made by the MRC.

The MRC grew tired of waiting for the University of Cambridge to make up its mind. In due course, the new LMB was officially opened in 1962 by Her Majesty Queen Elizabeth II, at a site close to Addenbrooke's Hospital, a few miles away from Cambridge.

Aversion to the use of the phrase 'molecular biology' was not restricted to Cambridge. I am told that at Yale University, the powers that be there preferred the designations Departments of Molecular Biophysics and Biochemistry.

Figure 8.6 *(a) The hut outside the Austin Wing of the Old Cavendish Laboratory c.1960.*

(Courtesy, Hans Boye/MRC-LMB)

(b) Max Perutz and some of the staff of the MRC Unit outside the hut in 1958. From the left: Larry Steinnauf, Dick Dickerson, Hilary Muirhead, Michael Rossmann, Ann Callis, Bror Strasberg, and his two unknown technicians. Standing are Leslie Barnett and Mary Pinkerton.

(Courtesy Hans Boye/MRC-LMB)

8.2.1.6 Perutz's activities for human rights

Max Perutz was deeply involved in the early activities of the *Network on International Human Rights of the Academic and Scholarly Societies*, of which Carol Corillon at the US National Academy of Sciences was a leading light. François Jacob (1920–2013), the French Nobel Prize-winning biologist, and Max Perutz worked closely together in the early days of the Network.

After Perutz's passing, I was invited by Torsten Wiesel (1924–), the Swedish–American Nobel Prize-winning neurophysiologist, to trace the trajectory of Perutz's life prior to the Max Perutz Memorial Lecture, given by the then President of Al Ouds University in Jerusalem, the eminent Palestinian scholar Sari Nusseibeh in the Royal Society, London, in May 2005.

In closing my tribute to Perutz at that meeting, having heard repeated mention in prior discussions of liberty, freedom, the pursuit of truth, and the elimination of injustice, I quoted the words of the Hindu mystic and poet Rabindranath Tagore (whom Max and Gisela, his wife, revered). Tagore's song 35 in his *Gitanjali* reads as follows:

> '*Where the mind is without fear and the head is held high;*
> *Where knowledge is free;*
> *Where the world has not been broken up into fragments by narrow domestic walls;*
> *Where words come out from the depth of truth;*
> *Where tireless striving stretches its arms towards perfection;*
> *Where the clear stream of reason has not lost its way into the dreary desert sand of dead habit;*
> *Where the mind is led forward by thee into ever-widening thought and action –*
> *Into that heaven of freedom, my Father, let my country awake.*'[32]

The current chair of the Network is Professor Martin Chalfie, the US Nobel Prize winner in 2006 for his work on green fluorescent proteins as markers for gene expression.

I also recalled that Max Perutz was utterly repulsed by the thought of the use of torture on political or other prisoners. He could be seen to cringe while talking about it. His revulsion of such practices was partly what animated him as a human rights activist. But he detested injustice of any kind and was dedicated to the eradication of ignorance. In Amsterdam, at the Royal Dutch Academy, he read a paper on '*By What Right Do We Invoke Human Rights?*' (This lecture was later published by the *American Philosophical Society* in its *Proceedings*, **1996**, *140*, 135). It is a closely reasoned history of the concept of human rights from the days of Aeschylus (458 BC) to the present. His response to the terrorist attack in New York City in 9/11 was to organize a petition intended for world leaders. He stated: '*Avoid military actions against innocent people. Military retaliation does not solve the problem of fanaticism, but instead fuels the anger by demanding counter revenge!*'

8.2.1.7 Max Perutz and Francis Crick

Frequently, both in private and publicly, Max Perutz used to say that Francis Crick was one of the ablest persons that he had ever encountered. He used to tell the story that, on the day he, Perutz, and John Kendrew accepted him as a research student, Crick was given a draft of a paper that Lawrence Bragg and Perutz had prepared for submission, dealing with the nature of the packing of components of molecules of haemoglobin in its crystalline state.

Perutz told Crick to take home the draft and to come and see him the following day so that, should he have any questions, they could be tackled together. By the time that Perutz arrived in his office, Crick had already deposited the text of the article, which had been annotated extensively in red ink by Crick. It was full of strong remarks, and very many statements like '*unjustified assumption*' and '*this does not follow*' were much in evidence.

Perutz and Bragg had made the assumption, from an examination of several tens of thousands of X-ray diffraction patterns and Patterson functions that Perutz had taken over a period of some ten years, that they could discern evidence for some linear regions of the haemoglobin molecules within the crystal of the material. As Perutz further analysed these criticisms by Crick, he slowly began to doubt the veracity of the conclusions that he and Bragg had made. This criticism by Crick, as it turned out, although regarded as unpopular at first by both Bragg and Perutz, was justified. The Perutz–Bragg approach in the use of the so-called Pattersons (see Chapter 2) was transformed by this incident. And, at the time, it made Perutz disconsolate.

Later on, when the validity of Crick's criticism was acknowledged, and later still, after Perutz had shown how the heavy-atom method could be used to resolve the phase problem in X-ray diffraction, Perutz took a relaxed and philosophical view of the whole incident. He used to say in his public lectures that, had he been a German, Japanese, or Indian professor and had he had a student from one of these countries (in the place of Crick), those students would not have had the courage to protest and criticize in the way that Crick had done so unceremoniously. National characteristics and traditions played a part in all this, said Perutz.

Very much later, well into the life of the LMB in the mid to late 1960s, Crick (and also Brenner) had no hesitation in criticizing publicly some of the thoughts of Perutz. At one of the first seminars attended by the future US Nobel Prize winner Michael Levitt in 1968, he heard Francis Crick interrupt the seminar speaker with the remarks, '*Peter, you are making this too complicated. Make it simpler. Explain it so that Max can understand it.*' This outburst shocked Levitt. Such rude behaviour towards the Head of the Laboratory and a Nobel Prize winner was totally unexpected. But I later learnt, from discussions with Richard Henderson and other members of the LMB, that Max quietly accepted such public outbursts. He rose above it all. He rather relished being regarded as one of the slowest members of an audience to appreciate the subtleties of a new argument. There was no shame in failure like that.

In fact, the story told by another US future Nobel Prize winner—Tom Steitz—bears out the pedagogic subtlety of the Perutz approach to his questioning of seminar speakers. The following are Steitz's remarks: '*During one lecture (in 1967) that involved a comparison of a process that occurs both in eucaryotes and procaryotes,*

Max asked "What is a eucaryote and what is a procaryote?"' Terms that were just beginning to be used. '*I was glad that Max asked the questions, since I had no idea what the terms meant.*' (I find it inconceivable that Perutz, at that time, did not himself know what these words meant; after all, Jim Watson's book '*The Molecular Biology of the Gene*' had used them and that had first been published in 1965. But this was Max's way of enlightening a junior and heterogeneous audience.)

Both Francis Crick and Sydney Brenner, each of whom had razor-sharp intellect and were extremely quick at picking up new information—in contrast to Max who took a more studious, steady approach to new knowledge—would sometimes ridicule Max's rather ponderous approach. In one of his descriptions of Max Perutz, Crick called him a 'plodder' who got to his goal in the end. But there was also enormous respect that Crick had for Perutz. In the obituary that Crick wrote after Perutz's death in 2002, he said of him that '*He was the still centre of the revolution that took place in molecular biology.*'

Throughout Perutz's life, he kept reminding people how very bright and sharp Crick was. According to Perutz, '*Crick and Watson shared the sublime arrogance of men who had rarely met their intellectual equals. Crick was tall, fair, dandishly dressed – and talked volubly, each phrase in his King's English strongly accented and punctuated by eruptions of jovial laughter that reverberated through the laboratory…to say that they (Crick and Watson) did not suffer fools gladly would be an understatement – Crick's comments would hit out like daggers at non sequiturs…Crick has a profound understanding of that hardest of the sciences, physics, without which the structure of DNA would never have been solved.*'

Perutz also could criticize Crick. When, for example, I asked Max what he thought of Crick's opinion that the origin of life on Earth was best ascribed to 'transpermia'—the arrival of living matter from a source in outer space—he said that he thought it was nonsense, '*And*', he added, '*I think Francis knows it's nonsense too.*'

8.2.2 The art of planning and managing a research laboratory

'*Every now and then*', Max Perutz wrote in one of his popular books,[26,33] '*I receive visits from earnest men and women armed with questionnaires and tape recorders who want to find out what made the LMB (where I work) so remarkably creative…I feel tempted to draw their attention to 15th Century Florence with a population of less than 50,000, from which emerged Brunelleschi, Donatello, Ghilberti, Masaccio, Botticelli, Leonardo and Michelangelo, and other great artists. Had my questioners investigated whether the rulers of Florence had created an interdisciplinary organization of painters, sculptors, architects, and poets to bring to life this flowering of great art?…My question is not as absurd as it seems, because creativity in science, as in the arts, cannot be organised. It arises spontaneously from individual talent. Well-run laboratories can foster it, but hierarchical organization, inflexible, bureaucratic rules, and mountains of futile paperwork can kill it. Discoveries cannot be planned; they pop up, like Puck, in unexpected corners.'*

In an age when the Paladins of accountability and the funding councils persist in preaching the necessity for all academic research centres to have a Mission Statement and Strategic Plans, it is prudent to recall how Perutz set about founding and running the extraordinarily successful LMB.[34] The principles he used were: Choose outstanding people and give them intellectual freedom; show genuine interest in everyone's work and give younger colleagues public credit; enlist skilled support staff who can design and build sophisticated and advanced new apparatus and instruments; facilitate the interchange of ideas, in the canteen as much as in seminars; have no secrecy; be in the laboratory most of the time and accessible to everybody where possible; and engender a happy environment where people's morale is kept high. Formally, the LMB was run by a Board, consisting initially of all the scientific staff who were already Fellows of the Royal Society, but, in Perutz's own words: '*The Board never directed the laboratory research, but tried to attract or keep talented young people, and give them a free hand. My job was to take an interest in their research, and to make sure they had the means to carry it out.*'

In an informative article published in *The Scientist*[35] in 1988, Perutz disclosed the principles he employed to create and manage his world-famous LMB. He said that, unlike Noah who gathered animals for his Ark, he did not select two mathematicians, two physicists, two biochemists, and two biologists and wait for their discoveries to hatch. Instead he let things grow naturally, like a tree, after first choosing outstanding people and giving them intellectual freedom. To set up a successful laboratory, what must be done, as a Director, according to Perutz, apart from granting intellectual freedom, is to show genuine interest in everyone's work.

It is pertinent to cite at this juncture some of the passages in the letter sent to Dr Richard Henderson shortly after Max Perutz died in February 2002[15] by the American Nobel Prize winner Tom Steitz, who, after his PhD at Harvard, came to work at the LMB in Cambridge: '*Max was excited by the success of others and always included the youngsters in his celebrations. When Dorothy Hodgkin, Guy Dodson and colleagues solved the structure of insulin in the late 1960s, Max decided to mount a Cambridge expedition to Oxford in order to help celebrate the great occasion.*'

In another passage in his letter, Steitz wrote: '*…not all Nobel prize-winning laboratory directors are alike. Joan* (Tom Steitz's eminent scientific wife) *and I decided to split the location of our first sabbatical from Yale between the Max Planck Institute in Göttingen and the LMB. In Germany…we were never introduced to the director. There was no encouragement of, or format for, inter-lab interactions and no chance to meet any of the Abteilung directors since they refused to eat in the general dining hall with the rest of the laboratory workers.*

Moving to Cambridge in the spring of 1977 was a breath of fresh air. Upon our first entry into the LMB canteen, Max came up to me and said "Let's have lunch together. Tell me what you have been doing." What a difference!'

Other eminent scientists throughout the world, including the US Medallist for Science Harry Gray of the California Institute for Technology, also recall exciting conversation, over lunch in the Athenaeum of that Institute, with Max Perutz whose infectious enthusiasm for new knowledge they still vividly remember. Hugh Huxley's view of the atmosphere and mode of operation of the LMB under Max Perutz as Chairman of the Board is expressed in the Preface to the book he edited in 2013 entitled '*Memories and Consequences*': '*In many ways, the lab was a kind of paradise for dedicated research scientists. No one had to do any teaching, or to write grant applications, or even to publish frequently. Expert practitioners in most of the relevant fields were readily at hand. Within reason, all supplies and equipment were either immediately available, or rapidly obtainable, via Michael Fuller* (who was in charge of the Stores). *The lab had sizeable mechanical and electronic workshops, where any novel equipment that the experimental groups needed and designed could be skilfully constructed and maintained...such arrangements followed the traditions of the Cavendish Laboratory.*'

Hallmarks of Max Perutz's management style were that he knew every scientist and technician in his laboratory and they knew him. And he was always available for discussions with them. As Director, he left the financial details to others. And this did cause a problem towards the end of his tenure as Head of the LMB. There was a financial crisis, and cuts had to be made in the number of staff that were given permanent posts.

Perutz often quoted the words of Sir Nevill Mott, another Nobel Prize winner, who had worked with Niels Bohr in his halcyon days in Copenhagen. Mott said,[35] '*We were in and out of each other's rooms all day, and so was Bohr. Nobody dreamt of keeping an idea to himself; our joy in life was to tell it to other people to get criticized and if possible accepted. Bohr himself, if he had a new idea, would tell it to the first person he could find...I learned from Bohr what physics was all about, that it was a social activity and that a teacher should be with his students.*'

Perutz certainly lived according to this ethic. What astonished so many about him was that he could create and manage a magnificently successful laboratory of molecular biology, and yet accomplish so very much in addition as a cultured, kindly compassionate human being. It is noteworthy that, well outside the realm of molecular biology—in, for example, a recent edition of the *London Business School Review* (2018)—Perutz's management style as the top individual is commended.

I can think of no better way of concluding this description of Perutz than by quoting his final paragraph when he spoke at Peterhouse on the occasion of his eightieth birthday. '*There are other things to be thankful for: to have lived in a country free from oppression and also from war – Falklands apart – these 49 years; to have worked among the British scientific community where you are judged, not by your origins, nor by your religion, nor by your politics, nor by your connection in high places, nor by your wealth, but solely by the quality of your work: to have enjoyed and to be still enjoying generous support for my work from both sides of the Atlantic, to be tolerated by my colleagues at the lab and here with affection and without being made to feel a burden, and finally for having received so many honours which in my youth I never expected to come my way, though I used to tease my son when he was little and when peerages were still hereditary, that one day I would become Lord Haemoglobin and he would inherit the title whether he wants to or not!*'

8.2.3 John Kendrew: a summary of his life and scientific legacy

In contrast to other giant contemporaries of his—notably Max Perutz, Dorothy Hodgkin, and Aaron Klug—John Kendrew has not yet had a full biography, although it is known that one will appear soon.[10]

When I wrote to him a few months before he passed away, he replied, in March 1997, '*I've written* nothing *autobiographical, but I enclose a detailed CV, which I hope you will find useful.*' His CV, which he compiled in December 1996, is reproduced in Appendix 1 to this chapter, and it conveys rich information about his achievements and recognition.

8.2.3.1 Early life

Kendrew was born in Oxford in 1917. His father W. G. Kendrew was a geographer who became University Reader in Climatology at the University of Oxford. His mother Evelyn Sandberg-Vavalà was a distinguished historian of early Renaissance Italian art. Although his parents separated when he was young, he had affectionate, though complicated, relations with both of them and inherited many of their own intellectual concerns—from his father, his interest in architecture, nature, and photography; from his mother, his interest in art and music. Florence, where his mother lived for most of her adult life, he called his 'second home'. He was educated at Dragon School, Oxford, and Clifton College, outside Bristol, whose science teachers—several of them FRS—he admired. After entering Trinity College, Cambridge, he pursued a degree in natural science and graduated with first class honours in Chemistry. For one year (1939–1940), he pursued research on chemical kinetics in physical chemistry with E. A. Moelwyn-Hughes (see Chapter 1 and Part II of this chapter, Section 8.4.2).

At the outbreak of war, he went to the Air Ministry as a Junior Scientific Officer working on radar. In 1940, he joined Operations Research with an honorary RAF rank. Thereafter, he worked primarily on anti-surface vessel warfare in the Middle East and South East Asia. He was exceptionally successful in this role, as is testified by the fact that he was one of the handful to have access to the ultra-Enigma codes.

Considerable details about Kendrew's life, both at Clifton and in his military service, are contained in the perceptive articles about him by K. C. Holmes[4] and Ross McKibbin.[36] But to gain a full appreciation of Kendrew's legacy, it is prudent first to recall some of the tributes paid to him at his Memorial Meeting in November 1997, organized by the Master and Fellows of his Cambridge College—Peterhouse (see Figure 8.7).

8.2.3.2 Reflections on Kendrew's achievements

We first recall some of the remarks made by his very close colleagues, both at the LMB and Peterhouse—Max Perutz and Sir Aaron Klug—before proceeding to the recollections of others who also interacted closely with him.

Figure 8.7 *John Kendrew in his days as Deputy Chairman of the LMB at Cambridge, looking down on a model of the low-resolution structure of myoglobin.*
(Courtesy MRC-LMB, Cambridge)

In his speech on 5 November 1997, Perutz began: *'John and I shared three great scientific adventures: founding the MRC Unit of Molecular Biology, solving the first protein structures and founding the European Organization of Molecular Biology... 14 years later, in the autumn of 1959, we find John secluded in the vast, bleak, window-less room of the Cavendish, where its obsolete cyclotron had just been dismantled, building up the first atomic model of a protein molecule. He created a towering forest of ⅛" steel rods on a wide wooden platform and marked the coordinates of the atoms derived from his X-ray crystallographic analysis on the rods with coloured meccano clips. His atoms were made of ⅛" brass rods representing the chemical bonds between them. One by one he clamped about 1300 of them to the rods and linked them all together until his model was complete. It became the Eighth Wonder of the World, and John was immensely proud of it. It is now on permanent exhibition in the Science Museum in South Kensington.'* (See Figure 8.8, panels (a) and (b).)

'John was decisive, but never pompous or overbearing. He had the Englishman's reserve, but he was the antithesis of the insular Englishman. He could be aloof, but he was urbane, cosmopolitan, fluent in French, German and Italian, and he was a devoted European. He was concerned that American universities quickly grasped the promise of

Figure 8.8 *(a) Kendrew's model, which Perutz described as the Eighth Wonder of the World. (Courtesy of the Science Museum) (b) John Kendrew seen through the model of the myoglobin molecule. (Courtesy of Science and Society Picture Library, London)*

molecular biology, while European ones ignored it. American postdocs could readily get fellowships to take them to Europe, but European ones had no funds to gain experience abroad. America had summer schools to spread the gospel, but Europe had none. With these concerns in mind, we founded, in 1963, the European Molecular Biology Organization, EMBO, with John as Secretary of its first Council. There I first saw John in action as a committee man, as he later became, diplomatic at achieving a consensus among divergent views and skilful at putting it into clear, unambiguous language. The EMBO fellowships and summer schools are still going strong and have had a decisive effect on molecular biology in Europe. However, from the very start of EMBO, John's aim was the creation of a European Molecular Biology Laboratory. This great laboratory was opened in Heidelberg in 1974 with John as its first director. It would never have come into being, but for John's determination and his astute organisational and diplomatic skill. It will remain as his lasting monument.'

Shortly after Kendrew's death, Max Perutz described to me (in a letter I still possess) that it was he (Perutz) who succeeded to persuade the Volkswagen Foundation to provide the substantial funds needed to initiate and sustain the EMBO Fellowship Scheme.

Aaron Klug first came across John Kendrew one night in Cambridge in 1953. They were fellow users of EDSAC, the first electronic, digital, and programmable computer in the Mathematics Laboratory at the University. Kendrew had set out to replace the tedious calculators necessary in analysing X-ray data, traditionally done by hand from tables or with the aid of primitive analogue machines (see Section 2.4). '*I remember well his sandwiches and thermos flask to fortify him against a long night.*' (See Figure 8.9.)

'*When I moved to Cambridge in 1962 to the MRC Lab, I also became, thanks to John, a teaching Fellow at Peterhouse. John was Director of Studies in Natural Sciences, and the sole scientist on the Fellowship of the College, which then numbered 14. In those days, one taught broadly; John taught most of Chemistry, and also History and Philosophy of Science. I had rooms on C staircase immediately opposite John's rooms. They were fairly spacious, and it was from here that he ran the office of the Journal of Molecular Biology, the first and, for many years, the leading journal in the subject.*

It was in this context that I had my first direct view of John's outstanding organisational ability. He kept in his pocket a large expandable notebook, a precursor of the Filofax (much later the symbol of yuppiedom). In this book he kept an extraordinary amount of information, easily accessible: the names of all undergraduates in Natural Science, their supervisors, their exam results; all the current papers submitted to J. Mol. Biol., and their status; the members of staff at LMB; scientific references and a host of other items. If he needed to know something, he took out the book, and there it was, in an instant.

At College meetings he did not speak a great deal, but when he did it was to good effect. He would listen while others spoke, and, even when discussion warmed up, as it sometimes did, he did not intervene. Usually only at the end, came a lucid and concise summing up of the debate, and his own judicious opinion.' Other Fellows of Peterhouse—Tony (later Sir Tony) Wrigley, who later became President of the British Academy, and Professor Jacques Heyman, Head of the School of

Figure 8.9 *John Kendrew.*
(Courtesy The Laboratory of Molecular Biology, MRC.)

Engineering at the University of Cambridge, and his colleague Professor Chris Calladine—on being asked recently their memories of John Kendrew, simply replied: '*exceptional mental clarity and urbanity*'.

According to Aaron Klug: '*It was this capacity for clear thinking, combined of course with high intelligence, that made John such an effective member or chairman of committees, and for which he was much in demand; for example, the Government committee on the future of CERN, or a trusteeship of the British Museum; and in many faraway places outside Britain. I remember* (said Klug) *once trying to reach him in the early '90s, to find he was at the World University in Tokyo (of which I had never heard). He loved travelling, and once told me that he felt there was something wrong if he hadn't been on an aeroplane in the last three weeks.*'

According to Aaron Klug: '*He will always be remembered on the analogy he once used between finding the structure of myoglobin and the first landfalls made by the early explorers of America, as one of the first two who had the first sight of a New World.*'

Although Hugh Huxley, John Kendrew's first PhD student, could not attend the Memorial Meeting held in Cambridge in November 1997, he was able to pay tribute to Kendrew at Heidelberg earlier that year. Some of his observations are cited below. But the reader ought to know that Kendrew's admiration of Hugh Huxley knew no limits (see Figure 1.2 of Chapter 1 and Figure 8.10).

Figure 8.10 *Hugh Huxley, John Kendrew's first ever PhD student, at his electron microscope.* *(Courtesy MRC-LMB)*

Only a few months before his passing, John Kendrew told me, with great excitement in his voice, that he had learnt that Huxley had been awarded the coveted Copley Medal of the Royal Society (its highest award). Previous awardees include Benjamin Franklin, Dmitri Mendeleev, Albert Einstein, and all the other individuals from the LMB seen in Figure 1.2 of Chapter 1—except John Kendrew.

John Kendrew proceededk to tell me that so skilful and able was his first-ever PhD student that he, Kendrew, and that student—Hugh Huxley—were each elected as Fellows of the Royal Society in the same year (1960), a very rare occurrence. As mentioned earlier, Hugh Huxley began to pursue the study of protein crystallography with Kendrew but soon became rather disenchanted by the almost endless computational labour it warranted. He soon switched fields and quickly established himself as a leading authority on the biophysics of muscle, which he investigated predominantly using electron microscopy (see Figure 8.10), and later kinetic, low-angle X-ray diffraction using synchrotron radiation. Huxley's early model (with collaborator Jean Hansen) of the myosin/actin interface has become a textbook feature for many years.

Figure 8.11 *View of Section of Old Court, Peterhouse, Cambridge, where Huxley met Kendrew on starting his PhD studies.*

(Courtesy Dr P. Pattenden)

Hugh Huxley first met John Kendrew in the latter's elegant rooms (Figure 8.11) in Peterhouse, Cambridge, in July 1948. '*I was his PhD student when he had not even quite taken his PhD himself.*'

'*To do research in Cambridge had been my great ambition since early high school, so at that moment it all seemed to be a magical world to me, and John seemed to fit into this ideal picture exactly, as the quiet, diffident Cambridge don, a man of action when necessary during the war, but now puzzling over the uninterpreted X-ray patterns from myoglobin crystals, enjoying the ironies and foibles of academic life and High Table, deeply appreciative of the intellectual pleasures of art and music, having great insight into all matters temporal and spiritual—an immediate, and long-term, hero to me if there was one.*

Just recall, those small rooms in the Cavendish, with Max and John peering at their X-ray patterns, Francis Crick in full flood, soon to be joined by Jim Watson, and then the two of them arguing excitedly over Linus Pauling's latest helix, or the news and gossip from Kings, London, and then in the science compound around us, Martin Ryle cycling out to his radio telescopes, Hoyle, Bondi and Gold dreaming up continuous creation—very stimulating even if wrong—Andrew Huxley and Alan Hodgkin working out their famous equation, Fred Sanger beginning to sequence proteins, and the first EDSAC computer whirring away in the Maths Lab. What a place to be! A Golden Age!

It was John who was instrumental in getting Jim Watson to Cambridge, and straightening out some Fellowship problems to keep him there, and I think it was his diplomatic intervention with Bragg which kept Francis Crick there when he had been particularly infuriating—that was before most people realised just how really clever he was, in spite of his brilliant exterior.

It was John's superb organizing ability—plus, of course, having picked the right problem and the right protein—which was essential to his and Max's success in showing how protein structures could be solved—finding good people to work on the chemistry of heavy atom labelling, organizing troupes of young ladies to densitometer thousands of films, organizing the first serious crystallographic computing, and finding out how to translate 3-dimensional density maps into polypeptide chains and the first protein structure, which began to reveal at last what a protein molecule was really like.

However, John had always also had a great interest in the organization and proper application of science. In the early 1960s, with a Nobel Prize to add to his credentials, he struck out in a number of different directions to implement these interests and ideas (this was in addition to founding the "Journal of Molecular Biology" a few years earlier, which he guided to a pre-eminent position among journals).

In those days—and for many years to come—we were all concerned about the danger of nuclear war, but mainly as protesters. John's policy was to go back into the defence system again, part time, first as a deputy to Solly Zuckerman (who was chief scientific advisor to the Ministry of Defence), and later to serve as Chairman of the Defence Advisory Council, in order to try to maintain some common sense and reason in the inner councils of a very dangerous world. I don't know exactly what he was doing there, and it's probably still secret, but I'm sure it involved figuring out sensible ways of doing things that people with disparate and apparently mutually exclusive views would find to be solutions to which they could give their willing support—and that worked out. It always struck me as quite remarkable that he should have received his knighthood for this work, rather than all his other scientific contributions at this time, which overlapped with founding the EMBL.'

According to Hugh Huxley, he felt that the early investigators of structural molecular biology in Cambridge in 1945 were keen to maintain the traditions of British science by organizing a first-class lab in the city, and by promoting a very active British involvement in scientific developments in Europe, by fostering collaboration in science.

'*The first step in this latter direction which John Kendrew and a number of others were responsible for was the setting up of an organization, EMBO, initially to provide Fellowships for European scientists to work in laboratories in other countries, to help spread more rapidly new ideas and techniques, and the new area of molecular biology. This was done first with support from the VW Foundation, and later with support from many governments.*' In this connection, Max Perutz's European contacts proved invaluable. '*This organization also maintained the idea of a European Molecular Biology Laboratory (EMBL), which first arose in 1963, and at which John worked tire-*

lessly with all his diplomatic skills for many years, to bring to fruition, before Governments finally became convinced, in the late '60s, that it was a desirable plan, and before a site was chosen and the lab designed and completed, in 1975.'

It is instructive, in the context of describing how EMBO was created, to consult the timely article that John Kendrew publicized in *Nature* in June 1968, entitled *EMBO and the Idea of a European Laboratory* where, inter alia, he recalls the special role that the Hungarian–American Leo Szilard, as early as 1962, and the Austrian–American Viki Weisskopf played as catalysts for the whole idea.

With John Kendrew as Director General, EMBL rapidly came into full-scale operation, with many very talented people working in Heidelberg and in the vitally important outstations that Kendrew had identified—the synchrotron facility at Hamburg and the neutron facility at Grenoble. These 'outstations' that Kendrew brought on board were a vitally important and perspicacious act, as he had realized that structural molecular biology would depend increasingly in future years on free access to synchrotron radiation and intense neutron sources.

The prestige, the promise, and the success of EMBL when John Kendrew stepped down as Director General in 1982 was universally acknowledged. And when, in November 2017, EMBL held a meeting at Heidelberg to celebrate the centenary of John Kendrew's birth, the outstanding all-round success of his initiatives were very apparent. Apart from its scientific excellence, its international character and diversity stand out. There are extensive opportunities for young creative thinkers and experimentalists, with ample opportunities for interdisciplinary collaboration. Twenty-three European countries are now members, with more likely to join, and Argentina and Australia are associate members. In addition, the six EMBL sites (see Figure 8.12) serve molecular biology in unique ways.

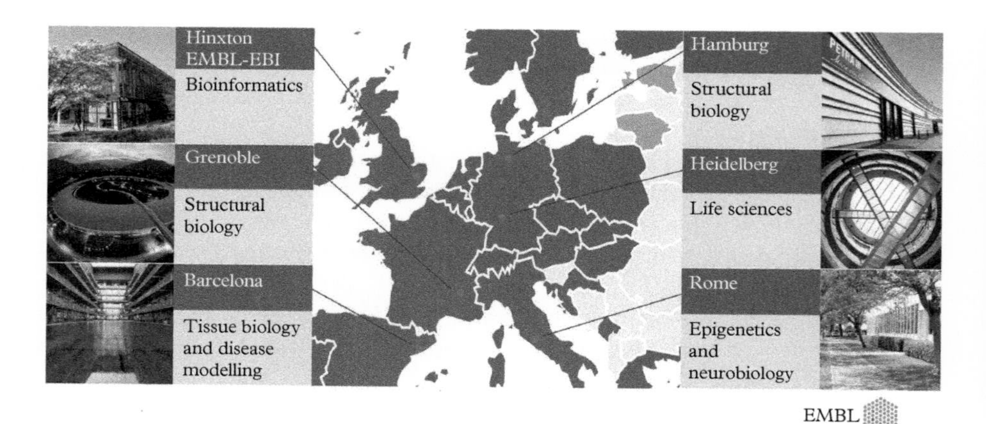

Figure 8.12 *The present-day 'outstations' related to the European Molecular Biology Laboratory (EMBL), Heidelberg. The EMBL now employs over 1700 people with over eighty nationalities. (Courtesy Director of EMBL. Copyright Tabea Rauscher)*

The Bioinformatics EMBL Centre in Hinxton, UK, has approximately twenty-five million web visits per day, and the structural biology centres at Hamburg and Grenoble have, between them, over 3000 people visit per annum. It is fitting that, just as the LMB in Cambridge has a Max Perutz Lecture Theatre, so at Heidelberg, there is a John Kendrew one and Hixton has a Francis Crick Auditorium.

In addition to his sterling work with Max Perutz in creating EMBO—Max Perutz was the founding chair of EMBO—and his own extraordinary skills in creating and leading EMBL, John Kendrew also participated on a wider international scene, in countless committees and advisory boards (see Appendix 1 to this chapter). Apart from his role as President of the International Organization of Pure and Applied Biophysics (1969–1972), he was Secretary General of the International Council of Scientific Unions (ICSU) from 1974 to 1980, then Vice-President and President from 1983 to 1988. Dr Julia Marton-Lefèvre, who served as Chief Executive Officer of ICSU, told me in a letter (October 1997) that it was John Kendrew's diplomatic skills that successfully negotiated the entry of both

Figure 8.13 *John Kendrew, Max Perutz, Cesar Milstein, and Fred Sanger pictured in Peterhouse on the occasion of Perutz's eightieth birthday in October 1994.*

(Courtesy JET Photographic and the Master and Fellows of Peterhouse)

mainland China and Taiwan into ICSU. She also pointed out that it was his great skills that led, in 1986, to the official launching of the International Geosphere-Biosphere Programme (IGBP), a study of global change.

The cultural depth of Kendrew's life is epitomized in another of the comments made by Julia Marton-Lefèvre: '*In the ICSU world, John's private passions for music and art found many sympathizers, and for a while when he was in office we even had an informal opera and art club, managing to add a little time to our official meetings to visit the world's art treasures and to take in an opera or two while we travelled the globe on ICSU business.*'

After he retired from being an outstanding President of St John's College, Oxford (1981–1987), John Kendrew returned often to Peterhouse. There were three major occasions that merit special mention: Max Perutz's eightieth birthday feast (see Figure 8.13) in autumn of 1994; John Kendrew's eightieth birthday feast in 1997 (see Figure 8.14); and a special dinner of the officers of the Royal Society and the US National Academy of Sciences, organized by Sir Michael

Figure 8.14 *Group present at the feast to celebrate John Kendrew's eightieth birthday in 1997. Perutz and Kendrew at the front, with Sir Michael Atiyah, second row, extreme right. Other molecular biologists, biochemists, and physiologists in this group are: Sophie Jackson, Andrew Lever, Tony Crowther, Chris Calladine, Tom Blundell, R. D. Keynes, M. Ferguson-Smith, Ken Holmes, and Fred Sanger.*

(Courtesy JET Photographic and the Master and Fellows of Peterhouse)

Atiyah (President of the Royal Society, 1990–1995) in 1993, all of which were very happy occasions.

Seated next to me at the third of these dinners was the then President of the US National Academy of Sciences Professor Bruce Alberts, who made a special mention, privately to me that night, of the occasion when John Kendrew lectured at Harvard (where Bruce Alberts was then a graduate student of the eminent US biological chemist Paul Doty). Alberts told me that, when Kendrew showed in the early 1960s his X-ray results on the structure of myoglobin and pointed out the individual amino acid residues in the polypeptide chain (see Figure 5.5 of Chapter 5), the intellectual impact was so startling that he, Alberts, still recalled the impact it made. (He repeated this reaction in a letter addressed to me in September 2018.) John Kendrew's lecture at the University of Pennsylvania in 1961 made a huge impression on a leading US protein scientist—Walter Kauzman of the University of Princeton. The opening words of his letter to Kendrew on 10 April 1961 says: '*It should not be necessary to let you know that your talk at the University of Pennsylvania made a tremendous impression. To say anything else would merely be a reflection of incompetence and blindness. You are obviously in a position to make tremendous contributions to our understanding of the finer details of protein structure and side chain interactions, and those of us who have been trying to approach this problem using the more indirect physical chemical methods might well have the feeling that we ought to close up shop.*'

Another eminent US scientist, now Emeritus Professor of Neurobiology, at Stanford University and recipient of the National Medal of Science in 2006—Lubert Stryer (who was a postdoctoral worker with John Kendrew at the LMB in 1962–1963)—also recalls the remarkable impact of Kendrew's lecture on myoglobin to the nascent Biophysics Program at Harvard University in 1960: '*His lectures were extraordinarily lucid and stimulating.*'

During his stay with John Kendrew at the LMB, Stryer (along with another full-time research worker of Kendrew's—Herman Watson) elucidated the mode of attachment of the azide ion to sperm whale myoglobin,[38] a significant advance in structural molecular biology, as it was the first use of the so-called difference Fourier approach to such problems.

Stryer's recollections of his time with John Kendrew are very revealing, in respect to both the latter's constructive attitude towards others and his scientific skills. In a letter I received from Professor Stryer (September 2018), he says: '*In March, John called me to his office to tell me that Arthur Kornberg (1918–2007)—the American biochemist who won the Nobel Prize in Physiology or Medicine for his discovery of the mechanisms in the biological synthesis of DNA—wrote that there was a faculty opening at Stanford. John asked whether he should write Arthur about me. "Sure" was my reply. A few weeks later, I received an invitation to visit Stanford and then an offer of an assistant professorship.*

I often looked at John's lab notebooks for information and inspiration [see Figure 8.15]. *They reflect a highly disciplined and focussed mind. He had a keen sense of what was important and decipherable.*'

- Myoglobin notebooks (Kendrew and collaborators)
  - Preliminary work [MSS. Eng. d. 2129-2134; MSS. Eng. e. 2345-2351]
  - Main myoglobin programme [MSS. Eng. d. 2135-2194]
  - Collaborators' notebooks [MS. Eng. c. 2398; MSS. Eng. d. 2195-2203; MSS. Eng. e. 2352-2358]
  - Atomic co-ordinates/amino-acid sequencing [MSS. Eng. d. 2204-2212]
  - Miscellaneous [MSS. Eng. d. 2213-2216]
- Myoglobin notes and data (Kendrew and collaborators)
  - Preliminary work [MSS. Eng. b. 2010-2014; MS. Eng. c. 2399]
  - Main myoglobin programme [MS. Eng. b. 2015; MS. Eng. c. 2400]
  - Collaborators' notes and data [MS. Eng. b. 2016]
  - Miscellaneous [MS. Eng. c. 2401]
- Myoglobin materials and apparatus (Kendrew and collaborators)
  - Supplies and specimens [MS. Eng. b. 2017]
  - Optical diffractometer [MS. Eng. b. 2017]
  - Microcamera [MS. Eng. b. 2017]
  - Densitometer [MS. Eng. b. 2017]
  - Computer time [MS. Eng. b. 2017]
- Myoglobin collaborators and staff
  - Individual files [MSS. Eng. c. 2402-2403]
  - Chronological files [MS. Eng. c. 2404]
- Myoglobin correspondence
  - Aspects of myoglobin [MSS. Eng. c. 2405-2406]
  - Atomic co-ordinates/amino-acid sequencing [MS. Eng. c. 2407]
  - Publications [MS. Eng. c. 2408]
- Myoglobin models
  - Skeletal model [MS. Eng. c. 2409]
  - Ball-and-spoke model [MS. Eng. c. 2409]
  - Science Museum, London [MS. Eng. c. 2409]
  - Correspondence [MS. Eng. c. 2410]
- Myoglobin miscellaneous
  - Pantographs [MS. Eng. d. 2217]

Figure 8.15 *A small section of the 397 boxes that constitute the Bodleian Library, University of Oxford,* 'Correspondence and Papers of Sir John Cowdrey Kendrew'. *This is a minute proportion of the scope and content of the Kendrew papers, as described on the Internet and that run to eight pages. (Copyright Bodleian Library, University of Oxford)*

Both Perutz and Kendrew elicited responses such as that experienced by Alberts and Stryer, when they travelled the world in the early 1960s to describe their structural advances. Most scientists knew that they were on course to win the Nobel Prize, no one more so than their mentor Lawrence Bragg, whose letter of recommendation—written at the Royal Institution—to the Nobel Committee is shown in Figure 8.16. Note that Bragg recommended that Perutz, Kendrew, and Hodgkin should be accorded the Nobel Prize in Physics.

The similarities in the excellent qualities of Perutz and Kendrew have been referred to earlier in this monograph: exceptionally gifted scientists, tenacity of purpose, extraordinary diplomatic, communicative, and negotiating skills, great visionaries, and generous leaders. They have left monuments that will stand forever in the scientific study of the living world. '*You asked whether I sensed any tension between Max and John*', wrote Lubert Stryer to me recently. '*The answer is a decisive no. The few times I saw them together in 1962–3, they were relaxed and smiling. I should also add that the people in John's lab, and in Max's, got along very well, both scientifically and socially.*'

K. Vetenskapsakademiens
Nobelkommitté
Inkom den 15.1.1960
13 bilagor, bästa ...

DAVY FARADAY RESEARCH
LABORATORY.

DIRECTOR: SIR LAWRENCE BRAGG, F.R.S.

TELEPHONE:
HYDE PARK 0669.

THE ROYAL INSTITUTION,
21, ALBEMARLE STREET,
LONDON, W.1.

9th January, 1960.

Dear Westgren,

I am making a recommendation to Professor Hulthen that the Nobel Physics Prize should be awarded in three shares to Perutz and Kendrew for their work on protein structure, and to Dorothy Crowfoot Hodgkin for her work on penicillin and vitamin B12, as outstanding examples of the highest achievements in the X-ray analysis of complex molecules. I enclose a copy of my letter to Professor Hulthen, hoping that this recommendation will also receive the support of your Chemistry Committee, though I consider it is primarily the concern of the Physics Committee since the molecules are analysed by what are essentially physical methods.

At the same time, I wish to ask your Chemistry Committee to consider the claims for consideration of the work which has established the structure of Deoxyribonucleic acid, DNA. We owe our knowledge of the atomic arrangement in this form of nucleic acid to:

J.D. Watson	Department of Biochemistry, Harvard University.
F.H.C. Crick, F.R.S.	Molecular Biology Research Unit, Cavendish Laboratory, Cambridge.
M.H.F. Wilkins, F.R.S.	Medical Research Council, Biophysics Research Unit, King's College, London.

The first analysis of the DNA structure was made by Watson and Crick at Cambridge in 1953, when Watson was a visitor to this country. The solution was their own single brilliant piece of work, but it was both partly inspired by Wilkins' patient researches over many years in getting fine diffraction pictures of DNA, and has been confirmed since by Wilkins, who has analysed the structure in much greater detail. In my opinion, these three researchers ought to be grouped together. Together they

have/

Figure 8.16 *Letter of recommendation by Sir Lawrence Bragg sent to the Nobel Committee from the Royal Institution.*

(Courtesy, Nobel Foundation)

J.D. WATSON, F.H.C. CRICK, F.R.S., M.H.F. WILKINS, F.R.S.

NUCLEIC ACID

- 2 -

have given us a detailed plan of the way in which the atoms are bonded in space. The picture on page 79 of Wilkins' paper (3) has now become widely familiar as an example of a complex biological structure.

The DNA solution has had so great an effect in inspiring research in many centres, especially in the United States, that it surely should be recognised by a Nobel award, either in Chemistry or in Medicine and Physiology. If your Committee considers that it falls in the latter category, will you please pass my letter to the appropriate Committee.

I attach a statement about the work on nucleic acid, and also of work on virus and other structures, together with copies of the more important papers. I have added a popular article by Perutz on "The Molecular Basis of Inheritance" which gives a general account of the investigation of DNA. I also include a report by Perutz on the work of the Medical Research Council Unit at Cambridge, in which a useful summary will be found.

I must make it clear that the recommendation to which I give priority is that of a Prize for Perutz, Kendrew and Hodgkin in my own field of Physics. I am adding a warm recommendation that Watson, Crick, and Wilkins should be considered for a Prize in Chemistry or Physiology, hoping that the recommendation may receive strong support from others.

Yours sincerely,

W L Bragg

W.L. Bragg.

Professor Arne Westgren,
The Nobel Committee for Chemistry,
Stockholm 50,
Sweden.

Figure 8.16 *Continued*

8.2.4 John Kendrew and Max Delbrück

There are several reasons why I have chosen to complete this section of Part I devoted to Perutz and Kendrew by focusing on Max Delbrück. Delbrück won the Nobel Prize, along with Luria and Hershey, in 1969 in physiology or medicine. Frequently, in my discussions with Max Perutz, I sensed that he greatly admired Delbrück. Perutz's son Robin has also told me that his father had great respect for the personal and intellectual qualities of Delbrück. (An outline of Delbrück's personal and intellectual life is given in Section 8.4.6 below.) I, myself, first came across Delbrück's name while reading Schrödinger's book *What is Life?*, wherein a mathematical model of the gene is headed '*Delbrück's Model*'.

Later, in my general reading, I learnt that Delbrück was an authority on so-called phages (bacterial viruses) and that he was the leader of the 'phage group'. Relatively recently—when I decided to write this monograph—I discovered that a Festschrift volume, entitled '*Phage and the Origins of Molecular Biology*' (edited by J. Cairns, G. S. Stent, and J. D. Watson), had been published in 1967 to mark Delbrück's sixtieth birthday. It was at that time that I came across John Kendrew's brilliant book review[37] of this volume, published in *Scientific American*. So full of insight is his analysis, and so lucidly is Kendrew's critique of this Festschrift, that I feel it merits further elaboration for two main reasons—first, the definition of what is meant by molecular biology, and, second, because it sheds much light on the ways in which scientific progress and revolutions occur.

Kendrew's review begins with a gentle and delicately worded chastisement of the editors and authors of the Festschrift. Nearly every contribution to this book traces the respective author's knowledge of the emergence of molecular biology. Kendrew agrees that the authors ' *...are celebrating an intellectual movement that, although only about 20 years old, has worked a major revolution in every branch of biology.*' Kendrew then recalls[37] the early X-ray studies of biological materials by W. T. Astbury and J. D. Bernal and their pupils, where the emphasis was on structure (see Chapter 3 especially). '*To anyone brought up in the British school of molecular biology...it is a little odd to find in nearly every contribution...the explicit or implicit assumption that molecular biology had its only real beginnings in the phage group, and that the central theme of the subject is biological information.*' It is relevant to recall here that, whenever, in my early days, I had conversations with E. G. Cox and Kathleen Lonsdale, who were contemporaries of Astbury and Bernal at the Davy-Faraday Research Laboratories (see Chapter 3), they frequently reminded me that Astbury especially, and later Bernal, were the progenitors of molecular biology.

Kendrew continues: '*It is true that the Watson-Crick double-helix structure of DNA has had an enormous influence on these workers*'—that is, the authors and editors of this Festschrift. '*Indeed, it has proved to be the central concept around which their thinking has developed. But the features of that structure that have been important to them...have been the one-dimensional nature of the information store and the role of specific pairs of nitrogenous bases in replication. Other kinds of biological structure, in particular the structures of proteins are hardly mentioned.*'

Kendrew then recalls that, in the early days, the two schools were almost entirely isolated from each other. On the one hand, there was Delbrück and Luria (who was J. D. Watson's teacher in Bloomington, Indiana) concerned mainly with the problem of interpreting the genetics of microorganisms at the molecular level in terms of a one-dimensional molecular information carrier that '*only by degrees emerged as the molecule of DNA*'. On the other hand, the British schools were developing methods of elucidating the three-dimensional structures '*of all kinds of biological macromolecules, but with a strong emphasis on proteins. For them the aim of interpreting function was a goal dimly discerned for the future, and they had little knowledge of, or interest in, the problems of genetics.*'

Fascinatingly, Kendrew notes that '*the schism was not entirely geographical in origin*'. Linus Pauling was US-based—indeed Delbrück and he were both at Caltech for a good while (see Chapter 7 for a brief account of their joint work). Pauling certainly '*brought the two schools in close geographical proximity, but it is questionable whether there was much intellectual relation between them*'.

In J. D. Watson's contribution to this Festschrift, mention is made of the fact that, in 1949, he had the impression that the phage group thought that Pauling's world and theirs did not have a great deal in common. '*These articles*', Kendrew observes, '*were written 15 years later, and it is surprising to find how little evidence they contain of a concern with structural problems even now. The first real link between the two schools was closed by the migration of Watson himself from the heart of the phage group by way of Copenhagen to Cambridge.*'

At Kendrew's Memorial Meeting in November 1997, Watson declared that the main reason why he came to Cambridge was because Luria had advised him to go and work with Kendrew. In his '*Double Helix*', Watson described how it was M. H. F. Wilkins' X-ray photograph which he saw in a conference in Italy earlier that provoked his interest in the structure of nucleic acids. And much later, Max Perutz thanked Watson publicly for importing to his and Kendrew's group the idea that the nucleic acids, not proteins, held the key to the secret of life.

'*It is curious*', so writes Kendrew,[37] '*that Watson, one of the few members of the phage school who was a real biologist was almost obsessed with the idea of a structural model, and that it was he who should have chosen this migration to a physics department.*' The result was the double-helix model of DNA, which, in turn, led to the phenomenally rapid development of numerous aspects of molecular genetics and structural biology.

In the final section of Kendrew's discerning book review, he focuses considerable attention on Delbrück's style of working and the profound influence that he exerted. His way of going to work was, according to Kendrew, as unorthodox as it was productive. '*At first he emerges not so much as a guide, but as a critic: we read how he scared his colleagues, how long seminars under his direction were an ordeal, how he told dozens of people confidentially after they had delivered a seminar that it was the worst he had ever heard. We read of his suspicions of chemistry, of his deprecation of biochemistry, of his disdainful indictment of phage work that did not come up to his standard...He is said to have been careless in treating history, and he was critical of the most exciting new results.*'

Kendrew recalls being in Caltech shortly after Watson and Crick had proposed in 1953 the double helix, full of excitement of seeing the model and understanding the replication hypothesis, and finding Delbrück '*breathing scepticism whether all this cleverness had anything to do with biology.*' Delbrück was not convinced, and he invented other explanations. Kendrew ends his masterly analysis with the following remarks: '*But of course Delbrück's influence, although sometimes superficially destructive was in fact of the most positive kind.*' Kendrew goes on to quote one of the authors of the Festschrift that: '*Delbrück's passionate rejection of vagueness in the building and testing of conceptual models has helped change radically the entire philosophy of biological research.*'

The result, according to Kendrew, was the most dramatic and rapid development in the whole of biology since Darwin and Mendel. '*It is described in this book in a way that not only illuminates the subject itself, but also has a much wider significance as an object lesson for the conditions under which scientific advances are made and intellectual movements gather momentum.*' (See Section 8.4.6 for a biographical sketch.)

8.2.5 A period of great uncertainty: the privatization of LMB?

During one of our walks, in 1993, Max Perutz told me of the great anxiety that reigned at the LMB at that time. This had arisen because it was felt that, through Government decree, the Laboratory might, in future, have to seek its own funding, rather than obtain it from the MRC. In the country at large, there was talk that the LMB might have to embark on a policy of seeking research funds from potential industrial collaborators, and it was felt that some of its most influential scientists (Sydney Brenner, for example) were in favour of seeing the LMB privatized. There were rumours that a large US multinational had an interest in taking over the Laboratory.

This atmosphere had gradually appeared ever since Lord Rothschild (1910–1990), as a Government advisor, had conceived and promulgated his so-called '*customer contractor*' relationship (first ventilated by Rothschild in the early 1970s, and later commended by Edward Heath, the Prime Minister, his successor Margaret Thatcher, and members of her Cabinet). I know from personal remarks made to me by Max Perutz in the early 1990s, that he was fearful of the future viability of the LMB, in view of the political climate in the UK. I also know that, at that time of uncertainty, he approached several of his former colleagues—all Nobel Prize winners—to solicit their support to avert the privatization or sale of the LMB and to seek its retention under the aegis of the MRC.

On 16 March 1993, an extremely powerfully worded letter appeared in *The Times* of London entitled '*Vital Importance of Pure Research*'. The signatories were: Francis Crick (then of the Salk Institute, California); J. D. Watson (Cold Spring Harbor, New York); Fred Sanger (a retired member of the LMB); George

Köhler (Max Planck Institute, Freiburg); and John Kendrew (who masterminded the epistle and submitted it from his home, The Old Guildhall in Linton, a suburb of Cambridge). This influential letter exhibited all the hallmarks of Kendrew's incisiveness, but all the signatories contributed to it, as I was subsequently informed.

The opening paragraph of the letter set the scene: '*Sir, The Government is about to define its policy for science and technology in a white paper. As Nobel laureates who have worked in the Medical Research Council Laboratory of Molecular Biology at Cambridge, we are concerned at suggestions that it should be hived off from the MRC and perhaps privatised, because any such move would jeopardise its continued world leadership in the biomedical field.*'

It then goes on to recall the enlightened policies pursued by the MRC and points out that most of the great advances that (at that time) have brought eight Nobel Prizes and many other awards were the outcome of years of exploring uncharted waters. Yet the work carried out by these signatories and their colleagues led to major changes in the intellectual basis of much of medicine, and some of the discoveries had benefited diagnosis and treatment.

Their next paragraph constituted a vital justification for long-term, pure research: '*However, at the outset the problems that we and others attacked were only known to be fundamental to biology; their relevance to practical medicine was not obvious and their commercial applications looked utopian. The MRC took a long-term view, often investing not so much in the research projects themselves as in the talents of the scientists concerned, whom they supported throughout the many lean years while the outcome of their research remained in doubt.*'

Then the authors of the letter reassured the reader that the MRC's long-term view did not imply security of funding, regardless of scientific quality. (Like all other work funded by the MRC, the LMB's research was reviewed at five-yearly intervals by panels of international experts and continued research was contingent on their reports.)

Finally, the wealth-creating, and other consequences, of work at the LMB was highlighted: '*Some of the scientific advances made in the laboratory have also created wealth in the form of new products, techniques and instruments, but these practical applications became apparent only after the fundamental scientific problems had been solved and could not have been foreseen. Commercial companies have to look for assured profits within a reasonable time and cannot risk investing in research in fields which are not fully formed and so may take many years to bear fruit. Moreover, the results might bring little direct financial reward even if they benefit medicine.*'

The key message proposed by Kendrew, Crick, and colleagues was that the LMB was unlikely to maintain its high quality in an environment which required it to seek its main funding from contracts. '*Its continued association with the MRC may pave the way for further advances of great benefit to medicine.*'

As is shown in Chapter 10, especially in its Appendix 2, the commercial rewards arising from major medical advances in the design of new drugs has amply vindicated the stand taken by the authors of this letter in 1993.

PART II

8.3 Dorothy Hodgkin (1910–1994)

Born in Cairo, where her father was working for the Egyptian Ministry of Education and her mother was interested in archaeology, in Egyptology, and in weaving, Dorothy Crowfoot, as she was before her marriage to Thomas Hodgkin in 1937, is arguably one of the most famous of British scientists—the only British woman to win the Nobel Prize in Chemistry.

As mentioned earlier (see Section 8.1), full accounts of her life and work have been chronicled by Ferry[1] and Dodson.[2] There are, however, some aspects of her unusual personality and stellar achievements which have not been previously discussed.

8.3.1 A summary of Dorothy Hodgkin's contributions to medical science through her structural biological triumphs

Earlier, we have seen that Dorothy Hodgkin, from the outset of her work as a PhD student with Bernal in Cambridge (see Section 1.3), made valuable contributions to our understanding of the nature of the constituents of the 'living' world. Her first efforts were concerned with the nature of steroids, especially of the sex hormone cholesterol, which she extended on returning to Oxford where she solved the structure of cholesterol iodide. We have also described how she solved the structure of penicillin (see Figure 3.16), and reference has also been made to her monumental work (that took over thirty years to complete) on the crystal structure of insulin. Many observers, however, believe that the most significant advance made by her and her colleagues in structural molecular biology was the determination of the detailed atomic nature of vitamin B_{12}, upon which we shall now focus.

Vitamin B_{12}, the most complicated of all members of the B vitamins, acts against pernicious anaemia. This disease is often called Addisonian pernicious anaemia, and it has very powerful effects on our metabolism and also on our behaviour. The body produces an antibody that attacks the protein responsible for extracting vitamin B_{12} from food (any animal product). Medical experts do not seem to know why this happens. This kind of anaemia is an example of an autoimmune disease. Vitamin B_{12} was discovered in the mid 1920s, and pharmaceutical companies ultimately produced pure samples of it in late 1940s. Hodgkin's work on vitamin B_{12} began in her laboratory when the Glaxo Laboratory in the UK gave her some beautiful-looking, deep red minute crystals. Like the pepsin crystals that she and Bernal studied in 1934,[39] these crystals diffracted X-rays very well. Because of her know-how concerning the relationship between density, unit cell dimension, and molecular weight (see equation [1] in Section 3.5.1), she was able to tell the

Glaxo scientists what the molecular weight of vitamin B_{12} was on the day she examined the crystals!

Aware that the material consisted of about a hundred atoms (but with no knowledge of the chemical formula), she realized that its analysis by X-ray diffraction would be a very formidable task. Her contemporaries opined that it would be a rather hopeless endeavour. But once she realized that vitamin B_{12} contained cobalt (a relatively heavy atom containing twenty-seven electrons), she grew confident that, ultimately, its structure would be solved.

A detailed account of all the subtleties of Hodgkin's subsequent arguments has been given by Dodson (see Figure 8.17) in his memoir.[2]

After a series of ups and downs in the crystallographic analysis of vitamin B_{12}, in which the American crystallographer K. N. Trueblood and Hodgkin's Oxford collaborators Jenny Pickworth, John Robertson, and others played crucial roles,[40–42] good progress was made. The full B_{12} structure was finally unveiled just in time for the International Union of Crystallography Congress in Paris in 1954 (see Figure 8.18). A. R. Todd's group in Cambridge were rivals of Dorothy in attacking the structure of the B_{12} molecule. With the perspective of time, it can now be seen that more reliable insight into its structure came from the Hodgkin, rather than the Todd, group. (An excellent account of the rivalry and rush for recognition on the part of Lord Todd is given in Georgina Ferry's book,[1] pp. 262–4.)

This breakthrough had an enormous impact. Lawrence Bragg described it as '*breaking the sound barrier*'. But what impressed chemists and biologically oriented scientists was the fact that the physical method of X-ray diffraction had resolved the structure of a molecule that was beyond the capacity of chemical analysis and the classical approaches of degradation and synthesis—just as W. H. Bragg in the 1930s, as Bernal a little later, and as John Kendrew and Max Perutz in the 1950s had emphasized (see Chapters 1 and 3).

Figure 8.17 *Photograph of Dorothy Hodgkin, alongside her New Zealand-born collaborator—the late Guy Dodson, taken c.1986.*

(Copyright Prof Eleanor Dodson)

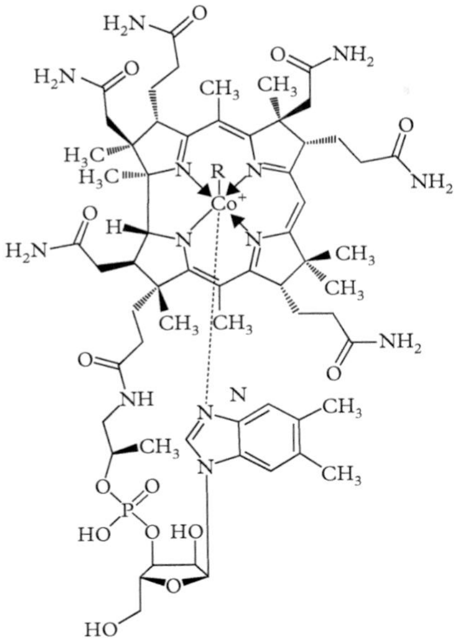

Figure 8.18 *The detailed atomic structure of vitamin B_{12}.*[41,42]
(Copyright The Royal Society)

In addition to elucidating the structure of vitamin B_{12}, this work of Hodgkin provided a basis for new synthetic (organic chemical) approaches. A. J. Eschenmoser in Zurich, in particular, regarded the so-called '*corrin ring*' at the core of the molecule, with its nine chiral centres, as '*the finest gift that X-ray analysis has so far bestowed on the organic chemistry of low molecular weight natural products*'.[42]

Hodgkin's triumph with vitamin B_{12} made it clear that X-ray analysis was the most effective tool for determining complicated molecular structures of the biological world. The added features of bond geometry, chirality, and ligand–solvent interaction also came as a bonus from her work. It is appropriate to recite here her credo, voiced in her Nobel lecture in 1964: '*The great advantage of X-ray analysis is its power to show some totally unexpected and surprising structure and to do so with absolute certainty.*'

Another scientific feature of the vitamin B_{12} structure, first discovered by Hodgkin, was the existence of the hitherto unknown Co–C covalent bond, which greatly excited the community of organometallic chemists. (It is to be recalled that, in her first paper, in 1934, with her Oxford University supervisor H. M. Powell,[43] Hodgkin was the first to discover the existence of a thallium–carbon bond, a fact also of great interest to the organometallic community.)

Hodgkin continued for another twenty years, with her numerous visitors from overseas and the UK, to pursue further studies of derivatives of vitamin B_{12} and to clarify its mode of action, medically. Along the way, she also solved, with her

DOROTHY CROWFOOT HODGKIN – IN MEMORIAM

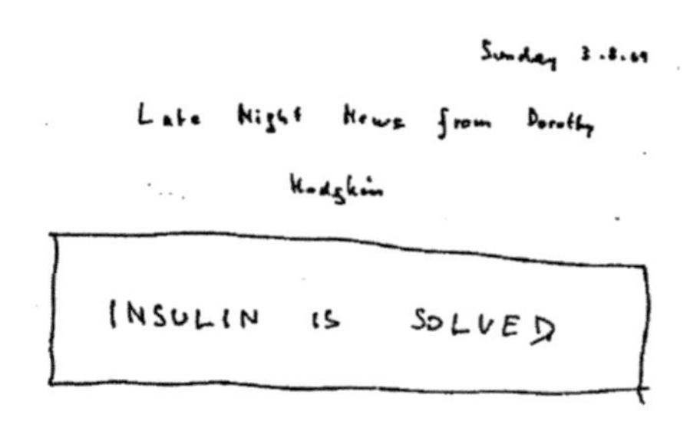

Figure 3. Poster at the Laboratory of Molecular Biology, Cambridge, announcing the great news.

is much to the Swedish crystallographer Gunner Hägg when I ran into him in a tram in Rome. He encouraged me to propose her, even though she had been proposed before. In fact, once there had been a newsleak that she was about to receive the Nobel Prize, but it proved false; Dorothy never mentioned that disappointment to me until long after. Anyway, it was easy to make out a good case for her; Bragg and Kendrew signed it with me, and to my immense pleasure it produced the desired result soon after.

'There are certain letters which I dread to open', Dorothy once told me, 'and when I saw one from Buckingham Palace I left it sealed, fearing that they wanted to make me *Dame* Dorothy'. I suppose it would have made her feel like a *femme formidable*, which she so happily is not. When she eventually opened the letter she was relieved that instead the Queen offered her the Order of Merit, which is a much greater honour and carries no title. She received it in private audience on the same day as Benjamin Britten. Once when they were both getting honorary degrees, Henry Moore said to her: 'It's really very good of them to give the OM to a simple chap like me.' I suspect that this remark echoed some of Dorothy's own feelings.

Figure 8.19 *Copy of the note posted up at the LMB, in great excitement by Max Perutz reporting Hodgkin's success with insulin (published by M. F. Perutz, in* Curr. Sci., **1997**, *72, 453). (Copyright MRC-LMB)*

collaborators, the structure of the important pharmaceutical morphine, and, as stated earlier, she finally solved the structure of insulin, after several decades of investigation that began in her laboratory in 1934.[44] This achievement delighted her friend Max Perutz, so much that he arranged a memorable visit of workers from the LMB to Hodgkin's laboratory in Oxford, a journey long-remembered by the participants[1] (see also Figure 8.19).

8.3.2 Dorothy Hodgkin's early years as a crystallographer

Both Dodson[2] and especially Ferry[1] have given full accounts of how Dorothy Hodgkin turned her interests in chemistry towards the study of crystals, and to using X-ray diffraction as a major tool for the investigation. Lectures that she attended as an undergraduate in Somerville College at the University of Oxford by Peter Debye, the impressive Dutch physicist, and especially those by J. D. Bernal, on the metallic state, each of which were laced with X-ray crystallography, greatly impressed her. So, after gaining first class honours in Part I of her undergraduate course, she willingly chose, for her Part II studies, a project under the supervision of the newly appointed University demonstrator in X-ray crystallography H. M. (Tiny) Powell. He chose for her the structural study of dimethyl thallium halides.[43] This was the first structural analysis of a metal–carbon covalent bond; it was twenty-seven years later that she discovered the first Co–C covalent bond in vitamin B_{12}. After her Part II, she joined Bernal in Cambridge, and her work there is described in Chapter 3.

On her return to Oxford, to her College post as a Research Fellow of Somerville College, she shared with Tiny Powell, her former supervisor, the X-ray laboratory in the old University Museum. As described by others,[1,2] this was a cavernous Gothic building and a somewhat gloomy and odd environment for a crystallographic laboratory. In late 1940s and 1950s, Hodgkin did a good deal of experimental work there—despite her duties as a supervisor and teacher of students from Somerville College and the demands of being a young mother. She attracted many able students, and excellent work was done by the team of crystallographers, headed by H. M. Powell in these years (see Figure 8.20).

In 1966, Tiny Powell, who was in charge of Chemical Crystallography, and Dorothy Hodgkin moved to the newly built Chemical Crystallography Laboratories

Figure 8.20 *The Chemical Crystallography Group in Oxford, 1949. Dorothy is sitting in the middle row, second from left; on her right is Frank Welch, the laboratory technician who spent his whole working life in this laboratory. David Sayre*[45] *is standing on the far right, and James Raynor is two places to his right. In the front row (from left to right) are Derek Holmes, Geoffrey Pitt, John Anthony (Tony) Jarvis, and Pauline Harrison (née Cowan).*[46]

(Courtesy University of Oxford. Photographer unknown)

in the Inorganic Chemistry Building. And in 1966, on W. L. Bragg's retirement at the Royal Institution, his right-hand man there David Phillips went to Oxford and established, with Louise Johnson (see Section 8.4.4), the Laboratory of Molecular Biophysics in the Zoology Department.

As stated by Guy Dodson,[2] as the two leading Oxford crystallographers, and colleagues for over forty years, Hodgkin and Powell's relations were not exceptionally friendly, even though they had similar political and social views and were sympathetic people. Powell's personality and other (especially linguistic) attributes are discussed in Section 8.4.3. Hodgkin and Powell had rather different approaches to life and research; they frequently failed to communicate.

8.3.3 Dorothy Hodgkin, David Phillips, and the Oxford Enzyme Group

When Lawrence Bragg moved to the Royal Institution in 1953, he endeavoured, at first, to convince Dorothy Hodgkin to join him in his intention to set up a protein structure group there. For domestic reasons alone, notwithstanding the non-ideal laboratory facilities that she had at her disposal in Oxford, it was quite impossible for her to accept. But she suggested that David C. Phillips (see Chapter 6) and J. D. Dunitz could be recruited. They were. Soon Dunitz left for a permanent post at the ETH, Zurich, and D. C. Phillips became the driving force of the soon-to-flourish protein crystallography group at the Davy-Faraday Research Laboratory. As described in Chapter 6, lysozyme's structure was solved there and its mode of action, now modified, elucidated.

In 1966, Bragg retired from the Royal Institution and for some time previously he had been concerned to find appropriate situations for members of his research group. As a result of proposals by Dorothy Hodgkin, Sir Hans Krebs, and John Pringle (Professor of Zoology), David Phillips was offered an ad hominem Professorship in Molecular Biophysics at the University of Oxford, with a Professorial Fellowship at Corpus Christi College. (The Oxford Chair, obtained at the age of forty-two, was his first tenured appointment, as he would sometimes remind anxious postdoctoral research assistants when he would stress the importance of doing good work with the expectation that the rest would follow.) With the active support of the MRC, led by Sir Harold Hemsworth, Lord (Howard) Florey of penicillin fame, who was President of the Royal Society (1945–1950), David Phillips was able to take with him the greater part of the lysozyme research group. Louise Johnson (see Section 8.4.4) joined the Oxford team in 1967.

The new team was viewed with suspicion by some biochemists, but given a warm welcome by chemists such as Jeremy Knowles[47] and Gordon Lowe. Their early years were enlivened by a sabbatical from Fred Richards,[48] Head of the Department of Molecular Biology and Biophysics at Yale University.

David Phillips, along with Sir Ewart Jones (Waynflete Professor of Organic Chemistry, in succession to Sir Robert Robertson at Oxford), Sir Rex Richards

Figure 8.21 *Some of the principal members of the Oxford Enzyme Group led by Dorothy Hodgkin.*

(Images courtesy of The Royal Society, Copyright Godfrey Argent, Royal Institution and Professor Eleanor Dodson)

(Head of the Physical Chemistry Laboratory, Oxford), Dorothy Hodgkin, and R. J. P. Williams (the eminent bioinorganic chemist at the Inorganic Chemistry Laboratory), Jeremy Knowles, and several others, were instrumental in forming the Oxford Enzyme Group in 1969 (see Figure 8.21).

The UK Science Research Council (SRC) responded (as described below) to an irate letter composed by Sir E. R. H. Jones and Sir Ronald Nyholm (of University College, London) who protested at a proposal to allocate more than forty per cent of the next ten years' SRC funds to a European 300 Ge Accelerator, the forerunner of the present facility at CERN. Jones and Nyholm wrote:[49] '*...other scientific fields which, cultivated and nurtured as nuclear physics has been in* *recent years, would yield still richer harvests...We should urgently be seeking opportunities of inventing comparable, and if possible, larger, sums in projects which offer some prospect of material advantage to the community and which at the same time serve to train useful scientists.'*

E. R. H. Jones was challenged by physicists to say what exciting new area of chemistry worthy of major initiative he had in mind; enzyme chemistry was his prime suggestion. He had been briefed by Jeremy Knowles, Dorothy Hodgkin, Bob (R. J. P.) Williams, and others. So the SRC relented and awarded money for the Oxford group.

The Oxford Enzyme Group was chaired by Sir Ewart Jones[49] from 1969 to 1984. Under Rex Richard's leadership, the development of nuclear magnetic resonance (NMR) spectroscopy, as applied to biological problems, flourished and was of great help to the Enzyme Group. R. J. P. Williams developed the new NMR technology with lysozyme to promote new ideas on protein dynamics. David Phillips took over as Chairman of the Enzyme Group from 1984 to 1988. A spirit of comradeship among the first twenty-two members from nine different departments was fostered by fortnightly dinner meetings during Term, held at one of the Colleges and financed by a donation to David Phillips from an industrial well-wisher who attached no conditions to how the funds were to be spent.

Full details of the detailed protein and other structures of prime biological interest pursued in the Phillips (and Oxford Enzyme) groups have been given by Louise Johnson.[46] He simulated David Stuart's studies on foot-and-mouth disease virus and promoted Janet Thornton's work[50] in protein folding and encouraged Greg Petsko's[51] studies of protein dynamics at low temperature. According to his former PhD student and later colleague Louise Johnson,[46] David Phillips in his later years at Oxford functioned as an enabler of science, encouraging others to flourish, qualities he had seen in Lawrence Bragg, Max Perutz, John Kendrew, and Dorothy Hodgkin. '*He was excellent at this...He demanded the highest standards in experimental work and expected students to be able to work independently. He was also pragmatic. One of his encouraging sayings was "Don't let the best be the enemy of the good"*.'[46] Sir Rex Richards,[52] one time Head of Physical Chemistry at Oxford, later the Vice-Chancellor of the University, contributed greatly to the Enzyme Group through his innovative work on NMR.

8.3.4 Other noteworthy aspects of Dorothy Hodgkin's personality and way of doing things

The memoir by Dodson[2] and the biography by Ferry[1] are again the best sources for becoming acquainted with Hodgkin's way of pursuing her science and of her reaction to society and world affairs. In Dodson's words,[2] Dorothy's approach made her open to the charges of naivety in political matters, instanced by her willingness to talk to individuals, whatever their history or associations, and by her reluctance to criticize comments stated publicly. Three visits to the USA, where she became friendly with Linus Pauling (see Figure 8.22), after the war widened

Figure 8.22 *Mutual admiration for one another characterized the relationship between Dorothy Hodgkin and Linus Pauling.*
(Courtesy Ava and Linus Pauling Archives, Oregon State University)

her knowledge of crystallography and science and of American attitudes. She assisted one of her key collaborators—Barbara Low (1920–2019)—in her pioneering work on penicillin to join Pauling's group at the California Institute of Technology, for instance. '*Dorothy Hodgkin's approach to the problems in international issues was the same as that which characterised her research. It was based on essential sympathy for people, a reluctance to condemn them and the belief that the best way to find solutions was through discussion. In the end she felt that one simply had to assume that people were honest and meant well.*'[2]

Having become a friend of Hodgkin from the early 1970s onwards (see Section 8.3.5), I was interested to hear from Max Perutz, Kathleen Lonsdale, W. L. Bragg, and her later colleagues (Judith Howard,[53] Guy and Eleanor Dodson[2]) of the way she used to greet visitors to her laboratory. From the beginning, these involved all the major crystallographers and later her own former students, often themselves distinguished. On these occasions, Dorothy Hodgkin would spread the news around the laboratory so that all were ready for a seminar and the discussions that might begin in the laboratory, but often drifted to pubs and her home. As Dodson said,[2] '*Such visits were frequent enough perhaps to explain her not organising any formal seminar programme.*'

Several of her co-workers have told me that she was always very excited whenever her former PhD supervisor J. D. (Sage) Bernal came to see her (see Figure 8.23). Throughout her life, her admiration of him never ceased.

The story is often told that when, in the mid 1940s, she had solved the structure of penicillin, she met J. D. Bernal on the steps of the Royal Society (then in Burlington House, London). And when she announced that she had elucidated the structure of penicillin, he retorted: '*This will earn you a Nobel Prize*', to which she responded, '*I would rather become a Fellow of this (Royal) Society.*' His response was: '*That will be much more difficult.*' It was not until 1946 that the first female

Figure 8.23 *A characteristic discussion between Hodgkin and Bernal taken towards the end of Bernal's life.*

(Photographer unknown)

Fellows of the Royal Society (Kathleen Lonsdale and Marjory Stephenson) were elected. Dorothy was elected FRS in 1947.

8.3.5 My first meeting with Dorothy Hodgkin

It was at J. M. Robertson's (see Section 3.5.4) seventieth birthday celebrations at the University of Glasgow that I first met Dorothy Hodgkin. It was also the occasion when I first met H. M. Powell (Dorothy Hodgkin's first research supervisor at the University of Oxford—see Section 8.4.3). I also met Jack Dunitz and Olga Kennard at this meeting. I had been invited to present a talk on dislocations in anthracene, a topic that I had described a year or so earlier, as my Corday Morgan Prize address at the Chemical Society Annual Meeting at the University of Southampton. In the audience at Southampton was Dr J. C. Speakman, a senior member of J. M. Robertson's outstanding school of chemical crystallography at Glasgow. He urged me to describe my work at Robertson's meeting. (Robertson and his associates had determined the detailed structure of anthracene from Fourier analysis of X-ray diffraction data with such accuracy that it was rumoured that Linus Pauling had taken their published C–C bond distances to estimate the degree of 'resonance' within the aromatic molecule.)

Before listening to an after-dinner talk by Dorothy Hodgkin on her work that had earned her the Nobel Prize, I had an interesting pre-dinner conversation with the multi-talented H. M. Powell, during the course of which I asked him when he had learnt Japanese. His reply surprised me: '*I have not learned Japanese.*' '*But*', I said, '*I vividly remember reading an article by you in Proc. Chem. Soc., in 1960, entitled "Japanese Chemical Writing – Read it yourself".*'[54] '*Yes*', he said, '*but that does not mean I can speak Japanese...I simply showed in that article how any chemist could work out the gist of a Japanese scientific article by following certain rules.*' (I recommend the reader to consult Powell's article; it conveys the extraordinary linguistic abilities of Powell, who, I subsequently learnt, mastered seventeen languages—see Section 8.4.3.)

I next encountered Dorothy Hodgkin at a Royal Society Meeting in March 1974. She attended my review lecture on '*Topography and Topology in Solid State Chemistry*'. In ending my talk, I cited a statement made years earlier by J. D. Bernal: '*Crystallization is death.*' As soon as I finished my talk, she immediately stood on her feet and said, '*I was there when Bernal made that remark.*' She then proceeded to allude to other aspects of my lecture that harmonized with Bernal's view. (At that time, I was unaware of how devoted and dedicated she was to her PhD supervisor J. D. Bernal.)

My next encounter with her was in Aberystwyth, where I was Professor of Chemistry, following an earlier visit to my Department by Dame Kathleen Lonsdale in 1970, where she addressed not only university students, but also invited members of the general public (doctors, schoolteachers, and others). Dame Kathleen stayed an extra day and became enamoured by my group's studies of photo-induced reactions in the organic solid state. (She communicated a paper by my colleagues and me to the Royal Society and recommended that my work be exhibited in a future summer Soiree.)

When, subsequently, Dorothy Hodgkin was invited to come to Aberystwyth, I promised that she could be driven from Oxford and back there by one of my research students, who was, at that time, based in the Department of Metallurgy at Oxford. She willingly obliged. Her visit was a resounding success. The packed lecture theatre contained townsfolk, as well as students. She described in almost mesmeric terms the way that she and her colleagues had arrived at the structures of penicillin and insulin. And she told us of her exciting visit to China.

Every time, almost, thereafter, whenever I saw her—she came frequently to listen to my lectures when I was Director of the Royal Institution—she would recall how, on being driven through mid Wales, she saw the sun shining on Cardigan Bay, with the sight of Aberystwyth in the distance.

8.4 Other Biographical Sketches

Several of the individuals mentioned earlier in this monograph exerted considerable influence on both Perutz and Kendrew, on Hodgkin, and on other scientists who contributed to the emergence and growth of structural molecular biology. Brief accounts given here are of some of the more interesting individuals concerned.

8.4.1 Walter Morley Fletcher (1873–1933)

The key person who was appointed first Secretary of the Medical Research Committee, when it was set up in 1914, as a consequence of Lloyd George's National Insurance Act of 1911, was Sir Walter Fletcher, FRS (see Figure 8.24).

Figure 8.24 *Sir Walter Morley Fletcher, FRS, first Secretary of the Medical Research Council. (Courtesy, Royal Society)*

He later became the first Secretary of the MRC in 1915. In this post, he was extremely effective and worked closely with the Prime Minister (Lloyd George) from 1916 onwards. He was largely instrumental in setting up the special conditions (described in Appendix 1 to Chapter 1) for long-term research that the MRC was able thereafter to pursue and support.

Brought up in Yorkshire of non-conformist stock—his father was an Inspector of Alkali Works in Liverpool and later in London—Walter Fletcher excelled as a student at Trinity College, Cambridge. While there, as a leading member of the Department of Physiology, he interacted closely with Gowland Hopkins, who was Professor of Biochemistry.

Animal respiration was the principal topic of Fletcher's researches at Cambridge. In 1907, he and Hopkins established the connection between lactic acid and muscle contraction. They showed that oxygen depletion causes an accumulation of lactic acid in the muscle, work that paved the way for the later discovery by A. V. Hill that a carbohydrate metabolic cycle supplies the energy for muscle contraction. In the opinion of the author of his *Biographical Memoir of Fellows of the Royal Society*, Fletcher's analyses of the respiration of muscle, as distinct from the general respiration of the body, was among the first of the modern quantitative studies that transformed biochemistry from a mere 'anatomical' identification of substances that can be extracted from tissues to a science describing the physiology of the chemical changes within the cell itself.

Shortly thereafter, as his Royal Society biographer recalls, Fletcher (like Kendrew long after him) became inclined '*...to abandon the laboratory and its slowly moving quest for knowledge*', and, in 1914, he took up the Secretaryship of the Medical Research Committee. For his sterling services during World War I, he received the honour of being made a Knight Commander of the Most Excellent Order of the British Empire (KBE) (a Gilbertian title when recited in full!). His FRS was awarded in 1914.

8.4.2 E. A. Moelwyn Hughes (1897–1978)

John Kendrew once told me that he was attracted to Emyr Alun Moelwyn Hughes (see Figure 8.25) as one of the most impressive members of staff at the University Chemical Laboratories, Cambridge, when he studied there as an undergraduate.

A graduate of the University of Liverpool, who later held a Fellowship at Magdalen College, Oxford, as the holder of one of the coveted 1851 Exhibitioners,

Figure 8.25 *Emyr Alun Moelwyn Hughes, lecturer in physical chemistry, Cambridge, with whom John Kendrew began to do research.*

Moelwyn Hughes and C. N. Hinshelwood did pioneering work on the kinetics of reactions in solution. He also collaborated with the eminent German chemist K. F. Bonhoeffer, in Frankfurt before moving to Cambridge, where he wrote one of the first comprehensive texts on physical chemistry. Linus Pauling is on record as having been much influenced as a young professor in California by Moelwyn Hughes' texts.

What particularly impressed young Kendrew and what influenced him in selecting Moelwyn Hughes as a research supervisor was the latter's combination of exceptional mathematical skills and extraordinary fluency in the written word. Moelwyn Hughes' verbal fluency, also, together with his wit, percipience and perspective, was unusually refreshing in its originality and impact. He often endowed molecules with human qualities, and he contemplated the beauty, magic, and mystery of their behaviour in the best traditions of the natural philosopher and poet. His physical chemistry texts were described—as Coleridge once said—by combining the best words in the best order. Some of his gems still trip off the tongue: '*Energy among molecules is like money among men: the rich are few, the poor numerous*'; '*Belief in the essential simplicity of things is one of the chemist's articles of faith.*'

Moelwyn Hughes' '*Introduction*' that he wrote for the *Everyman* edition of *The Sceptical Chemist* is still much admired. In asking when the discovery of the inverse relationship between pressure and volume was made and who Robert Boyle was, he provided a lyrical reply: '*He flourished in the seventeenth century, that turbulent time of pestilence and fire so amply described by Evelyn and Pepys, when Bunyan wrote in Bedford jail and Penn left England's shores, when Milton sang his "Paradise Lost' and Wren built London's churches, when Britain's monarch was overthrown and Cromwell made Protector.*' Little wonder that young undergraduates, Kendrew among them, were attracted to the scholarly stature of this research supervisor, with whom, in the space of a few months since embarking on his research, Kendrew authored a Royal Society publication on the mutarotation of sugars.

8.4.3 H. M. (Tiny) Powell

Tiny Powell (see Figure 8.26)—the designation arises because he was only 5 ft 2 in, i.e. 1.57 m tall—was, according to his contemporaries at Oxford, a secretive man possessed of genuine curiosity. Apparently, he was a keen observer of life and a sympathetic and amusing commentator on it, and he wrote beautifully, as may be judged from his extraordinary article,[54] published in 1960, entitled '*Japanese Chemical Writing – Read it yourself*'.

His scientific legacy will forever be associated with his pioneering work on clathrates, which are compounds in which molecules of one component are physically trapped within the crystal structures of another. It was Davy, and later Faraday, at the Royal Institution who discovered and first described clathrates, when it was found in the early nineteenth century that chlorine could be encapsulated in a cage of crystalline water. But it was Powell who put the subject properly into perspective. Nowadays it is believed that methane clathrate ($CH_4{\cdot}5.75H_2O$)

Figure 8.26 *Self-made sketch of H. M. Powell (1906–1991) (Keith A. McLauchlan, Biographical Memoirs of Fellows of the Royal Society,* **2000***, 46, 425).*

or ($4CH_4 \cdot 23H_2O$), also called methane hydrate, hydromethane, methane ice, fire ice, and natural gas hydrate, occurs extensively on ocean beds. (Efforts are currently being made in Japan, Russia, and elsewhere to extract methane from its clathrate.)

Earlier we recalled the relationship that existed between Dorothy Hodgkin, his first research student, and Tiny Powell. In brief, they did not enjoy a very happy relationship, and this fact was well known to their Oxford contemporaries.

The most extraordinary aspect of Tiny Powell's academic qualities, verified by a colleague of his at Hertford College—Professor Keith McLauchlan—where they were Fellows was his linguistic skills. Apart from Japanese, Chinese, Russian, and most European languages (including Romanian and Welsh), he also mastered Mongolian.

To illustrate his mastery of languages—he could read, write, speak, or understand seventeen in all—we repeat a paragraph from his 1960 paper in *Proc. Chem. Soc.*:[54] '*The writing in the Japanese papers is less credible than the giraffe's neck, yet the complexity itself makes some of the content more accessible for Western people who have no Japanese. Parts of the paper may contain mathematical symbols as used everywhere with both Greek and Roman alphabets to denote mathematical and physical constants, variables, and operations. Chemical formulae are as* H_2O*. Identifying numbers to formulae are Roman, as in the Journal, but Arabic numerals are used for tables. Some Chinese numerals occur. They are used on title pages and page headlines where they may give the volume and part number, though pages are usually numbered in Arabic. Chinese numerals may be used in names of some simple substances such as car-*

bon disulphide. Four other kinds of writing may follow each other in seemingly erratic alternation in a single sentence of the main text. They include some of the thousands of Chinese characters either in their Chinese dictionary forms or in official Japanese modifications.'

8.4.4 Louise Johnson (1940–2012)

A key member of the Phillips–North team that solved the three-dimensional structure of lysozyme (see Chapter 6), Louise Johnson (see Figure 8.27) graduated in physics at University College, London, and decided to apply for a PhD at the Davy-Faraday Research Laboratory where Perutz and Kendrew were Honorary Readers and Bragg the Director. Assigned to David C. Phillips, she first worked on the determination of the structure of the sugar molecule N-acetylglucosamine, before moving on to investigate substrate binding in the enzyme lysozyme. After her PhD was awarded, she worked with Fredrick M. Richards at Yale University, and on returning to the UK, she was appointed Demonstrator in the Department of Zoology at Oxford University, at which David Phillips had earlier been appointed Professor of Molecular Biophysics.

Figure 8.27 *Professor Dame Louise Johnson.*
(Courtesy Mr Umar Salam)

At Oxford, she solved the structure of the protein glycogen phosphorylase, which is found in muscle and is responsible for mobilizing the energy store of glycogen to provide fuel to sustain muscle contraction. Continuing her research towards understanding the molecular basis of biological properties and of catalytic mechanism, she became a major user, with her team, of the Synchrotron Radiation Source at Daresbury. From 1990 to 2007, she was the David Phillips Professor of Molecular Biophysics, University of Oxford, where she was also a Fellow of Somerville College. Together with Tom Blundell, she wrote an influential textbook on protein crystallography. From 2003 to 2008, she was Director of Life Sciences at Diamond Light Source, the UK national synchrotron facility at Harwell, Oxfordshire.

The recipient of many honours, including honorary doctorates from the universities of St Andrews, Bath, Cambridge, and Imperial College, London. In 2003, she was made Dame Commander of the British Empire (DBE). She married the Nobel Prize winner Abdus Salam in 1965 and had one son and a daughter. She was an outstanding lecturer, researcher, and a compassionate and charming human being, who was an ardent supporter of the Third World Academy of Science.

8.4.5 Archer J. P. Martin (1910–2002)

Martin (see Figure 8.28) was the first member of Peterhouse to win a Nobel Prize, which he shared with the former Trinity College, Cambridge scholar R. J. M. Synge (1914–1994) in 1952 for their development of a technique that proved indispensable to workers studying molecular biology, especially proteins and nucleic acids, namely partition chromatography.

As a teenager in Bedford school, he was fascinated by the subject of separation of closely similar materials. As a student in Peterhouse, he graduated with a lower-second degree in 1936; then he worked alongside C. P. Snow[55] in the Department of Physical Chemistry, Cambridge, where, shortly thereafter, he was advised by J. B. S. Haldane to switch to a research post at the University's Dunn Nutritional Laboratory where he helped separate vitamin E from vegetable oils. He then went to the Wool Industries Research Association in Leeds (1938–1946) where he sought ways of elucidating the structure of the keratin proteins of which wool is composed. It was there that he collaborated with Synge and the two of them did their seminal work (described in Chapter 3—see Sections 3.3.2 and 3.3.3). After a brief spell at the Boots Drug Company in Nottingham, he went in 1948 to the MRC Laboratory at the Lister Institute of London, the precursor of the National Institute of Medical Research and the Francis Crick Institute. Martin and Synge's technique of partition chromatography is very simple, inexpensive, rapid, and very efficient. The process entails placing a drop of the mixture to be separated at the edge of a sheet of filter paper that holds the stationary phase (often water); the moving phase is another solvent, typically a mixture such as chloroform and butanol. The solvents soak into the paper by capillarity, taking with them the

Figure 8.28 *Archer J. P. Martin, first ever Nobel Prize winner from Peterhouse.* *(Courtesy Royal Society)*

components of the mixture to be analysed to different distances, depending on their respective solubilities. The identity of the various components (e.g. amino acids in hydrolysed wool) are discerned by comparing the positions of the spots with a reference chart (see Figure 8.29).

To gauge the transformative magnitude of Martin's contribution to molecular biology, it is relevant to quote James Lovelock's words (of Gaia fame[56]) in his Royal Society biographical memoir of Martin. '*We judge the worth of a scientist by the benefits he or she brings to science and society; by this measure Archer Martin was outstanding, and rightfully his contribution was recognised with a Nobel Prize. Scientific instruments and instrumental methods now come almost entirely from commercial sources and we take them for granted and often have little idea of how they work. Archer Martin was of a different time when scientists would often devise their own new instruments, which usually they fully understood, and then they would use them to explore the world. The chromatographic methods and instruments Martin devised were at least as crucial in the genesis and development of molecular biology as were those from X-ray crystallography.*'

Erwin Chargaff (1905–2002), the Austrian–American biochemist, also used the Martin–Synge paper chromatography technique to determine the composition

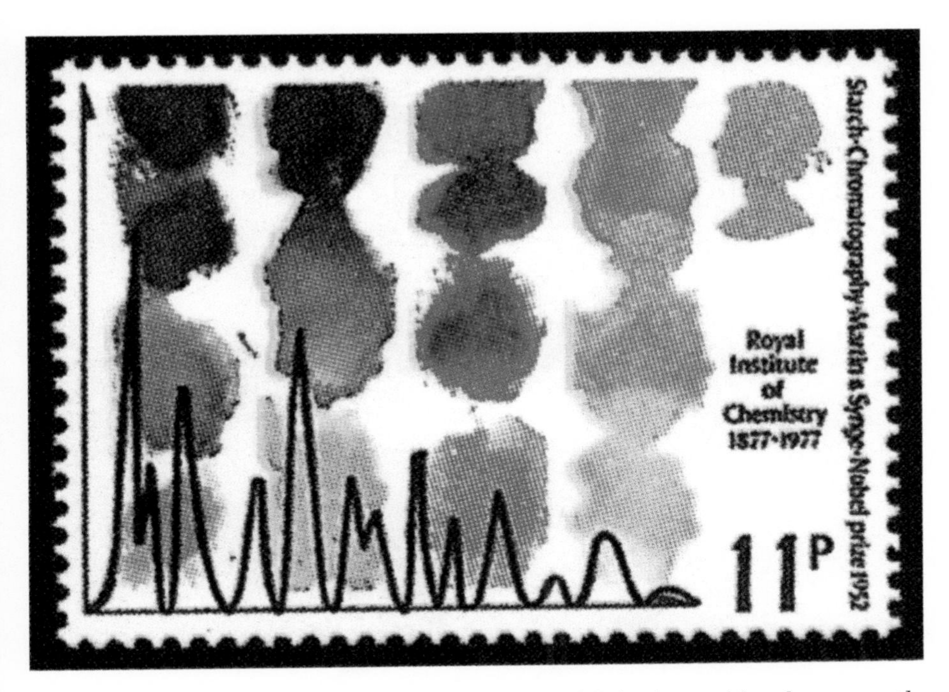

Figure 8.29 *British postage stamp celebrating the Nobel Prize for partition chromatography awarded to A. J. P. Martin and R. L. M. Synge.*

(Copyright General Post Office)

of DNA. He found its composition to be constant within a species, but to differ widely between species. This led him to conclude that there were as many different types of DNA as there were different species (see Appendix 1 to Chapter 9). He also found, by partition chromatography, that the number of adenine (A) and guanine (G) components was always equal to the number of cytosine (C) and thymine (T) components. Moreover, he found that the number of adenine bases is always equal to the number of thymine ones, and that the number of guanine bases equals the number of cytosine ones. This determination by Chargaff in 1950 was of crucial importance in the determination of the double-helix structure of DNA (see Figure 8.30). Another important molecular biological use of partition chromatography was made by Vernon Ingram, as described in Chapter 5, when he used the technique, in conjunction with paper electrophoresis, to establish the precise nature of the first identified molecular disease, namely sickle-cell anaemia (see Section 5.4.1).

While he was at the Lister Institute, supported by the MRC, Martin had the good fortune to team up with the young Tony (A. T.) James.[57] James and Martin were the principal innovators of another powerful analytical technique—gas chromatography—which has been as transformative an influence in chemistry and

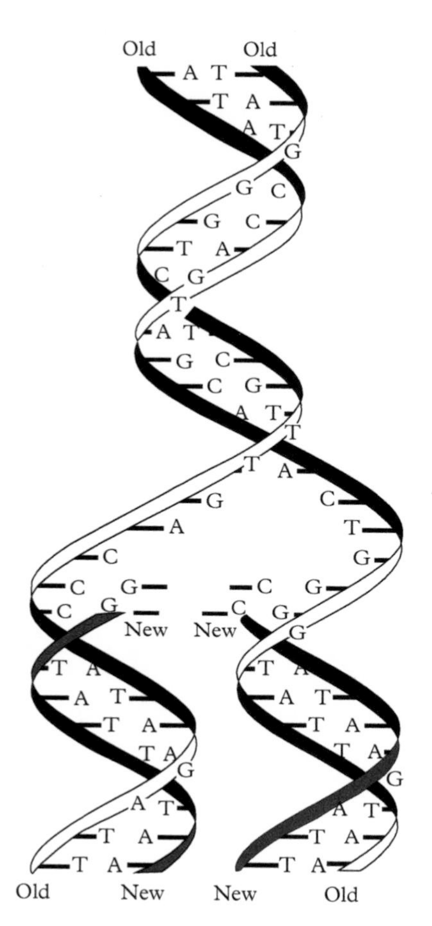

Figure 8.30 *The well-known double-helix structure of DNA.*
(Courtesy MRC-LMB)

industry as partition chromatography has been in molecular biology. Indeed, many experts[58] felt that the introduction of gas chromatography merited another Nobel Prize.

Martin's dyslexia hampered him in several ways. According to Lovelock: '*Dyslexics find the linear world of sequential expression, so necessary for written examinations or scientific papers, difficult, even baffling.*' For Martin, therefore, the fluent dispatchful ways in which Tony James did things with him was a godsend. In his entire scientific career, Martin wrote only ninety or so scientific papers.

After retiring from his posts in the UK, he took up a professorship in a US university for several years. But he was ultimately asked to leave because he did not publish enough papers.

8.4.6 Max Delbrück (1906–1981)

Delbrück (see Figure 8.31), according to Bill Hayes, the author[59] of his biographical memoirs in the Royal Society (of which he was elected a Foreign Member in 1967), was '*one of the outstanding natural scientists of our time. A man of rare intellectual ability, clarity of thought and perceptions, he excelled in theoretical physics, biology and philosophy, and possessed a deep knowledge and appreciation of the arts. His dedication to truth, and his intolerance of half-truths and intellectual pretension, were sometimes expressed with a disturbing frankness and abruptness of manner, often construed as arrogance by those who did not know him well. His disclaimer, "I don't believe a word of it", when told of some new experimental result or hypothesis, became famous among his colleagues. In fact, Max was very gregarious and had a rich vein of friendship and affection in his nature which he was always ready to share with others of all ages.*'

In Section 8.2.4, examples are given of the way in which Delbrück operated among his colleagues and followers. Many of these are not particularly flattering and they suggest an intemperateness on the part of Delbrück that many of his students could not tolerate. I was recently told by my friend Jack Dunitz, who knew both Delbrück and Pauling well, from his sojourn at Caltech, that an outstanding young chemist—Martin Karplus,[60] Nobel Laureate in chemistry in

Figure 8.31 *Max Delbrück.*
(Courtesy, Archives of California Institute of Technology, via H. B. Gray)

2013 '*for the development of multiscale models for complex chemical systems*'—went to work as Delbrück's graduate student after graduating at Harvard. But, after a month or so, Karplus felt that Delbrück's reaction to his contributions to seminars was one of ridicule. Karplus then switched to work as Pauling's student.

As described in Section 8.2.4, others were made to feel humiliated by Delbrück's criticisms. But, as stated earlier, Max Perutz was a great admirer of Delbrück, and Max never accorded such respect (in my experience of him) to anyone who lacked sincerity, integrity, and fundamental human decency. It is therefore gratifying to learn from the memoir[59] by Bill Hayes (1913–1994), the famous Irish geneticist, the admirable qualities that Delbrück possessed.

According to Hayes, Delbrück was a born leader whose Socratic influence on those who worked with him was enormous, whose rare praise was something to be coveted, and whose criticism was generally welcomed with respect. He was often proved wrong, but when this occurred, he was ready to admit it.

Scientifically, he was the foremost pioneer—after Bernal's early work (see Section 3.5)—in the study of bacterial viruses (i.e. bacteriophage). Early he noted that infection of a bacterium by a single particle is followed, about thirty minutes later, by rupture of the cell and liberation of more than a hundred progeny particles. Delbrück, along with his co-investigators Salvador Luria and Alfred Hershey, was awarded the Nobel Prize in 1969. He and Pauling, who was the dominant partner, wrote a very influential note on the nature of intermolecular forces operative in biological processes,[61] in which they argued that it is complementarity, not identity, that holds sway in biological processes,[61] a fact elaborated in Chapter 7.

Delbrück was born in Berlin, the son of a professor of military history. His mother was the granddaughter of the famous German organic chemist Justus von Liebig of Giessen. His mentor, as a young scientist, was Karl Friedrich Bonhoeffer, with whom E. A. Moelwyn Hughes (see Section 8.4.2) had collaborated at the University of Frankfurt in the early 1930s. Delbrück studied theoretical physics in Göttingen where he gained his doctorate in 1930. He later spent two years with John Lennard-Jones at the University of Bristol before spending a somewhat longer period with Niels Bohr in Denmark. While there, each of them pondered whether biological processes could ever be fully understood in terms of the laws of chemistry and physics.

Delbrück became an American citizen, and he did most of his work at Vanderbilt University and at Caltech. As an addendum to this portrait of Delbrück, it is instructive to consult Appendix 2 to this chapter, which contains a paragraph about Delbrück, written by Professor Jack Dunitz, now of ETH, Zurich, a colleague of Delbrück and Pauling in Caltech in the 1950s.

8.4.7 William (Bill) Coates (1919–1993)

Sir Lawrence Bragg, and his immediate successor Sir George (later Lord) Porter, made Bill Coates (see Figure 8.32), a technician at the Royal Institution, one of

Figure 8.32 *William (Bill) Coates, technician supreme at the Royal Institution and the Davy-Faraday Research Laboratory. In his hands, he is holding an original of the miners' safety lamp invented by Sir Humphry Davy.*

(Photograph presented to author by W. Coates, 1987)

the most admired and widely recognized scientific personalities by the general public in the UK from the mid 1950s to the mid 1980s. Millions of children and television viewers saw him in action (either live or on the screen) in the theatre of the Royal Institution. His skills as an improviser, dextrous manipulator, and, at times, human guinea pig or acrobat were exceptional. But beneath the professional, unpretentious showmanship, there was the shrewd, inventive technician who, in record time, could translate inchoate or incompletely formulated ideas by a new, or visiting, lecturer into an exhilarating spectacle.

Born in the East End of London, educated at a local grammar school, Bill Coates, like so many of his generation, joined the armed services in 1939. He entered as a Private and left as a Captain. He served in Norway and other parts of Europe with the Parachute Regiment and was involved in D-Day operations. After the War, he worked for a short period as technical assistant at Charing Cross Medical School before being recruited to the Royal Institution by Sir Eric Rideal, its Director, in 1948. Although he served under five directors in all, he was particularly closely associated for a period of thirteen years with Sir

Lawrence Bragg (and his Deputy Professor Ronald King) and for twenty years with Sir George (now Lord) Porter. The others were E. N. da C. Andrade and myself.

In Coates' early days as a technician at the Royal Institution, he was part of the world-class team assembled by Lawrence Bragg, consisting of David C. Phillips, Tony North, U. Arndt, R. Poljak, and L. Johnson (see Chapter 6), to pursue X-ray studies of proteins. He was largely in charge of building (with U. Arndt and Phillips) the rotating-anode X-ray sources and the linear X-ray diffractometer. Phillips used to say that Bill Coates was not only a jack of all trades, but also a master of most.

In the early 1950s, however, his career changed when, on the advice of Professor Ronald King, he turned his attention to lecture demonstrations, so much a feature of the Royal Institution from the days of Sir Humphry Davy and the incomparable Michael Faraday. But it was the arrival of Bragg from the Cavendish Chair in Cambridge in 1953 that marked the real turning point in Coates' career.

Bill Coates' remarkable skills, which were brilliantly harnessed and further developed by Bragg's successor George Porter and by the numerous creative scientists with whom he interacted at Friday Evening Discourses, were recognized by the Clothworkers' Company[62] which, conscious of the importance of conveying scientific advances to a lay audience, generously provided funds to establish at the Royal Institution the post of Clothworkers' Lecturer and Lectures Superintendent, of which Coates was the first occupant.

Max Perutz, and countless others who delivered Friday Evening Discourses in his day at the Royal Institution, were immensely impressed by his seemingly endless ingenuity in devising appropriate practical demonstrations.

Bill Coates had a great fund of stories encompassing his days as a parachutist, near-misses while handling circuits carrying hundreds of amps, temperamental X-ray sources, or capriciously explosive gas mixtures. But the one that used to bring him out in a sweat was his recollection of the occasion in 1965 when a glittering array of Nobel Laureates came to the Royal Institution to celebrate the fiftieth anniversary of Bragg's Nobel Prize (see Section 6.6). Bragg's (gold) Nobel medal was on display in the library. But in the preparation of the exhibits, Coates had laid down the medal on a drop of mercury and so it gained an unsightly stain. The bullion merchants Johnson–Matthey were hurriedly contacted by phone, and they prescribed the exact temperature of the heat treatment required to drive off the mercury. Coates claimed that he lost several years of his life before the medal emerged in its pristine glory from the oven. '*I never told Sir Lawrence what had happened.*'

Coates, with his engaging cock-robin demeanour and 'can do' approach, always exuding vitality and irrepressible enthusiasm, was himself a popular lecturer, particularly with teachers and schoolchildren. He was also an authority on the way in which Faraday carried out lecture demonstrations, as I was to appreciate when he helped me several years ago to re-enact Faraday's famous lecture experiments on platinum first given at the Royal Institution in 1861.[63]

8.4.8 Dai Rees (1936–)

Born in England of Welsh parents, Rees (see Figure 8.33) started his research career in the Department of Chemistry at the University College of North Wales, Bangor, of Stanley Peat, an eminent carbohydrate chemist from the school of Sir Norman Haworth, FRS, Nobel Laureate. Like many scientific academics of the time, Peat had been drawn into scientific advisory functions in support of the 1939–1945 World War effort. Japan's armies had overrun a number of countries in the Far East to cut off the supply of agar needed for penicillin production for the treatment of war injuries. Agar was manufactured in cottage industries by extraction of certain species of seaweeds peculiar to those regions. Related species from European waters did yield agar-related materials, but these lacked key properties. The British Government assembled a panel of experts to advise how local materials might be manipulated to be made useful. The Committee failed to arrive at any helpful conclusion, and for the perfectionist Peat, this failure rankled. Years later, however, resources became available in his Department to return to the challenge. Thus, the research career of the newly graduated David Allan Rees was launched with an assignment to look again at the agar problem.

Figure 8.33 *Sir Dai Rees, ex-Chief Executive Officer at the MRC.*

It turned out that the special properties of agar were conferred by an exotic sugar component (3,6-anhydrogalactose), which allowed the linear chain to coil into a double helix (like DNA, but parallel-, rather than anti-parallel-, stranded). Some of the theoretical insights by Crick into the fibre diffraction patterns of DNA helped in understanding these structures. Rees continued this work in the Department of Sir Edmund Hirst, FRS, in Edinburgh. This research formed the core of the work recognized by the award of the Colworth Medal of the Biochemical Society (1970) and the Carbohydrate Award of the Chemical Society (1970). It gradually emerged that the new understanding was also relevant to a wide range of other problems with industrial products based on gels and structured liquids. These occurred in food technology, cleaning products, and elsewhere. Hence, there arose an ever increasing interaction with industrial R&D, which eventually led to the relocation of the academic research team from Edinburgh to become an in-house entity at Unilever Research Laboratories.

Dai Rees was asked to take increasing responsibilities within the wider Unilever Research, eventually for overall strategic management of biological research. Common interests with the MRC in cell biology led to an invitation from the MRC that Rees should move to a jointly funded appointment in which he would combine his Unilever position with MRC responsibilities as Associate Director of the Cell Biophysics Unit at King's College, London.

Having now become willy-nilly part of the MRC's management structure, Rees was appointed to a new Council sub-committee to determine the future of the National Institute of Medical Research (NIMR), following the imminent retirement of the incumbent Director. What was to be its role in the new economic climate? Thatcher's government was demanding more financial accountability and indeed return from publicly funded research activities. Competition for the shrinking pot of research funding was becoming fiercer. Based on the sub-committee's recommendations, the Council proposed a leaner and more focused structure and invited Rees to the Directorship to flesh out and implement the aspirations. In due course, these led to the founding of the MRC Collaborative Centre on the NIMR site. This had MRC-wide responsibilities for initiating projects with industry, to realize the potential of MRC science for prosperity and health of the nation. Financial returns were shared according to pre-agreed patterns with inventor scientists and their MRC institutions. Examples of successful outcomes are described in Chapter 10 of this book.

Sir James Gowans, the CEO of the MRC now decided to retire, and Rees was appointed as his successor. To Rees, it seemed that the challenges remained the same as those confronted by Sir Walter Fletcher and Lloyd George when they agreed that the health of the nation required that the most sophisticated modern science be partnered with experience in the health services to develop the scientific foundations for ever improving care; the MRC was to be and remains the instrument to deliver this. Societal changes now demand faster and more cost-effective delivery and more linkage to economic development. For these, it was necessary to bring industry into the relationship. The learning process for this had already begun through initiatives of individuals and teams, and a start made on co-ordination of effort across the Council, for example, through the Collaborative Centre.

Rees, who was the Colworth Medallist of the Biochemical Society and winner of the Carbohydrate Award of the Royal Society of Chemistry, has received numerous Honorary Doctorates. He was knighted in 1993 and was President of the European Science Foundation from 1993 to 1999. His role in gaining royalties for MRC/LMB discoveries was crucial, as described in Chapter 10.

APPENDIX 1

A CV Compiled by John Kendrew, December 1996

Sir John Kendrew

Born: 24 March 1917, Oxford

British

Education

1923–1930	Dragon School, Oxford
1930–1936	Clifton College, Bristol
1936–1939	Scholar of Trinity College, Cambridge, later Senior Scholar Natural Sciences Tripos Pt I (Chemistry, Physics, Mathematics, Biochemistry), 1st Class; Pt II (Chemistry), 1st class
1939	BA
1942	MA
1939–1940	Research Student, Department of Physical Chemistry, Cambridge (Reaction Kinetics)
1946–1949	Research Student, Cavendish Laboratory, Cambridge (Protein Structure)
1949	PhD
1962	ScD

Positions Occupied

1940	Appointed Junior Scientific Officer, Air Ministry Research Establishment (later Telecommunications Research Establishment, Ministry of Aircraft Production)
1940	Operational Research Officer, RAF Coastal Command (on the staff of Sir Robert Watson-Watt)
1941	Promoted to Scientific Officer
1941–1943	Deputy Officer-in-Charge, Operational Research Section, RAF Middle East (Cairo); Honorary Commission as Squadron Leader
1942	Promoted to Senior Scientific Officer (Acting)

1943	Promoted to Principal Scientific Officer (Acting)
1944–1945	Officer-in-Charge, Operational Research Section, and Scientific Adviser to the Honorary Commission as Wing Commander
1946–1982	Staff member of Medical Research Council; Deputy Chairman of MRC Laboratory of Molecular Biology, Cambridge until 1974 (thereafter seconded to EMBL, see below)
1947–1975	Fellow of Peterhouse, Cambridge
1954–1968	Reader at the Davy-Faraday Research Laboratory, Royal Institution, London
1960–1963	Deputy Chief Scientific Adviser, Ministry of Defence
1971–1974	Project Leader, European Molecular Biology Laboratory (EMBL), Heidelberg
1975–1982	Director-General, EMBL, Heidelberg
1981–1987	President, St John's College, Oxford

Part-time and Honorary Positions

1959–1962	Member of Council, Biophysical Society (USA)
1963–1971	Member of Council
1975–1977	Member of Council
1969–1974	Secretary-General, European Molecular Biology Organization
1964	Chairman, British Biophysical Society
1964–1967	Member of BBC Scientific Advisory Group
1964–	Governor of the Weizmann Institute, Rehovot, Israel
1964–	Governor of Clifton College
1964–1969	Vice-President, International Union of Pure and Applied Biophysics
1969–1972	President (as above)
1972–1975	Honorary Vice-President (as above)
1964–1972	Member, Council for Scientific Policy, Department of Education and Science
1970–1972	Deputy Chairman (as above)
1965–1967	Member of Council, The Royal Society
1965–1967	Vice-President, Institute of Biology
1969–1974	Member of the Defence Scientific Advisory Council
1970–1972	Chairman, International Scientific Relations Committee, Council for Scientific Policy
1970–1974	Secretary-General, European Molecular Biology Conference
1971–1974	Chairman, Defence Scientific Advisory Council
1973–1983	Member of Academic Advisory Council, University College, Buckingham
1973–1981	Chairman of the Scientific Council, Laboratory of Molecular Embryology, Naples

1974–1979	Trustee of the British Museum
1974–1979 1981–1987	Member, Kuratorium, Max-Planck-Institut für Kemphysik, Heidelberg Member, Kuratorium, Max-Planck-Institut für Kemphysik, Heidelberg
1974–1979	Member, Scientific Advisory Board, International Institute of Cellular and Molecular Pathology, Brussels
1974–1980	Secretary-General, International Council of Scientific Unions
1975–1980	Trustee, International Foundation for Science, Stockholm
1975–1979	Member, Board of Advisers, Basel Institute for Immunology
1976–1984	Chairman, Natural Sciences Advisory Committee, UK National Commission for UNESCO
1980–1994	Member, Fachbeirat, Max-Planck-Institute für Zellbiologie, Ladenburg
1980–1986	Member, Council of the United Nations University
1983–1985	Chairman, Council of the United Nations University
1980–1985	Member, Board of Directors (and its Scientific Committee), Basel Institute for Immunology
1981	Member, Advisory Panel on Science, Technology and Society, UNESCO
1981–1985	President, Confederation of Scientific and Technological Organizations for Development (CISTOD)
1982	Honorary Professor, University of Heidelberg
1982–1983	First Vice-President, International Council of Scientific Unions (ICSU)
1983–1988	President, ICSU
1984–1985	Chairman, High Energy Particle Physics Review Group, Department of Education and Science
1985–1992	Chairman, Board of Governors, Joint Research Centre, European Economic Communities
1987	Patron, Glynn Research Foundation
1988–1990	Past President, ICSU
1989	Emeritus Member, Academia Europaea
1990	Honorary Vice-President, Research Defence Society
1992	Chairman, Board of Honorary Advisers, Glynn Research Foundation
1993	Vice-President, Alzheimer Research Trust
1995	Member, Independent World Commission on the Oceans

Honours

1943	Mentioned in Despatches
1960	Fellow of the Royal Society

1962	Nobel Prize in Chemistry
1963	Commander of the British Empire
1965	Royal Medal of the Royal Society
1972	Honorary Fellow, Trinity College, Cambridge
1974	Knight Bachelor
1974	President, British Association for the Advancement of Science
1975	Honorary Fellow, Peterhouse, Cambridge
1980	Order of the Madara Horseman, 1st degree, Bulgaria
1987	Honorary Fellow, St John's College, Oxford
1988	William Procter Prize for Scientific Achievement, Sigma Xi
1991	Forum Engelberg Prize

Honorary and Foreign Memberships

American Society of Biological Chemists (1962); American Academy of Arts and Sciences (1964); Leopoldina Academy (1965); Institute of Biology (1966); Max-Planck-Gesellschaft (1969); Weizmann Institute (1969); National Academy of Science, USA (1972); Romanian Union of Medical Sciences (1977); Heidelberg Academy of Sciences (1978); Bulgarian Academy of Sciences (1979); Royal Irish Academy (1981); British Biophysical Society (1982); Indian National Science Academy (1989); Chilean Academy of Sciences (1992); Academy of Creative Endeavours, Moscow (1993)

Honorary Degrees

ScD Keele (1968); ScD Reading (1968); D Univ Stirling (1974); Doctor *honoris causa*, Medical University of Pécs, Hungary (1975); DSc Exeter (1982); DSc Buckingham (1983); Doctor *honoris causa*, Madrid (1987); Doctor *honoris causa*, Siena (1991); Doctor *honoris causa*, University of Chile (1992)

Lectureships

Herbert Spencer, Oxford (1965); Hans Sloane, Belfast (1966); Crookshank, Faculty of Radiologists (1967); Robbins, Pomona, USA (1968); John R. Bloor, Rochester, USA (1968); Procter, International Society of Leather Trades Chemists (1969); Fison, Guy's Hospital (1971); Monsignor de Brùn, Galway (1979); Saha Memorial, Calcutta (1980); 1st Sir Florizel Glasspole, Kingston, Jamaica (1989)

Selected Publications (various original publications in scientific journals)

1949	(ed. with F. J. W. Roughton) Haemoglobin: a Symposium based on a Conference held at Cambridge, June 1948, in memory of Sir Joseph Barcroft (Butterworth)
1959–1987	Founder and Editor-in-Chief, *Journal of Molecular Biology* (Academic Press)
1966	*The Thread of Life* (Bell)
1994	Editor-in-Chief, *The Encyclopaedia of Molecular Biology* (Blackwell Science)

APPENDIX 2

Statement About Delbrück Compiled by J. D. Dunitz, July 2018

I have never regarded Delbrück as one of the fathers of molecular biology. In fact, in the days that I knew him at Caltech, it was my impression that he was rather hostile to the ideas behind molecular biology. As I recall, he sat beside me at the lecture where Pauling first publicly announced his stable hydrogen-bonded model structures for polypeptide chains. Pauling had a feeling for drama. On the table in front of him stood bulky columnar objects shrouded in cloth, which naturally excited the curiosity of those in the packed auditorium. Only after describing in detail the structural principles behind the models did he turn to the table and unveil the molecular models with a characteristic theatrical gesture. There were the two structures—the three-residue and the five-residue spirals—later dubbed the α- and δ-helices! I was immediately converted, a believer right from the start. In contrast, Max made no secret of his scepticism and especially his disapproval of Pauling's manner of presentation, and asked if I thought there was anything of value in these models. I believe I may have disappointed him when I told him that in my opinion the models were based on sound structural principles and were very likely to represent important building blocks of actual proteins.

REFERENCES

1. G. Ferry, '*Max Perutz and the Secret of Life*', Chatto and Windus, London, **2007**.
2. G. Dodson, *Biographical Memoirs of Fellows of the Royal Society, D. M. C. Hodgkin*, **2008**, *48*, 3.
3. D. M. Blow, *Biographical Memoirs of Fellows of the Royal Society, M. F. Perutz*, **2004**, *50*, 227.
4. K. C. Holmes, *Biographical Memoirs of Fellows of the Royal Society, J. C. Kendrew*, **2001**, *47*, 311.
5. R. E. Dickerson, '*Present at the Flood: How Structural Molecular Biology Came About*', Sinauer Associates, Sunderland, MA, **2005**.
6. M. Rossmann, *J. Mol. Biol.*, **2017**, *429*, 2601.
7. S. de Chadarevian, '*Designs for Life: Molecular Biology after World War II*', Cambridge University Press, Cambridge, **2002**. See also S. de Chadarevian, *Protein Sci.*, **2018**, *27*, 1136.
8. R. Olby, '*Francis Crick: Hunter of Life's Secrets*', Cold Spring Harbor Laboratory Press, Cold Spring Harbor, NY, **2008**.
9. H. F. Judson, '*The Eighth Day of Creation: Makers of the Revolution in Biology*', Penguin, London, **1979**.
10. P. M. Wassarmann, *J. Mol. Biol.*, **2017**, *429*, 2594.
11. A. B. Pippard, in '*The Legacy of Sir Lawrence Bragg: Selections and Reflections*' (eds. J. M. Thomas and Sir David Phillips), Science Reviews Ltd, Northwood, **1990**, p. 97.
12. M. F. Perutz, *Protein Sci.*, **1997**, *6*, 2684.
13. M. G. Rossmann, private communication, 10 July **2018**.

14. V. Perutz (ed), '*What a Time I Am Having: Selected Letters of Max Perutz*', Cold Spring Harbor Laboratory Press, Cold Spring Harbor, NY, **2009**.
15. T. Steitz, Letter to R. Henderson on the death of Max Perutz in '*A Nobel Fellow on Every Floor*' (ed. J. Finch), MRC/LMB, Cambridge, **2008**, p. 332.
16. Origo, an English-born biographer who lived in Italy (1902–1988) and wrote '*The Merchant of Prato*' (1957) on the life and commercial operations of Di Marco Datini.
17. Ruth Draper was an American actress, dramatist and noted diseuse who specialized in character-driven monologues.
18. The Gog-Magog Hills are a range of low chalk hills, extending for several miles to the south east of Cambridge.
19. M. Faraday, '*On the magnetization of light and the illumination of magnetic lines of force*', *Phil. Trans. R. Soc.*, **1846**, pp. 1–62.
20. J. M Thomas, '*Michael Faraday and the Royal Institution: The Genius of Man and Place*', Taylor and Francis, New York, NY, **1991**.
21. See ref [20], p. 75. The final paragraph of Faraday's paper '*On the possible relation of gravity to electricity*' (*Phil. Trans. R. Soc.*, **1851**, pp. 1–122) has an immortal ring: '*Here end my trials for the present. The results are negative; they do not shake my strong feeling of an existence of a relation between gravity and electricity, though they gave no proof that such a relation exists.*'
22. Robin Perutz is Professor of Inorganic Chemistry at the University of York, UK. His son Timothy is Professor of Mathematics at the University of Texas, Austin. Vivien Perutz, Max Perutz's daughter, is an art historian.
23. H. Trevor-Roper (1914–2003), otherwise known as Lord Dacre, was Regius Professor of Modern History at Oxford and later, Master of Peterhouse, Cambridge, where he became very friendly with Max Perutz. Widely recognized as one of the greatest prose stylists in the English Language; also described by his contemporaries as '*the greatest letter writer of our age*'.
24. A. J. P. Taylor, a famous Oxford historian who was Trevor-Roper's contemporary and rival. He and Trevor-Roper were candidates for the Regius Chair of History at the University of Oxford. He is quoted as saying: '*When I read a passage by Trevor-Roper, tears of envy fill my eyes*'.
25. J. M. Thomas, '*Max Perutz: chemist, molecular biologist, human rights activist*', *Notes and Records of the Royal Society*, **2006**, *60*, 59.
26. M. F. Perutz, Preface of '*I Wish I'd Made You Angry Earlier*', expanded version. Cold Spring Harbor Publishers, Cold Spring Harbor, NY, **2003** (with nine new essays by the author and an *Appreciation* by John Meurig Thomas).
27. Brendan P. Simms is now Professor of the History of International Relations in the Department of Politics and International Studies at the University of Cambridge and a Fellow of Peterhouse.
28. M. F. Perutz, in '*The Legacy of Sir Lawrence Bragg*' (eds. J. M. Thomas and Sir David Phillips), **1990**, p. 67.
29. T. S. Kuhn, '*The Structure of Scientific Revolutions*', University of Chicago Press, Chicago, IL, **1970**.
30. Dr Hasnain is now the Max Perutz Professor of Molecular Biophysics at the University of Liverpool.
31. M. F. Perutz, S. S. Hasnain, P. J. Duke, J. L. Sessler, and J. E. Hahn, *Nature*, **1982**, *295*, 535.
32. J. M. Thomas, *Notes and Records of the Royal Society*, **2006**, *60*, 59.

33. See Preface to ref. [26], p. ix.
34. Cited in J. M. Thomas, *Angew. Chemie. Int. Ed.*, **2002**, *41*, 3155.
35. M. F. Perutz, *The Scientist*, 8 August **1988**.
36. R. McKibbin, remarks made at the Memorial Meeting at St John's College, Oxford, **1998**.
37. J. C. Kendrew, *Scientific American*, **1967**, *216*, 141.
38. L. Stryer, J. C. Kendrew, and H. C. Watson, *J. Mol. Biol.*, **1964**, *8*, 96.
39. D. Crowfoot and J. D. Bernal, *Nature*, **1934**, *133*, 794.
40. K. N. Trueblood, in '*Structural Studies on Molecules of Biological Interest*' (eds. G. G. Dodson, J. Glusker, and D. Sayre), Oxford University Press, Oxford, **1981**, p. 87.
41. J. Pickworth, J. M. Robertson, R. J. Prosen, R. H. Sparks, K. N. Trueblood, and D. C. Hodgkin, *Proc. Roy. Soc. A*, **1959**, *306*, 352.
42. A. J. Eschenmoser, *Pure Appl. Chem.*, **1963**, *7*, 297.
43. D. H. Crowfoot and H. M. Powell, *Z. Kristallogr.*, **1934**, *87*, 370.
44. M. J. Adams, T. L. Blundell, E. J. Dodson, *et al.*, *Nature*, **1969**, *224*, 491.
45. It is to be noted that there were three female co-workers in the Hodgkin–Powell group, an unusually high number for British university science research groups in that era. Note that D. Sayre, born in New York City, did his doctorate with Dorothy Hodgkin. He later was one of the IBM scientists who created the FORTRAN computer language. Later still, he pioneered electron beam lithography as a means of creating X-ray lenses.
46. L. N. Johnson, *Biographical Memoirs of Fellows of the Royal Society*, D. C. Phillips, **2000**, *46*, 377.
47. Jeremy R. Knowles (1935–2008). As a Fellow of Wadham College and University Lecturer at Oxford, he was a leading member of the Oxford Enzyme Group. He later became Professor and Dean of the Faculty of Arts and Science at Harvard University. He made many seminal contributions to enzymology.
48. Fred Richards and his wife sailed their own yacht across the Atlantic on their way to sabbatical leave in Oxford!
49. J. H. Jones, *Biographical Memoirs of Fellows of the Royal Society, Sir Ewart Ray Herbert Jones*, **2003**, *49*, 263.
50. Janet M. Thornton (b. 1949) graduated in physics at Nottingham University. After completing her PhD at the National Institute of Medical Research, she joined David Phillips's group in Oxford. Now Dame Janet Thornton, she was Director of the European Bioinformatics Institute (EBI). She has contributed greatly to our understanding of the structure of proteins. As the Director of EBI, she was responsible for strategic developments relating to the impact of life sciences data on medical science.
51. Greg Petsko (b. 1948) is now at Weill Cornell Medical College at Harvard and a Professor Emeritus at Branders University. His current research interests are in understanding the biochemical and structural biological basis of neurological diseases like Alzheimer's disease and Parkinson's disease. He has made key contributions to the field of protein crystallography, enzymology, and neuroscience.
52. Sir Rex Richards (b. 1922). Former Head of the Physical Chemistry Laboratory and former Vice-Chancellor of the University of Oxford. Distinguished for his work on nuclear magnetic resonance.
53. Judith A. K. Howard, a distinguished Professor at the University of Durham and former Vice-President of the Royal Society.
54. H. M. Powell, *Proc. Chem. Soc.*, **1960**, *138*.

55. C. P. Snow (1905–1980), novelist and physical chemist, renowned especially for his 1958 lecture on '*The Two Cultures*'.
56. James Lovelock (1919–), FRS, an independent scientist and environmentalist, renowned for proposing the Gaia hypothesis.
57. A. T. James, FRS, graduated with distinction in chemistry from University College, London, in 1942. He accompanied most of the members of the Department of Chemistry during the War in 1943 to the University College of Wales, Aberystwyth, where he was Student President. Later in his life, when we became friends and he was then Head of a Unilever Laboratory in Bedford, he told me that A. J. P. Martin, with whom he developed gas chromatography, was the most ingenious scientist he ever met.
58. E. Heilbronner and F. A. Miller, '*A Philatelic Ramble through Chemistry*', Wiley-VCH, Weinheim, **1998**.
59. W. Hayes, *Biographical Memoirs of Fellows of the Royal Society*, **1982**, *28*, 59.
60. Martin Karplus, private communication, August 2018—see also M. Karplus on '*Spinach on the Ceiling: A Theoretical Chemist's Return to Biology*', *Annu. Rev. Biophys. Biomed. Struct.*, **2006**, *35* (see p. 15 in particular).
61. The livery companies of the City of London comprise London's ancient and modern trade associations and guilds. They play a significant part in the life of the city by providing funds and other charitable support to various worthy causes. The Worshipful Company of Clothworkers was granted a Royal Charter in 1528.
62. L. Pauling and M. Delbrück, *Science*, **1940**, *92*, 77.
63. G. Stent, *Science*, **1968**, *160*, 390.
64. J. M. Thomas, The RSC Faraday-Prize Lecture of 1989 on '*Platinum: A Re-enactment of Faraday's Famous 1861 Lecture*', *Chem. Commun.*, **2017**, *53*, 9185.

9

Contributions of Cambridge College Life to Structural Biology: Peterhouse as an Exemplar

9.1 Introduction

For a fifteen-year period from 1982, when Aaron Klug was the single recipient of the Nobel Prize in Chemistry, four Fellows of Peterhouse held Nobel Prizes for their contributions to structural molecular biology. Perutz and Kendrew were joint recipients in 1962, and Archer J. P. Martin's Nobel Prize came in 1952.

We shall see in this chapter how an amalgam of factors—astute recruitment, personal choice, and a fortunate concatenation of circumstances—together can lead a single College—Peterhouse—to be a vital centre in the pursuit and florescence of structural biology. Similar stories could be related pertaining to other Colleges and subjects. For example, St John's College, Cambridge, attracted the following contemporary giant founders of quantum mechanics: P. A. M. Dirac, N. F. Mott, and D. R. Hartree. They were all active in that College at the same time as J. D. Cockcroft. In King's College, Cambridge, Maynard Keynes collaborated effectively with junior colleagues there—Richard (later Lord) Kahn, who contributed significantly to Keynes' magnum opus '*The General Theory of Employment, Interest and Money*', and, arguably, one of the cleverest, ever, economists—Frank Ramsey. And ever since Isaac Newton studied at Trinity College, it has attracted a brilliant succession of mathematicians, including Fields Medallists such as Sir Michael Atiyah, Sir Timothy Gowers, Alan Baker, and Richard Borcherds.

9.2 Aaron Klug and his Peterhouse/LMB Colleagues and Collaborators

Late in 1993, towards the time of Aaron Klug's retirement as a Fellow of Peterhouse, the photograph shown in Figure 9.1 was taken.

Figure 9.1 *The three Nobel Prize winners—Perutz, Kendrew, and Klug—are accompanied by three other members of Peterhouse who were involved in teaching undergraduates, as well as pursuing research at the LMB: Drs R. A. Crowther, J. T. Finch, and P. J. G. Butler. (Photograph first published in* Notes and Records of the Royal Society, *2000,* 54, *369, under the title* 'Peterhouse, The Royal Society and Molecular Biology'*)*

All the six structural biologists shown here were members of Peterhouse—four as full Fellows, two (Finch and Butler) supervised undergraduates at the College—and participated regularly in the lively discussions pursued in the Kelvin Club, the students' science society.

9.3 A Brief Outline of Aaron Klug's Life

Aaron Klug, who died as this chapter was being written (November 2018), was born in Lithuania, the son of a saddler, cattle breeder, and occasional journalist. His parents and their Jewish family found life in the Soviet empire increasingly difficult. So when Aaron was two, the Klug family fled to Durban, South Africa. A full account of his precocious childhood and teenage years has been given in a recent biography by Holmes.[1] He entered the University of Witwatersrand in Johannesburg aged fifteen to study medicine. He slowly abandoned this subject, and he majored in physics, mathematics, and chemistry. He then pursued an MSc degree in X-ray crystallography at the University of Cape Town with R. W. James, one of W. L. Bragg's former colleagues at the University of Manchester in the 1930s. In 1949, he was awarded one of the prestigious 1851 Scholarships[2] to study for a PhD in Cambridge where the Cavendish Professor Lawrence Bragg assigned him to study, under D. R. Hartree, theoretical problems associated with the phase transitions occurring during the cooling of steel. This metallurgical problem introduced him to solid-state physics and gave him a grounding in computing. It also acquainted him with the phenomenon of martensitic transformation, a diffusion-less phase transition, which later helped him in his structural biology.[3]

From the Cavendish Laboratory, he migrated to J. D. Bernal's Department of Physics in Birkbeck College, University of London, with a grant from the Nuffield Foundation. There he had the good fortune to meet Rosalind Franklin—they first met on the stairs of the overcrowded building, in Torrington Square. She was working on viruses, and he quickly became a strong collaborator of hers, especially in view of his ability to interpret the X-ray diffraction patterns that she so expertly recorded (see below).

John Finch was a PhD student of Rosalind Franklin, as was K. C. Holmes,[1] at that time. They both accompanied Klug to the Laboratory of Molecular Biology (LMB) when Max Perutz invited him to come to Cambridge in 1962. Sadly, Rosalind Franklin had died in 1958.

Aaron Klug told me on numerous occasions that he felt forever indebted to Bragg, to Perutz, and to Kendrew. Of Bragg, he spoke of the gentle and kindly way in which he was treated by him.[4] Of Perutz, he was grateful for the invitation to join the LMB where, for forty years or so, as I outline below, he did his brilliant work in so many aspects of structural and theoretical molecular biology. And of Kendrew, whom he had first met in 1950 as a PhD student, he was eternally grateful for introducing him to Peterhouse, as a successor as Director of Studies in the Natural Sciences (from 1962 to 1994) (see Section 8.2.3).

Much of the remainder of this chapter deals with work initiated and/or pursued by Klug and his collaborators. More details are given below, but we note here that, apart from his Nobel Prize in 1982, he was knighted in 1988, he was Director of the LMB until 1996, taking over from Sydney Brenner in 1986, and he was President of the Royal Society from 1995 to 2000, and its Copley medallist—its highest honour—in 1985. He was also pivotally involved in the establishment of the Sanger Institute (now the Wellcome Trust, Genome Campus, Hixton, Cambridge) and in the appointment of his Nobel Laureate colleague Sir John Sulston as its first Director. He was also instrumental in enabling the LMB to earn royalties for its work on monoclonal antibodies, as described in Chapter 10.

9.3.1 Klug's personality and ability and his interaction with others

To all those who knew him, Klug was a rather shy individual, but endowed with a prodigious retentive memory and a degree of intellectual curiosity among the greatest I (and numerous others) have ever encountered. He was, from an early age, an autodidact. He read extensively—not only most aspects of the fundamental sciences, but also about ancient civilizations, literature, religious texts, poetry, philosophy, history, and political discourses. I have never known anyone who possessed a more omnivorous intellectual appetite. He could talk authoritatively about vast swathes of knowledge. When I first got to know him (when, along with other electron microscopists, we attended a Nobel Symposium in Stockholm in 1979[5]), he was an avid reader of the BBC weekly publication called *The Listener*.[6] He read each issue from cover to cover, while also pursuing the life of one of the most active molecular biologists in the world.

Stories abound about his extraordinary range of cultural, scientific, and historical interests. One of his successors as Director of the LMB—Sir Hugh Pelham, said that, while he was working alongside Aaron Klug on a difficult problem, the conversation, led by Klug, turned to T. S. Eliot's iconic masterpiece '*The Love Song of J. Alfred Prufrock*'. On another occasion, during a dinner in Klug's home in 1980, Sir Huw Wheldon (the Director General of the BBC, whom the Klugs knew from their period in London working at Birkbeck College) expressed utter astonishment at how much Klug knew about the articles published in *The Listener*.

Whenever I spoke to him about aspects of fundamental physics, he would quote what he had learnt from the Soviet Nobel Prize winner genius Lev Landau (1908–1968),[7] and he usually recalled, almost verbatim, what he had learnt from Landau's writings and numerous theoretical innovations. Once, after returning from a visit to the Lawrence Berkeley Laboratory in California, I told Aaron that E. O. Lawrence was the first physicist to embark, in the mid 1930s, on what was then and is now called *Big Physics*. Aaron quickly corrected me by drawing to my attention that John Cockcroft and Ernest Walton had established *Big Physics* in Rutherford's Cavendish Laboratory in Cambridge when they carried out their Nobel Prize-winning experiment in 1932 on the splitting of the atom, thus performing the first artificial nuclear disintegration. Whether the Cockcroft–Walton experiments constitutes the beginning of *Big Physics* is an open question.

In executing his duties as a Fellow of Peterhouse, he sat for many years on the Research Fellowship Committee. This was an annual exercise that entailed scrutinizing anything from a hundred to 200 research projects proposed by young (fresh PhD) candidates and the associated referees' reports and testimonials of their former supervisors. His colleagues on that Committee never ceased to be amazed by how much Aaron Klug knew—of subjects well outside his own research fields—but also his phenomenal memory of what the referee had said in previous years about candidates for Research Fellowships. My predecessor as Master of Peterhouse the Rev Prof Henry Chadwick, a very scholarly theologian, said of Aaron Klug that he knew more about ancient philosophy than most professors in the humanities. As a young man, Aaron Klug studied some Egyptology and learnt to decipher some hieroglyphs.

When Klug died, the President of the Royal Society—the Nobel Prize winner Venki Ramakrishnan—said of him that '*he was a towering giant of 20th century molecular biology*'. At his funeral, reference was made to his favourite verse from the Old Testament, '*What doth the Lord require of thee? To do justly, to love mercy, to walk humbly with your God*' (Micah 6:8). Aaron Klug was not deeply religious. He cherished Judaism and honoured its traditions to the extent of going occasionally to the synagogue.

Aaron Klug also had the intrinsic gift of being able to pick winners. For example, when he set about combining his life as a major leader of research at the LMB with teaching duties for undergraduates in many branches of natural science—mathematics, physics, biology, and chemistry principally—he recruited the 'right' people within his ambit to do so. R. A. (Tony) Crowther (later elected FRS and who, some scientists felt, should have shared the Nobel Prize with Klug and David DeRosier) was a mathematics graduate and he joined the LMB principally to mastermind sophisticated computational programmes. But, gradually, with encouragement from Aaron, he learnt his electron microscopy from John Finch (later FRS), with whom Aaron had collaborated from his Birkbeck days. Likewise, Aaron recruited Jo Butler as an expert practitioner from the Department of Biochemistry in Cambridge. And, largely as a result of discussions at mealtimes in Peterhouse, the eminent structural engineer Chris Calladine, FRS, was recruited by Aaron to investigate how the self-assembly of identical protein subunits could construct the corkscrew-like flagellar filaments that act as propellers in the swimming of bacteria. The solution turned out to be that there is a bi-stable interface between neighbouring subunits which endows the filaments with the ability to construct a helical form, which moreover may switch into members of a family of different helical forms under appropriate environmental conditions.[8,9]

Klug himself was very happy to supervise physicists, materials scientists, biochemists, and physical and inorganic chemists, but he was uneasy about supervising organic chemists, especially those studying natural products. Astutely, he chose a young Australian member of the Cambridge Department of Chemistry—Dr Andrew B. Holmes (later elected FRS)—to do that work and he served as supervisor at Peterhouse for twenty years under Klug's general direction. In this choice, Klug again excelled. Holmes had worked with Eschenmoser at ETH, Zurich on the total synthesis of vitamin B_{12}. Later, Holmes did such outstanding work, both as a natural product chemist and as an inventive polymer chemist, that he became Professor in Cambridge, FRS, and a Royal Medallist of the Royal Society. Later, he returned to the University of Melbourne to take up an Australian Research Council Federation Fellowship at the University of Melbourne. He then became President of the Australian Academy of Science and a holder of the Order of Australia.

Klug, like Kendrew, his predecessor at both the LMB and Peterhouse, also had an enormous range of contacts with scientists all over the world, especially with those in the United States (see Section 9.3.2).

9.3.2 Three eminent Stanford University molecular biologists at the LMB

The links between Peterhouse and the LMB were always strong, as adumbrated earlier, and Figure 9.2 shows three eminent molecular biologists who spent some time in Peterhouse. Dr Lubert Stryer, Emeritus Professor of Cell Biology at Stanford University School of Medicine, was a postdoctoral scientist working in Sir John Kendrew's group at the LMB in the period 1962–1963, and he visited the college socially on a number of occasions. He is renowned for his multiple achievements, which include work on the primary stage of amplification in vision and in the elucidation of the so-called G protein cascade[10] that generates a neural signal in visual excitation. He is a co-inventor of the DNA chip and received the US National Medal of Science in 2007 from President G. W. Bush.

John Kendrew was approached in 1963 by Arthur Kornberg, US Nobel Prize winner (who discovered the mechanism for the biological synthesis of DNA), and asked if there was a suitable candidate at the LMB for the post of Assistant Professorship in Stanford. Dr Stryer, Kendrew's colleague, was encouraged by Kendrew to apply. He duly was selected; and, in the fullness of time, he recruited Arthur Kornberg's distinguished son Roger, who was a close colleague of Aaron Klug (and who won the Nobel Prize in Chemistry in 2006 for the molecular basis

Figure 9.2 *Professor Lubert Stryer flanked by Professors Michael Levitt and Roger Kornberg, three structural molecular biologists at Stanford University, all three alumni of the LMB.*

of eukaryotic transcription) (see also Section 9.6.1), to a Professorship at the Department of Structural Biology in Stanford. Roger Kornberg lived in Peterhouse College accommodation, while at the LMB. In due course, Michael Levitt, a close collaborator at the LMB of Aaron Klug and a graduate student at Peterhouse, was also appointed to the distinguished Stanford Faculty. Levitt, whose initial work was greatly encouraged at its outset by John Kendrew, is one of the founders of computational structural biology,[11] for which he was awarded a Nobel Prize in 2013, in association with Martin Karplus and Arieh Warshel, two other distinguished alumni of the LMB.

9.4 Klug's Views on Lawrence Bragg

The detailed views of Klug in regard to John Kendrew, as well as his respect and admiration for Max Perutz, have been given in Chapter 8. Here we focus on his views on Bragg. When, in 1990, Sir David Phillips and I invited Aaron Klug to contribute to our monograph on '*The Legacy of Sir Lawrence Bragg*', he responded enthusiastically, as he held Bragg in the highest regard ever since he had pursued an MSc study with R. W. James, one of Bragg's former colleagues at the University of Manchester in the 1930s. [R. W. James (1891–1964) was a most unusual individual, having been an Antarctic explorer in the famous Shackleton expedition that was stranded on Elephant Island in 1914–1916.][12]

James disclosed to Klug many of Bragg's attributes, but Klug also came to know Bragg once he joined the Cavendish Laboratory in Cambridge as a PhD student in 1949. In Cape Town, James told Klug of his experiences working with the impressive Peter Debye as a visitor at the University of Leipzig in 1931, when one of Debye's professional colleagues was Heisenberg. The difference in approach between British and Continental scientists was illustrated by James when he described '*the impressive learning in theoretical physics and mathematics not only of the leaders, as might have been expected, but also of all the junior workers*'.[4] Cambridge undergraduates in physics had hardly been exposed to any formal teaching in mathematics at that time (1931), and James had learnt it only as it was needed in various parts of physics.

James had gone to work with Debye to investigate the influence of temperature on the scattering of X-rays by molecules in the gas phase. He recounted to Aaron Klug how, for example, Debye could write down, without notes, the quantum–mechanical Hamiltonian for the electromagnetic field. The lack of rigorous training among their English contemporaries had not gone unnoticed by their German counterparts—'*Tell me*,' said one of James's colleagues, emboldened after several months of acquaintance, '*How does Bragg discover things? He doesn't know anything!*' Klug also made the following remarks:[4] '*I would imagine that the men in Leipzig must have been further astonished some three years later when Bragg and Williams*[13] *produced their theory of order–disorder in alloys, the beginning of the subject of cooperative phenomena in physics. It seems unlikely to me that Bragg would have known the hybrid formalism of thermodynamics and statistical mechanics employed (that is surely due to Williams), but his hand is unmistakable. In the first presentation of the theory, the energy of interaction between an atom and its neighbours is assumed to depend upon the degree of order in the whole assembly. Elsewhere Bragg made the analogy that the energy keeping any given member of a society in the right place is dependent on the degree to which its neighbours are in their right place, that is "public opinion". The theory surely reflects Bragg's imaginative insight into the physics of the phenomenon.*'

Klug certainly is right both in assigning the imaginative analogy of public opinion, so typical of the mastery of metaphor (see Chapter 7), to W. L. Bragg. And he is also right in attributing the fundamental thinking behind the now famous Bragg–Williams order–disorder phenomenon to Williams. As P. M. S. Blackett said in his memoir[14,15], '*Williams was one of the most brilliant physicists of the age*'. He worked with extraordinary success alongside Chadwick, Rutherford, Bragg, and Niels Bohr.[14]

Klug, in his article,[4] also acknowledges the kind cooperation and encouragement by Bragg given to him, John Finch, and Rosalind Franklin, when they were granted permission, in 1954–1956, to use the high-power rotating-anode X-ray

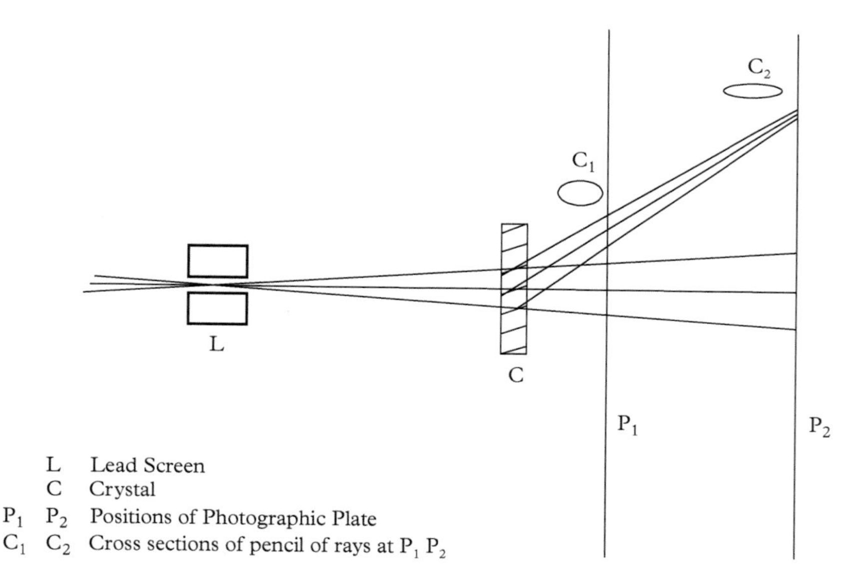

Figure 9.3 *Change of shape of the X-ray reflexions as the photographic plate was moved away from the crystal. Reflexions that were round when the plate was near the crystal became drawn out in the horizontal direction further away. Bragg pointed out that reflexions by the lattice planes of an incident core of X-rays of continuously varying wavelength would come to a focus in the vertical direction but would spread out in the horizontal direction.*[18]

(First published in J. M. Thomas and D. C. Phillips (Eds), 'The Legacy of Lawrence Bragg', *Science Reviews Ltd., 1991)*

tubes constructed early on at the Royal Institution (see Section 9.5 and ref [17]) for their work on poliomyelitis virus crystals. In another section of his article, Klug describes how Bragg, on discovering that Kendrew's different heavy atom derivatives of myoglobin (see Chapter 5) could solve the phase problem (see Section 2.4.3), he, Bragg, '*wept with joy*'. Finally, Klug makes admiring reference to Bragg's initial paper in 1912, read to the Cambridge Philosophical Society:[16–17] '*Bragg's Law is known to all students, but how many learn that what helped him to arrive at the notion of X-rays being reflected by a crystal was his observation of the change in shape of the diffracted spots, as the distance between the crystal and photographic plate was varied in his experiments. This is described in Bragg's first and great paper submitted in November 1912* [see Figure 9.3]. *I used to give a copy of it to my new research students or associates to bring to life a great scientist at work. I remember Bragg's observations when, in 1972, our crystals of tRNA gave X-ray spots which had different shapes in different regions of the film. The thin plate-like crystals were being bent by contact with the glass wall of the capillary tube, and reflecting as from a curved mirror. So, sixty years later, Bragg's great paper still had a resonance.*'

9.5 A Summary of the Contributions to Structural Biology by Aaron Klug and his Collaborators

When Klug arrived in 1962 at Cambridge, he brought with him not only both Finch and Holmes, but also an American virologist—Don Caspar (who had been directed towards England by Isidor Fankuchen, a former student of Bernal—see Section 3.5.1). All joined the LMB at the same time (see Figure 9.4).
Gradually, Klug established himself as a world authority in several distinct aspects of molecular biology;[1,19,20] and, in some of his principal interests, John Finch, FRS, as well as R. A. Crowther, FRS, together with Jo Butler, were his key co-investigators at the LMB. The recent biography by his former co-worker Holmes[1] gives full details of all the major advances (scientific and administrative) that Klug achieved at the LMB and elsewhere.

As stated earlier, he had started to study viruses with Rosalind Franklin using X-rays at Birkbeck College in Bernal's Department. With her and John Finch, he published several papers on the structure of turnip yellow mosaic virus, tobacco mosaic virus (TMV), poliomyelitis virus, and papilloma–polyoma (human wart) virus, some of which he had investigated earlier with Caspar.[4]

In the Franklin–Klug–Finch project[21,22] on crystals of polio virus (received from workers at the University of California, Berkeley), problems of safety arose in working with this pathogen. In fact, it was Sir Lawrence Bragg, at the Royal Institution, who allowed John Finch to use the uniquely powerful X-ray set-up there—the facility initiated in the days of Bernal and Lonsdale, using a rotating anode source of X-rays (see Section 3.5). John Finch and his colleagues working on the polio virus at the Davy-Faraday Research Laboratory were vaccinated

Figure 9.4 *Photo taken at a Madrid meeting, April 1956, from left to right: Francis Crick, Don Caspar, Aaron Klug, Rosalind Franklin, Odile Crick, and John Kendrew.*
(Photo courtesy D. L. D. Caspar)

beforehand against polio with the newly available Salk vaccine. As is stated in Finch's biographical memoirs,[17] '*X-ray exposures were long, sometime overnight and as someone had to be in attendance, John remembered the night-time Royal Institution as an eerie place, with busts of past Directors lit from the streetlights outside.*'

It is outside the scope of this monograph to summarize all the key facts pertaining to viruses, since they were first investigated by Bernal. We focus only on a few. There are the rod-like ones and the spherical ones. TMV, for example, is rod-shaped, with a diameter of 150 Å and a length of about 2800 Å, and it has a molecular weight of 4×10^7 daltons. Like other plant viruses, it contains both RNA (six per cent) and protein (94 per cent). The near-spherical viruses often have an icosahedral symmetry—just as in a football (see Figure 9.5).

Dr Jo Butler joined Aaron Klug's group as a biochemist in the virus group where he made significant contributions to the nature of spherical viruses, as well as the rod-like TMV. It was during this time, especially from the period of the

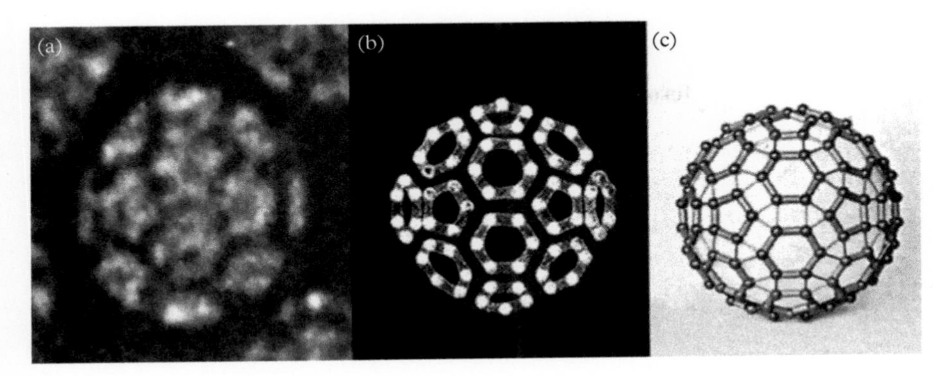

Figure 9.5 *(a) A negatively stained electron micrograph of a single particle of turnip yellow mosaic virus viewed over a holey grid (the protein is white). The particle is aligned looking along a twofold axis. In this view, the front and back surface features are the same, which makes the image easier to interpret. (b) An idealized drawing derived from the inspection of hundreds of images, as shown in (a). The fivefold and sixfold groupings of the virus coat protein can be seen. There are twelve fivefold and twenty sixfold rings, accounting for the thirty-two knobs seen in low-resolution electron micrographs. (c) A ball-and-spoke model based on the drawing in (b) that was used for generating shadowgraphs at various tilts to be compared with images from virus particles tilted in the electron microscope.*

(Reprinted from Finch, J. T. and Klug, A., J. Mol. Biol., 15, *344–364,* **1966**, *with permission from Elsevier)*

landmark paper published by DeRosier and Klug[23] on the use of electron microscopy in structural biology (see below), that R. A. Crowther made vital contributions to the electron microscopic determination of virus structure.[24–26] He, along with Finch, Klug, and other co-workers, demonstrated how to retrieve the detailed structure of spherical viruses from electron micrographs.

In view of the pioneering nature of DeRosier and Klug's[23] work on the reconstruction of three-dimensional structures from two-dimensional electron micrographs, and the present-day major activity in electron cryo-microscopy in structural biology—discussed further in Chapter 10 (see Section 10.6.1)—the essence of electron microscopy as a structure-determining technique is outlined in Section 9.5.1.

9.5.1 Essence of the electron microscopic approach pioneered by Klug for the determination of atomically resolved biological structures

The fundamental difference between electron microscopy and X-ray diffraction as techniques for the retrieval of structural information is that, in the former, because electrons can be focused by appropriate magnetic lenses, a structural image is obtained, whereas in the latter, since X-rays cannot be readily focused, the X-ray diffraction pattern has to be converted to the structural image (by

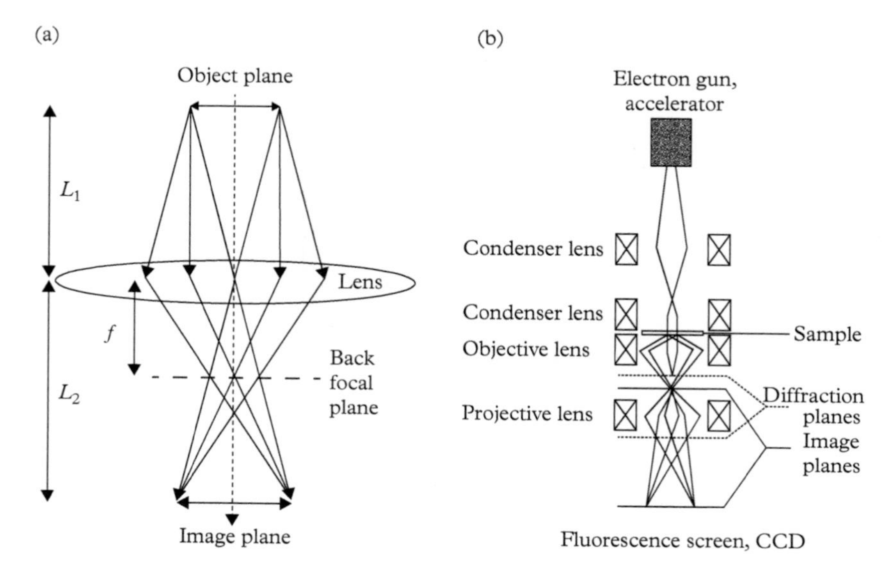

Figure 9.6 *The essential similarity between (a) optical and (b) electron microscopy.*

Fourier transformation) using both the amplitude and the phase of each diffraction spot, as explained in Chapter 2.

Figure 9.6 depicts the essential features of image formation in an electron microscope and, also for heuristic purposes, in an ordinary light microscope.

It is important to appreciate that high-resolution electron microscopes have a very considerable depth of focus, making the image a two-dimensional superposition of different levels in the three-dimensional structure. The focus cannot be adjusted to different levels within the object, and so three-dimensional structures are difficult to analyse. Klug, DeRosier, Crowther, and Finch, as well as later workers at the LMB, notably Henderson and Unwin,[27] and especially Huxley,[28] realized all this early on.[29] They were able to determine, as a consequence, molecular biological structures by recognizing that more than one view is generally needed to see a continuous object in three dimensions.

Klug worked out a tomographic approach to reconstructing a three-dimensional image from a series of two-dimensional ones recorded by electron microscopy.[30] He did this by what is called Fourier reconstruction. The essence of the method can, however, be readily appreciated by reference to the similar method of back projection, summarized in Figure 9.7, much used by Baumeister and co-workers[31] and by Midgley and his.[32]

In addition to Crowther, Finch, and Butler, Klug collaborated with the Peterhouse structural engineer C. R. Calladine, FRS, who made important contributions to the behaviour of bacterial flagella,[33] which are constructed from molecules of protein known as flagellin by a process of self-assembly. The flagella are rigid, corkscrew-shaped, slender filaments, which are rotary motors.

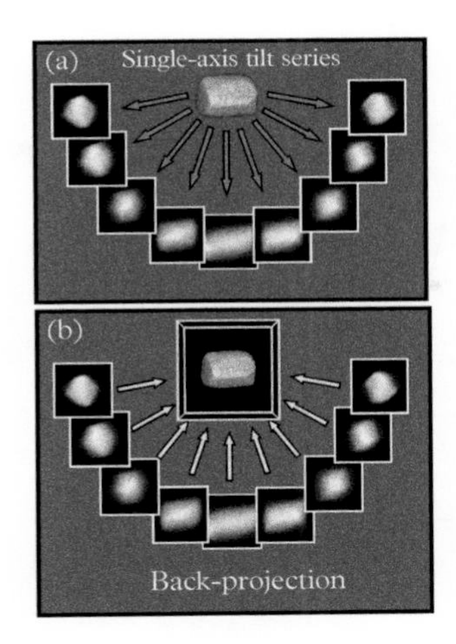

Figure 9.7 *Schematic diagram illustrating the principles of tomographic reconstruction using the so-called back-projection method. In (a), a series of images is recorded at successive tilts. These images are back-projected in (b) along their original tilt directions into a three-dimensional object space. The overlap of all the back-projections defines the reconstructed object.*[32]

(Courtesy Royal Society of Chemistry)

9.6 Klug's Other Contributions to Structural Biology

Several scientists who have worked alongside him, or who know the depth of learning and ability of Aaron Klug, have been heard to refer to him as the cleverest person that they have ever met.[34] Perusal of Holmes' biography of Klug[1] reinforces this view; Rosalind Franklin, for example, was extremely impressed by his ability. Quite apart from his profound studies of viruses and the other work he initiated with Crowther, whereby they identified filamentous lesions found in Alzheimer's disease, he also made very many other important advances in molecular biology. These are now very briefly described.

In 1985, he identified[36] so-called zinc fingers (see Figure 9.8). These are small protein structural motifs that are characterized by the co-ordination of one or more zinc (Zn^{2+}) ions in order to stabilize the protein fold.[37] These zinc fingers have subsequently been found to be ubiquitous in the living world and may be found in some three per cent of the genes of the human genome. They are protein modules for nucleic acid recognition and are now also extremely useful in therapeutic work.

He also contributed, with Roger Kornberg, Jean Thomas, John Finch, Tim Richmond, Daniella Rhodes, Jo Butler, and other colleagues in Cambridge, to study the structure of chromatin—the complex that condenses long DNA using a set of small, basic (positively charged) proteins, called histones, such that, in mammalian cells, for example, two metres of DNA fits into a nucleus of about ten microns in diameter. This packaging plays a crucial role in the control of gene expression, in DNA replication, and in protection of DNA from damage.

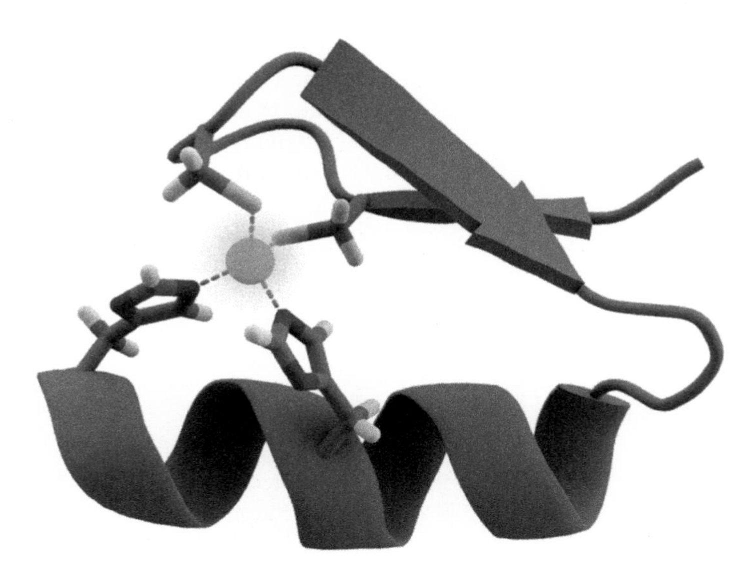

Figure 9.8 *Pictorial representation of the essence of a zinc finger, in which a Zn^{2+} ion stabilizes a particular section of a protein.*

(Reproduced from http://www.scistyle.com under creative commons license 4.0, drawn by Thomas Splettstoesser)

9.6.1 The nucleosome

In 1975, Jean Thomas (by now Dame Jean, is on the academic staff of the Biochemistry Department as Professor of Macromolecular Chemistry and former Vice-President of the Royal Society, having earlier been, as a Beit Memorial Fellow, at the LMB) and Roger Kornberg (postdoctoral fellow at the LMB, where he had come to learn crystallography from Aaron Klug) provided key evidence on histones and their association which led to Kornberg's formulation of a new, and now proven, model for chromatin structure. DNA is wrapped around successive discrete histone complexes (octamers), in a repeating structure, giving the 'beads-on-a-string' model of chromatin. The beads, the structural subunits, were later termed nucleosomes (by Pierre Chambon and co-workers in Strasbourg).

The protein core of the nucleosome—the histone octamer (two copies of each of four types of histone) was characterized, in 1977, by Jean Thomas and Jo Butler.[38] This material was found by John Finch to form regular aggregates that were shown by electron microscopy to be hollow tubes, similar in appearance to TMV, which were suitable for image reconstruction analysis (see Section 9.5.1). In collaboration with Aaron Klug and others, this led to the publication, in 1980, of a low-resolution structure of the protein core, as a disc shape, and a proposal that DNA would fit into a left-handed helical groove on the surface, forming a low-pitch super helix—the first model to be proposed for the nucleosome.[39] A few years later, so-called 'trimmed' nucleosomes ('core particles') were crystallized, and X-ray crystallography by Tim Richmond (who had joined the Klug group) and colleagues led to direct visualization of the arrangement of DNA and histones, at 7 Å resolution, and confirmed all the features of the earlier[39] low-resolution model. Thirteen years later, Tim Richmond (who had, by then, moved to the ETH in Zurich) and colleagues published the 2.8 Å resolution structure of the nucleosome (see Figure 9.9).

Meanwhile, Klug and Finch, alongside Jean Thomas and Jo Butler who pursued hydrodynamic studies, also addressed the nature of the next level of structure into which the string of beads is folded. (This is a highly enigmatic problem to tackle, and work is ongoing in many laboratories.)

9.6.2 *t*RNA

The existence of *t*RNA was first hypothesized by Francis Crick, based on the assumption that there must exist an adaptor molecule capable of mediating the translation of the RNA alphabet into the protein alphabet. Although the mechanistics of RNA, DNA, ribosomes, and, as stated in the Preface, the whole interlinked processes that involve protein production from DNA, lie outside the scope of this book, it is necessary to make reference to Klug *et al.*'s work on the discovery of the structure of *t*RNA. Intense rivalry developed between the group of Alexander Rich at the Massachusetts Institute of Technology (MIT) and Klug's

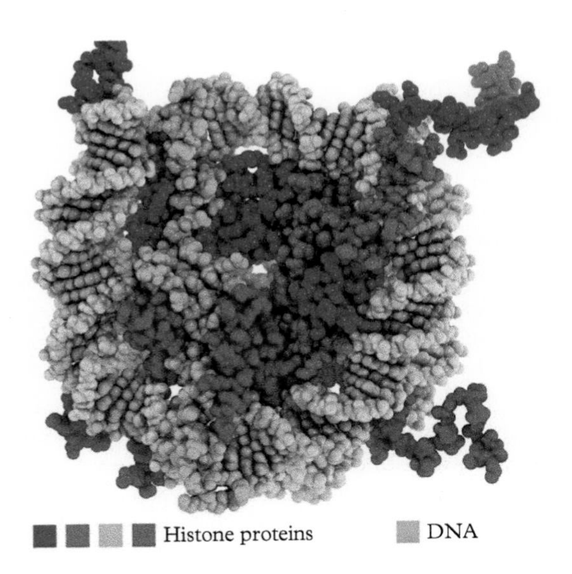

Figure 9.9 *Depiction of the nucleosome core particle. This is the association of DNA and proteins in the core, which holds many water molecules and various anions. (See C. A. Davey, D. F. Sargent, K. Luger, A. W. Maeder, and T. J. Richmond,* J. Mol. Biol., *2002,* 319, *1097.)*

(Courtesy of Professor Song Tan)

group at the LMB on this particular topic. In Rich's group, Kim Sung-Hou played a leading role in the determination of the structure.[40] At the LMB,[41] there were even thoughts (in Klug's mind) of improper practice attributed to the work of others. But this was later proven to be untrue.

9.7 Recent Structural Biological Work by Members of Peterhouse

Six current members of Peterhouse are currently engaged in molecular biological research of one kind or another: Prof Ben Luisi, Prof Sophie Jackson, Prof Andrew Lever, Dr Graham Christie, Dr R. A. Crowther (see Figure 9.1), and Dr C. R. Calladine. The last two are now retired from their full-time work in, respectively, the LMB and the Department of Engineering. Earlier, they worked very closely with Aaron Klug and his associates.

Professor Sophie Jackson, a chemist, is *inter alia* focusing on the enigma of knots in α-helices. What is their biological function? (See Figure 9.10.)

Figure 9.10 *Professor Jackson is holding a 5_2-knot—this is the knot that is formed by the enzyme ubiquitin C terminal hydrolase (see ref [42]).*

(Courtesy S. E. Jackson)

In Figure 9.11 is shown one of the relatively small molecular machines that Luisi and colleagues have been studying recently[43]—the architectonically impressive structure, teeming with α-helices. The outer and inner membranes in this figure are bacterial ones through which this assembly forms a channel. It would have been impossible to determine this structure by X-ray diffraction methods. But electron cryo-microscopy has yielded its intricate structural details, as described fully in Chapter 10.

Prof Calladine has made many other fundamental contributions to our understanding of the packing of DNA under various circumstances—see the comments of Klug[8] and Finch.[19] He has contributed greatly to our understanding of why various types of DNA double helix have been found, and also explained how and why DNA twists and curves. His impressive monograph,[44] now in its third edition and co-authored with collaborators Ben Luisi, H. R. Drew, and A. A. Travers (the latter two also LMB alumni), has been much praised.

As mentioned earlier (see Section 9.5.1), Crowther's electron microscopic work with Aaron Klug[45] led to the identification of the so-called tau protein (see Chapter 10), the paired helical filament in Alzheimer's disease. Very recently, he has used electron cryo-microscopy in his studies of another neurodegenerative condition—Pick's disease,[46] and this work is described in Section 10.6.1.

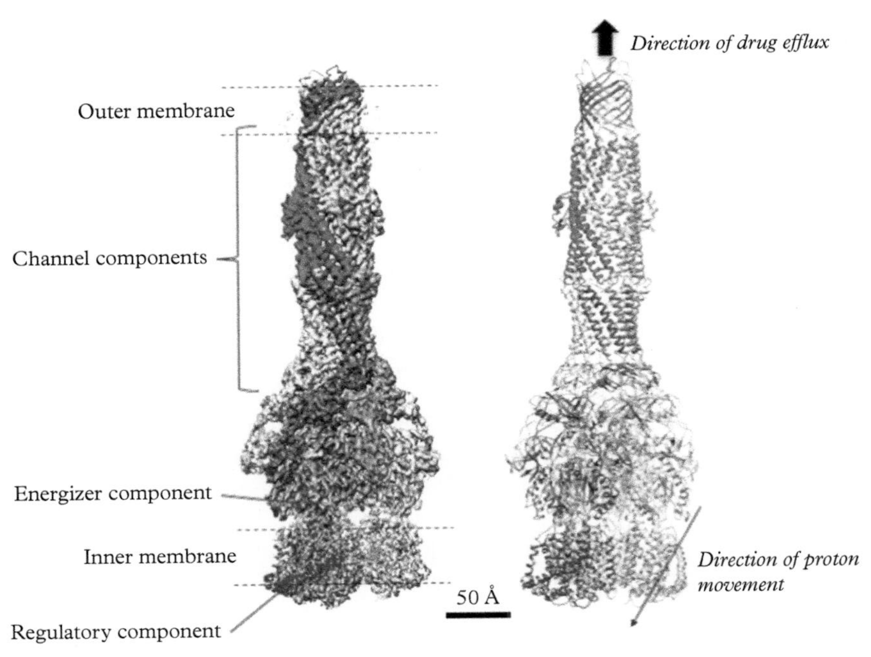

Figure 9.11 *This intrinsic molecular assembly of proteins (called TolC) is an energy-driven machine to pump drugs and other molecules out of* Escherichia coli *and several other bacteria. (Courtesy Professor B. F. Luisi)*

Professor Andrew Lever of the Clinical School of the University is deeply implicated in developing novel therapies to prevent virus recognition. Lastly, Dr Graham Christie, a member of the Department of Chemical Engineering and Biotechnology, pursues a fascinating series of studies in the large field of bacterial spores.[47] Much of the structural molecular biology pursued by Dr Graham Christie's group is aimed at gaining insights into the proteins and molecular mechanisms that underpin the process of spore generation, and also into the numerous proteins that constitute the protective outermost layers of the spore.[47] Progress in this study is stimulated by the prospect that it may help overcome problems associated with healthcare, food, veterinary medicine, and biosecurity.

As an outsider, looking into the current, wide-ranging fields of molecular biology, I never cease to marvel at the myriad of inter-dependent chemical processes that sustain the sheer variety and extraordinary prodigality of living things.

9.8 CODA

In completing this chapter, where I have singled out, for special mention, interactions between scientists in one college—Peterhouse, I feel it is necessary to amplify briefly how the uniquely effective ways in which Cambridge college life have contributed so remarkably to excellence in structural molecular biology.

First, there are the magical and inspiring traditions of Cambridge itself and the intellectual giants in numerous fields who have worked in its university and its colleges. Max Perutz was deeply aware and appreciative of this fact, as is described in Section 8.2.1. And so were Aaron Klug and Hugh Huxley (see Section 8.2.3.2), when they contemplated the extraordinary experience in interacting and following on from John Kendrew at Peterhouse.

Second, following numerous conversations I had with the son and daughter of Lawrence Bragg (Stephen and Patience), I am aware of how much their father was inspired by the greatness and world-changing achievements of his predecessors as Cavendish Professors at Cambridge: J. Clerk Maxwell, Lord Rayleigh, J. J. Thomson, and Lord Rutherford.

Shortly before he died in 2002, Max Perutz, writing to Queen Elizabeth II, said: '*I have had 65 years of productive research in Cambridge, the most wonderful centre of learning in the world, and I could never have accomplished what I did anywhere else.*'

Other scientists, far removed from molecular biology, have also expressed the view that, being present in a university such as Cambridge, they were inspired, as were their students and colleagues, to address worthwhile targets of research. Take, for example, the late Dennis Sciama (1926–1999), a distinguished astrophysicist and cosmologist, though not as well-known as Bragg, Perutz, Kendrew, and Hodgkin. He was proud to proclaim that he had been the beneficiary of listening to lectures, in Cambridge as a student, on philosophy from Wittgenstein and on quantum mechanics from Dirac, who became his PhD supervisor. Sciama's two most noteworthy PhD students are now regarded as the most famous contemporary cosmologists: the late Stephen Hawking and Martin Rees. Sciama also profoundly influenced another renowned cosmologist, Roger Penrose.

Eminent US scientists have told me that students and researches in Cambridge, conscious of the legacies of Newton, Darwin, Clerk Maxwell, Rutherford, G. H. Hardy, Milton, Keynes, Dirac, Sir James Frazer, Sir Gowland Hopkins, and others, are constantly reminded that one's research should address problems that are truly important, not just fashionable. They feel it imperative to attack worthwhile problems.

One of my friends recently said that the LMB, which still harvests results of supreme structural biological importance (see Chapter 10), is one of the most remarkable institutions in history, thanks largely to the ethos engendered by Perutz, Kendrew, and their eminent successors.

Apart from its recent record in structural molecular biology—five of its members have been awarded Nobel Prizes in this field—Peterhouse has, in previous

William Hopkins, FRS
1793–1866

G. G. Stokes, F. Galton, A. Cayley,
Lord Kelvin, P. G. Tait, J. C. Maxwell, E. J. Routh

Edward J. Routh, FRS
1831–1907

Lord Rayleigh, J. J. Thomson, G. Darwin,
J. M. Larmor, A. N. Whitehead

Figure 9.12 *Two former Fellows of Peterhouse who were extraordinary successful teachers of mathematics. Underneath each photograph are given the names of the eminent students taught by them. (Note that Whitehead became the mathematics tutor of Bertrand Russell.)*

(Courtesy JET Photographic and Master and Fellows of Peterhouse)

eras, excelled in other subjects, notably history and mathematics. Its record as a hot-bed of mathematic excellence was well known to Aaron Klug. He was aware that, for a period of some sixty years from *c.*1840, two prodigiously successful Fellows of Peterhouse—William Hopkins, FRS (1793–1866) and Edward Routh, FRS (1831–1907), his student—were the foremost teachers of mathematics in Cambridge (see Figure 9.12). Between them, they tutored some thirty Senior Wranglers (in mathematics). Among their distinguished progeny were: Francis Galton, G. G. Stokes, A. Cayley, Wm. Thomson (Lord Kelvin), P. G. Tait, J. C. Maxwell, John W. Strutt (Lord Rayleigh), J. J. Thomson, A. N. Whitehead, and J. Larmor. Figure 9.13 shows the portrait of P. G. Tait that nowadays inspires Peterhouse members who enter its main library (where it is now displayed). Tait wrote an important text with his fellow Petrean Lord Kelvin on thermodynamics, of which he was a major authority. Tait is also acknowledged as one of the principal founders of topology and knot theory.

Figure 9.13 *Portrait of Peter Guthrie Tait, FRS (1831–1901), painted by Sir George Reid. For the thirty-one years that Aaron Klug taught undergraduates at Peterhouse, this portrait was prominently placed behind his desk,* 'so as to inspire my students'.

(By kind permission of JET Photographic and Master and Fellows of Peterhouse)

APPENDIX 1

The Visit of Erwin Chargaff to Peterhouse in 1952

In the constituent Colleges of places like the University of Oxford and University of Cambridge especially, not only students, but also Fellows and supervisors at these Colleges derive great intellectual stimulus through the cross-fertilization that takes place from the mingling that occurs at mealtimes with others from different disciplines and different countries.

In addition to intra- and inter-college relationships between individuals from various fields, there is also the great intellectual stimulus that arises from the stream of influential visitors who are attracted to Cambridge. One famous example, which has received much subsequent publicity—in, for example, J. D. Watson's '*The Double Helix*'—is the visit of Professor Erwin Chargaff of Columbia University.

The following story was told by Max Perutz in a talk he gave at the University of Toronto in 1994.[48] He mentioned that '*Our colleague, John Kendrew, happened to be friendly with Erwin Chargaff (1905–2002), the Hungarian-born biochemist at Columbia University, who was working on the chemistry of DNA. One day, Chargaff visited Cambridge and Kendrew invited him to dinner in his college together with Watson and Crick. There, Chargaff drew their attention to a paper he had recently published in an obscure Swiss journal, "Experientia", showing that in the DNA from several different sources the ratios of adenine (A) to thymine (T) and of guanine (G) to cytosine(C) were always near unity.*

It was a vital clue. If DNA was made of two halves, Chargaff's result suggested that A in one chain is linked to a T in its opposite chain, and G in one chain is always linked to C in its opposite chain. The chemical formula of the bases suggested hydrogen bonds…' The rest is history and has often been told—see refs [49] and [50]. It led almost immediately to the solution of the double-helical structure of DNA by Crick and Watson.

A good deal of confusion and debate about this famous confrontation between Chargaff and Crick and Watson has been in circulation among the Cambridge scientists who knew Chargaff—namely, Kendrew and Perutz—and there is much said about him in Watson's popular book '*The Double Helix*'—see also the account by Bretscher and Mitchison.[49] It was John Kendrew, a friend of Chargaff, who arranged a dinner in his room at Peterhouse, to which Crick and Watson were invited to join them later for drinks. The story I heard from Kendrew was that when Chargaff described his rules—adenine to thymine 1:1 and guanine to cytosine 1:1—both Crick and Watson were startled, and Chargaff almost ridiculed them for not reading the literature. In Watson's book '*The Double Helix*', however, Watson says that he knew of those observations by Chargaff earlier. He also disclosed that Francis Crick was hopelessly confused about them and even by the individual nature of the molecules involved in the nucleotides.[51]

Whatever the truth is, Chargaff was not impressed by either the attitudes or the model-building skills of Crick or Watson and their achievement in deriving a convincing picture of the double-helical structure of DNA. Chargaff, it must be remembered, was a deeply cultured man. Like Max Perutz (see Section 8.2.1.2), he was the beneficiary of the Central European Bildungsbürger ethos, which meant familiarity with philosophy, literature, and history. He was unimpressed by the manner and talk of both Watson and Crick. He often, subsequently, ridiculed their achievements and accused them of practising biochemistry without a licence.

APPENDIX 2

Bernal and Picasso

Like so many of their contemporaries, Bernal and Picasso were ardent pacifists and they protested publicly about the disastrous dangers of war, especially with the arrival of bombing of civilian populations, as exemplified by the horrors of Guernica. Other intellectuals who aligned themselves with Bernal and Picasso were Joseph Needham, Dorothy Hodgkin, Aldous Huxley, and P. M. S. Blackett.

An international Peace Meeting was arranged in November 1950 in Sheffield, England. It was to be attended by Bernal, Picasso, and many foreign delegates. The UK Government aborted the Sheffield meeting. So Bernal threw a party at his home in London. Picasso, therefore, stayed in Bernal's home in Torrington Square, London. During the night, he drew the 'cartoon' shown in Figure 9.14. This is a copy of the original, now owned by the Wellcome Trust, made by Bernal's biographer Andrew Brown.[52]

Figure 9.14 shows the drawing made by Pablo Picasso on the wall of J. D. Bernal's home in Torrington Square, London.

Figure 9.14 *A copy made of the mural by Picasso biographer Andrew Brown.*
(Copyright Wellcome Trust)

REFERENCES

1. K. C. Holmes, '*Aaron Klug: A Long Way from Durban*', Cambridge University Press, Cambridge, **2017**.
2. The Royal Commission for the 1851 Exhibition was so successful that it has awarded for over a hundred years the 1851 Scholarships to promising researchers from all over the British Commonwealth. Other recipients have included Ernst Rutherford, John Cockcroft, and Sydney Brenner.
3. (a) J. M. Thomas, *Nature*, **1981**, *289*, 633.
   (b) J. W. Christian, '*Theory of Transformation in Metals and Alloys*', Pergamon Press, Oxford, **1975**.
4. A. Klug, in '*The Legacy of Sir Lawrence Bragg*' (eds. J. M. Thomas and Sir David Phillips), Science Reviews Ltd, London, **1991**, p. 129.
5. J. M. Thomas, *Nature*, **1979**, *281*, 523.
6. '*The Listener*' started in 1929 and was discontinued in 1991. It had culturally profound articles, such as the Reith Lectures and others on English romantic poetry, psychology, and literature, all of which interested Klug.
7. P. L. Kapitza and E. M. Lifshitz, *Biographical Memoirs of Fellows of the Royal Society, L. D. Landau*, **1964**, *15*, 140.
8. A. Klug, '*Chris Calladine and biological structures: a personal account*', in '*New Approaches to Structural Mechanics, Shells, and Biological Structures*' (eds. H. R. Drew and S. Pellegrine), Kluwers Publishers, Dordrecht, **2002**, p. 43.
9. C. R. Calladine, H. R. Drew, B. F. Luisi, and A. A. Travers, '*Understanding DNA: The Molecule and How it Works*', third edition, Elsevier, Amsterdam, **2014**.
10. G-proteins, also known as guanine nucleotide-binding proteins, are a family of proteins that act as molecular switches inside cells and are involved in transmitting signals from a cell to its interior.
11. M. Levitt, *Nat. Struct. Biol.*, **2001**, *8*, 392.
12. P. P. Ewald, '*R. W. James*', *Physics Today*, **1965**, *18*, 154.
13. E. J. Williams and W. L. Bragg, *Proc. Roy. Soc. A*, **1934**, *145*, 699.
14. P. M. S. Blackett, *Biographical Memoirs of Fellows of the Royal Society, E. J. Williams*, **1947**, *5*, 326.
15. Williams was 42 years of age, and the Professor of Physics at the University College of Wales, Aberystwyth, when he died. One of his experiments there involved confirming Yukawa's concept and existence of the pi-meson in nuclear physics. Earlier, Williams had worked in Bohr's Institute in Copenhagen, and during World War II, working alongside Blackett, he made vital contributions to operational research.
16. W. L. Bragg, *Proc. Camb. Phil. Soc.*, **1912**, *17*, 43; *Nature*, **1912**, *90*, 402.
17. R. A. Crowther and K. C. Holmes, *Biographical Memoirs of Fellows of the Royal Society, J. T. Finch*, **2019**, *66*, 183.
18. M. F. Perutz, see ref [4], p. 71.
19. J. T. Finch, '*A Nobel on Every Floor*', MRC-LMB, '*A History of the Medical Record*', Council Laboratory of Molecular Biology, Cambridge, **2008**.
20. H. E. Huxley, '*Memories and Consequences: Visiting Scientists at the MRC Laboratory of Molecular Biology*', MRC-LMB, Cambridge, **2013**.
21. R. E. Franklin, D. L. D. Caspar, and A. Klug, '*Plant Pathology! Problems and Progress (1908–58)*', Golden Jubilee Volume of the American Phytopathology Society, University of Wisconsin Press, Madison, WI, **1959**.

22. A. Klug, J. T. Finch, and Rosalind E. Franklin, *Nature,* **1957**, *179*, 683. See also J. T. Finch and A. Klug, *Nature,* **1959**, *183*, 1709.
23. D. J. DeRosier and A. Klug, *Nature,* **1968**, *217*, 130.
24. R. A. Crowther, *Phil. Trans. R. Soc. A,* **1971**, *261*, 221.
25. R. A. Crowther's classic paper[24] on procedures for determining the three-dimensional structure of spherical viruses from electron microscopy was chosen by the Royal Society as one of its landmark publications in the last 350 years.
26. R. A. Crowther, L. A. Amos, J. T. Finch, D. J. Rossiter, and A. Klug, *Nature,* **1970**, *226*, 421.
27. R. Henderson and P. N. T. Unwin, *Nature,* **1975**, *257*, 26.
28. H. E. Huxley (see Figure 8.10, Chapter 8), even earlier. Huxley was a pioneer in the use of the electron microscope long before Klug and Finch began their studies.
29. K. C. Holmes and A. Weeds, *Biographical Memoirs of Fellows of the Royal Society, H. E. Huxley*, **2017**, *63*, 309.
30. A. Klug, Nobel Lecture, 1982, *Angew. Chemie. Int. Ed.*, **1983**, *22*, 565.
31. W. Baumeister, *Curr. Opin. Struct. Biol.*, **2002**, *12*, 679.
32. P. A. Midgley, E. P. W. Ward, A. B. Hangria, and J. M. Thomas, *Chem. Soc. Rev.*, **2007**, *36*, 1477.
33. C. R. Calladine, *Nature,* **1975**, *255*, 121.
34. This is the view of the eminent MIT chemist and recipient of the US National Medal of Science Stephen Lippard.[35]
35. S. Lippard. See ref [19], p. 305.
36. A. Klug and D. Rhodes, Cold Spring Harbor '*Symposium on Quantitative Biology*', **1987**, *52*, 473.
37. A. Klug, The Robert A. Welch Foundation Conf Ref XXXIX. *The Molecular Basis of Heredity*, **1985**, 133.
38. J. O. Thomas and P. J. Butler, *J. Mol. Biol.*, **1977**, *116*, 769.
39. A. Klug, D. Rhodes, J. Smith, J. T. Finch, and J. O. Thomas, *Nature,* **1980**, *287*, 509. See also D. Kornberg, *Science,* **1974**, *184*, 868.
40. K. Sung-Hou, G. I Quigley, F. L. Suddath, and A. Rich, *Science,* **1973**, *179*, 285.
41. J. E. Ladner, A. Jack, J. D. Robertus, and A. Klug, *Proc. Natl. Acad. Sci. U. S. A.*, **1975**, *72*, 4414.
42. S. A. Jackson, A. Susma, and C. Mitchelatti, *Curr. Opin. Struct. Biol.*, **2017**, *42*, 6.
43. Z. Wang, G. Fan, C. F. Hryc, *et al.*, *eLife,* **2017**, *6*, e24905.
44. C. R. Calladine, H. R. Drew, B. F. Luisi, and A. A. Travers, '*Understanding DNA: The Molecule and How it Works*', third edition, Elsevier, Amsterdam, **2014**.
45. M. Goedert, A. Klug, and R. A. Crowther, *J. Alzheimers Dis.*, **2006**, *9*, 195.
46. B. Falcon, W. Zhang, A. G. Murzin, *et al.*, *Nature,* **2018**, *561*, 137.
47. F. I. Üstok, D. Y. Chirgadze, and G. Christie, *Proteins,* **2015**, *83*, 1787.
48. M. F. Perutz, 'Living molecules' in '*Science and Society*' (ed. M. Moskovits), House of Ananse Press, Concord, **1995**, p. 1245.
49. M. S. Bretscher and G. Mitchison, *Biographical Memoirs of Fellows of the Royal Society, F. H. C. Crick,* **2017**, *63*, 159.
50. See R. Olby, '*Francis Crick: Hunter of Life's Secrets*', Cold Spring Harbor Press, Cold Spring Harbor, NY, **2008**, Chapter 7.
51. J. D. Watson, '*The Double Helix*', W. A. Norton & Co., New York, NY, **1980**, p. 78.
52. Andrew Brown, '*J. D. Bernal: The Sage of Science*', Oxford University Press, Oxford, **2005**.

10

The Summing Up: The Astonishing Successes of the LMB and the Dawn of a New Structural Biological Era

10.1 Introduction

The principal theme that permeates this monograph is the fructifying interactions between the four protagonists—Bragg, Perutz, Kendrew, and Hodgkin, and the vital contributions made by them and several of their contemporaries and predecessors—Astbury, Bernal, Pauling, Klug, Phillips, and especially von Laue, who first demonstrated the reality of the phenomenon of X-ray diffraction. Without a scintilla of doubt, however, it was Lawrence Bragg—not W. H. Bragg AND W. L. Bragg (see Section 7.5.2)—who first realized the immense potential of X-ray diffraction as a new means of investigating, at near-atomic resolution, the nature of solid matter. It was he who produced Bragg's Law; and it was he, and a little later, Dorothy Hodgkin, who fully realized the potential of the idea contained in his father's famous Bakerian Lecture to the Royal Society in 1915, that Fourier's analysis could, in principle, enable X-ray diffraction to reveal the inner structure and atomic constituents of crystalline materials.

Prior to the determination of the structures of biologically important molecules at near-atomic resolution by Perutz and Kendrew, no one knew what enzymes and other proteins looked like, and therefore how such biological macromolecules functioned. Whereas it was their predecessors (who influenced them greatly)—Bernal and Crowfoot (Hodgkin) in 1934—who first demonstrated that X-ray diffraction could, in principle, achieve insights into the structure of an enzyme (such as pepsin) at close to atomic resolution, it was the structural advances registered through the deployment of X-ray diffraction by Kendrew and Perutz at the close of the 1950s that constituted a major landmark in structural biology. Their work, more than that of any other group, spawned the subject of structural molecular biology, the very name and birth of which was, at one time, challenged (even ridiculed).

True, W. H. Bragg, J. D. Bernal, and W. T. Astbury had already perceived (in the mid 1930s at the Royal Institution) that X-ray crystallography held the key to understanding the nature and function of 'living molecules'. It often surprises many modern molecular biologists to learn that it was in a basement laboratory in Mayfair, central London, used earlier by Michael Faraday at the Davy-Faraday Research Laboratory (DFRL), and later in the Department of Mineralogy and Petrography at the University of Cambridge and the Department of Textile Physics at the University of Leeds, that some of the first concerted efforts were made to elucidate the structure of the molecules of the living world.

All four protagonists of X-ray crystallography—W. L. Bragg, Perutz, Kendrew, and Hodgkin—in the early stages of their scientific adventures, as well as later, had connections with the DFRL, as did their later crucial collaborators Arndt and D. C. Phillips (see Sections 6.5 and 6.6).

10.2 The Relaxed Pre-World War II and Immediate Post-War Years, and the Free Exchange of Ideas

It is both endearing and heart-warming to reflect that the exchange of ideas, samples, and results, as well as the cooperation among the protagonists and their contemporaries, during the genesis of structural molecular biology in the 1930s and 1940s was conducted—with rare exception—in an atmosphere of exciting, selfless collaboration, rather than that of feverish (sometimes cut-throat) competition that tends nowadays to prevail. The qualities of human decency, generosity of spirit, and sense of exhilarating adventure in the pursuit of knowledge for its own sake infused the activities of the early investigators. Recall the words of both Kendrew[1] and Earl Mountbatten[2] (see Section 1.3), when describing the attitude of J. D. Bernal: '*He had an infectious delight in new ideas, whether his own or another's; the question of credit did not arise…Other people's results gave him as much pleasure as his own*'.[3]

And again, describing Bernal:[2] '*…perhaps his most pleasant quality was his generosity. He never minded slaving away at other people's ideas, helping to decide what could or could not be done…*'

We recall also that Bernal was himself the recipient of much spontaneous generosity. Thus, samples of the enzyme pepsin were sent to him by an altruistic friend in Sweden. And in his early days in the Department of Mineralogy and Petrography at Cambridge, he was deluged with samples of sterols, viruses, amino acids, and a host of other 'living' molecules (see Section 3.5.1) from colleagues in sister departments at Cambridge, as well as from organic chemists and biochemists from several British, German, and Swiss universities.

Ideas and excitement flowed freely between the geographically separated participants. The joy of discovery is apparent when one consults the early publications of pioneering structural molecular biologists. In Chapter 3, reference was made to one of Astbury's papers,[3] in which he and his colleagues, on realizing (while studying the clotting of blood) that fibrinogen and fibrin are myosin-like and were therefore members of the keratin group, felt impelled to compose a beautifully worded letter to *Nature*[3] where, as mentioned in Chapter 3, they expressed their delight at discovering the commonality of the protein structures in distinct manifestations of the living world. That letter ended with the magical passage, so rare in present-day scientific literature: '*...all spring from the same peculiar shape of molecules and are therefore probably all adaptations of a single root idea, we seem to glimpse one of the great co-ordinating facts in the lineage of biological molecules.*'

Perutz, in describing how to manage a successful scientific laboratory,[4] mentions that openness and mutual excitement in the success of others were not only a feature of Niels Bohr's prodigiously successful Mecca for physicists in Copenhagen in the 1930s, but also one that he himself succeeded in creating in the early days of the Laboratory of Molecular Biology (LMB). Visitors to his laboratory, especially from the United States, as may be gleaned from numerous testimonies in the compilation by Hugh Huxley[5] at the fiftieth anniversary of the founding of the LMB, commented approvingly on the way Max Perutz would organize visits for his colleagues to the DFRL to witness the revolutionary work of the Phillips–North team on the structure of lysozyme or to Dorothy Hodgkin's laboratory in Oxford to taste and share the exhilaration of the achievement in (her) solution of the structure of insulin.

The present Director of the Center for Molecular and Cellular Dynamics at Harvard Medical School and Head of the Laboratory of Molecular Medicine at Boston Children's Hospital Stephen C. Harrison recalled[5] the excitement that overtook him in 1965 when Max Perutz organized a visit to the DFRL at the Royal Institution to see David Phillips' model of the lysozyme structure determined there. Harrison also remarked that while '*most of us stared at the model, Francis (Crick) seemed more interested in ogling Miss Johnson.*' Miss Johnson later became Dame Louise Johnson (see Section 8.4.4). Harrison also said,[5] '*I was excited to see β-strands and disulfide bonds.*' These had not hitherto been detected by X-ray diffraction.

Another distinguished US molecular cell and developmental biologist Harry E. Noller of the University of California, Santa Cruz, recalls the trip that Max Perutz organized to see the work on lysozyme at the DFRL. In his chapter in Huxley's book,[5] Noller re-lives the description of how David Phillips revealed the sequence of amino acids in the lysozyme chain that appeared (in technicolor) from the high roof of the lecture theatre—see Louise Johnson's '*magical evening*' described in Section 6.5.1—and how, by ingenious lighting effects on the three-dimensional model, '*we all saw the blue lights blink on the interior of the enzyme and red lights on the surface...we were stunned.*'

As Ramakrishnan has recently described,[6] nowadays the human messiness of science and the corrupting influence of prizes are all too apparent, and rivalries and jealousies abound. This is in contrast to the atmosphere that pervaded the LMB in 1964 when the distinguished US biochemist Ignacio Tinoco (1915–2016) of the University of California, Berkeley, worked there. He reflected[7] on the free discussions that he witnessed there: '*All sorts of problems were discussed, some profound, but most trivial. However, what came through in the conversations was that everybody was excited about their work and wanted to share their excitement, or their frustrations. There did not seem to be fear that anyone "would steal their idea" or that somebody else would point out how stupid they were for not knowing an obvious fact. You would be told that a critical control had been omitted, or that a more clear-cut result would result if the experiment were done differently, but it was clear that the goal was to produce better science, not to put you down. The attitude in the LMB seemed to be: it is us against the world. We competed strongly against the world (and usually win), but within the lab we are only collaborators.*'

Later, in his article, Tinoco addresses some of the less attractive aspects of current molecular biological and other research that have also been discussed by Ramakrishnan:[6] '*I still try to continue the environment that I learned at the LMB where the researchers in my group don't compete, but collaborate. I observe, unhappily, that in top-ranked American universities students and postdocs, even in the same group, compete with each other. A student is afraid to mention an idea for fear that it will be presented to the professor by a fellow student as their idea. Everybody is too busy to help a newcomer. Everybody is too busy to have coffee or lunch together, and nobody drinks tea. The professor has lunch in the office while writing grants, and the student's lucky to talk to him or her once a week. Scientific ability is measured quantitatively by the amount of funding obtained, and by first or last author papers in Nature and Science…the ultimate measure of your contribution to science is your H-index of cited publications. Whether your most highly-cited paper is wrong or irrelevant.*' It was pointed out to me recently by Ramakrishnan that Fred Sanger's H-index was 18. (A double Nobel Prize winner.) These days, this low value would not secure him a lectureship in most universities. (The H-index of Bernal, Astbury, W. L. Bragg, and Kendrew were, respectively, 29, 30, 37, and 29. Max Perutz's and Pauling's were 81 and 100. They each continued publishing well past the age of eighty.)

Professor David Eisenberg, now Head of the Institute of Genomics and Proteomics, UCLA, recalled that his sabbatical at the LMB in 1985 was one of the most stimulating and enjoyable of his career: '*The whole atmosphere was geared for discovery: an absence of red tape, an abundance of resources, a crowd of intelligent and curious colleagues, and a designed social system featuring a high concentration of brain power in a small canteen at coffee, lunch and tea every work day.*'

10.3 A Summary of the Key Milestones Reached at the LMB

Even before the Perutz–Kendrew collaboration began to succeed, substantial significant advances had been made, using the simple principles of X-ray diffraction, to elucidate the nature of some important aspects of 'living molecules'. Thus, Bernal, when he read the details of the structures proposed by the two German organic chemists (who were later to win Nobel Prizes) Wieland and Windaus (see Section 3.5.2) for the framework structure of steroids, could indisputably reject their validity. Using straightforward unit cell measurements by X-rays, as well as knowledge of the observed space groups (see Chapter 2) of crystalline steroidal compounds, he could incontrovertibly assert that the Wieland–Windaus proposed structures were erroneous (see Figure 3.13 of Chapter 3).

Furthermore, well before Perutz and Kendrew embarked upon their work, Bernal and Crowfoot had demonstrated the ease with which, from the accurate measurement of the density of a crystal and a knowledge of its unit cell volume, the molecular weight of complicated organic and biochemically relevant molecules could be accurately determined (see Section 2.3)—see also the Appendix to Chapter 2. This was an early triumph for simple X-ray crystallography, which helped further the early scientific career of Dorothy Hodgkin. It also enabled her, in record time, to determine the molecular weight of vitamin B_{12} from her simple crystallographic measurements.

10.3.1 1953: *annus mirabilis*

Few would dispute that the single most revolutionary advance in structural molecular biology to emerge from the Perutz–Kendrew school was the double-helix structure of DNA in 1953. This was, in a sense, a miraculous year for structural molecular biology, since the structure of DNA promised, in principle, how to explain heredity and replication. Max Perutz showed, also in 1953, that, contrary to previous expectations, the heavy atom phasing technique would indeed work on protein crystals and enabled their nature to be elucidated by X-ray diffraction.

In 1953 also, John Kendrew began organizing the equipment, assistants, and computing to do X-ray diffraction at high resolution—a very major undertaking. And although Hugh Huxley was, by then, at the Massachusetts Institute of Technology (MIT), he, too, made major advances in his electron microscopic study of muscle cross-sections and confirmed his earlier (Cambridge) work on separate filaments of actin and myosin.

Because so much has already been written—and is available elsewhere—about DNA and also about the work commenced in 1957 by Crick and Brenner on establishing the non-overlapping triplet code, and many other features of the translation machinery of DNA via RNA into proteins, hardly any coverage has been given in this monograph to the structural subtleties of RNA and DNA and the whole escalation of molecular biological phenomena, especially genetics, that flow from them. It is a common belief, though I contend a fallacious one, that molecular biology as a subject is essentially synonymous with all the structural, biological, and medical properties of DNA. But this is patently not true, for there are enormous areas of biology where proteins still hold sway, notwithstanding the pivotal importance of DNA and ribosomes. And the early preoccupation by Bragg, Bernal, and others—especially those like Hodgkin, Perutz, and Kendrew whom they influenced the most—with proteins still remains highly relevant in view of the ubiquity of proteins in the corpus of modern biology.

10.3.2 Milestones reached since 1957

To continue on the roll-call of significant milestones registered by the Cambridge LMB, we note that in:

- 1957—the realization, unambiguously established by the experimental work of Vernon Ingram, that a single amino acid change causes sickle-cell anaemia (one of the first molecular diseases to be identified—see Chapter 7);
- 1959—John Kendrew and colleagues produced the first atomic resolution map of a protein (myoglobin);
- 1959—Perutz and his team were able to map out the structure of haemoglobin;[8]
- 1968—Aaron Klug and his colleagues showed how a series of two-dimensional electron micrographs could be reconstructed to yield a three-dimensional picture of the specimen (such as viruses) under investigation;
- 1975—monoclonal antibody methodology was invented by Milstein and Köhler (see Appendix 1 to this chapter); Klug *et al.* published the structure of *t*RNA;
- 1975—Unwin and Henderson, using electron microscopy, determined the first structure of a membrane protein (bacteriorhodopsin), a feat that earned them the Aminoff Prize of the Royal Swedish Academy of Sciences;
- 1977—Sanger worked out a new method that enabled him to sequence the nucleotides in DNA;
- 1985—Aaron Klug discovered zinc fingers, an important DNA-binding motif;
- 1986—as a consequence of the early work of Milstein and Köhler, Greg Winter's group at the LMB produced the first specimens of humanized antibodies;
- 1987—the MRC confocal microscope first went into commercial production;
- 1987—Klug, Crowther, and Finch analysed by electron microscopy the map of the filaments found in Alzheimer's disease and identified them as paired helical filaments (phf) that were subsequently identified by M. Goedert as tau proteins;
- 1988—the first patient was treated with the humanized antibody known as Campath-1, work that flowed from discoveries by Greg Winter;
- 1989—the first LMB spin-out company Cell Tech was set up by Brenner and facilitated by the change of policy of the MRC, engineered by its Chief Executive Sir Dai Rees (see Section 8.4.8);

- 1997—Walker and colleagues solved, by X-ray diffraction, an important sub-complex of the mitochondrial enzyme known as ATPase.[9] Walker was awarded the Nobel Prize in 1997. The origins and propagation of life on earth depends on developing an ability to generate adenosine triphosphate (ATP). Each human being makes and expends about 60 kg of this 'fuel' every day of its life;
- 2000—using synchrotron radiation for his X-ray diffraction studies, Ramakrishnan solved the structure of the so-called 30S ribosomal subunit, and subsequently (2000–2009) solved the structure of the entire 70S ribosome, work that earned Ramakrishnan the Nobel Prize in 2009;
- 2014—electron cryo-microscopy came into its own at the LMB, led by the work of Henderson and his colleagues. Henderson was awarded the Nobel Prize in
- 2017—'*for developing cryo-electron microscopy for the high-resolution structure determination of biomolecules in solution*';
- 2018—award of the Nobel Prize to Greg Winter for the phage display of peptides, which led to the development of monoclonal antibodies as powerful new drugs.

For reasons explained earlier, I have omitted mention of such milestones as mutants of the nematode worm and the Nobel Prize-winning work of Sydney Brenner and John Sulston in molecular genetics.

The new MRC Laboratory of Molecular Biology in Cambridge was opened in 2013, the cost of the building having been met, as described in Section 10.5, by the royalties generated from the commercial successes that flowed from the work of Winter and Neuberger.

10.4 The Academic Successes of the LMB

At present, January 2019, twelve Nobel Prizes have been awarded to scientists who were permanently employed at the LMB, the most recent being to Greg Winter (in Chemistry, 2018) for his pioneering work on monoclonal antibodies and their use as powerful new drugs. This is further testimony to the atmosphere and attitudes that pervade the laboratory and which were the direct legacy of Perutz and Kendrew, especially the former. Several accounts and analyses have been given elsewhere for the notable success of the Laboratory—see, in particular, de Chadarevian,[10] Finch,[11] and Huxley.[5] The collection of forty-one articles that were written by former visiting scientists, principally from the United States, but also from Australia, Canada, Japan, Venezuela, and Germany, all convey the special nature of the LMB and how important a part it played in their development as successful molecular biologists. Eleven LMB alumni have received Nobel Prizes, and very many others have had bestowed upon them other distinctions, such as US Medals of Science, and membership of the foremost academies of the world.

10.5 The Commercial Successes of the LMB

The LMB has become outstandingly successful in a commercial sense and in fostering the transfer of technology for the common good. Illuminating accounts of the manner in which research at the LMB led to commercial exploitation have been given by John Finch[11] and, more recently, by de Chadarevian.[10,12] As a prelude to summarizing these accounts, it is prudent to recall how attitudes to commercial exploitation and entrepreneurial activity have changed over the years since the inception of the MRC. In the Royal Charter establishing the MRC, there was no mention of industrial relationships.

For a considerable time thereafter, well into Max Perutz's period as the head of the LMB, research was pursued for its own good, and few, if any, of the personnel were interested in entrepreneurial activities. Fred Sanger, for example, did not patent his method of sequencing the bases in DNA, for which he was awarded the Nobel Prize in 1980. Perutz himself often told me that, like Rutherford and Faraday—two of his scientific heroes—he was proud not to have filed a single patent.

Rather dramatic, and sobering, accounts have been given by both de Chadavarian[10,12] and Finch[11] of the unfortunate relationship that developed (see ref [10]) between César Milstein—one of the discoverers of monoclonal antibodies—and the National Research and Development Corporation (NRDC) of the UK, the public body set up by the Government to exploit new inventions for the national interest.

In de Chadarevian's article,[13] entitled '*The Making of an Entrepreneurial Science Biotechnology in Britain, 1975–1995*', a nuanced and detailed account is given of how successive Directors of the LMB, notably Sydney Brenner, and especially his successor Aaron Klug, and key individuals such as Milstein and his successor as one of the prime authorities and creators of monoclonal antibodies, Greg Winter, gradually led the way to the extraordinary commercial successes of the LMB. Accounts are given by de Chadarevian[12] of the financial benefits of work done at the LMB on monoclonal antibodies. Two large pharmaceutical firms—AstraZeneca and GlaxoSmithKline—acquired companies (such as Cambridge Antibody Technology and Domantis founded at the LMB) in 2006 (see Figure 10.1). The MRC, as well as the LMB and the scientists involved, received significant revenues from share sales and royalties from the two companies. Between 2000 and 2006, the sale of human antibodies alone resulted in revenues for the MRC of £127 million. The gross annual income apportioned to the LMB in the period 1995–2008 was comparable to the entire commercial income of all UK universities, as well as on a par with that of much larger US academic institutions such as MIT.[13] Monoclonal antibodies now make up a third of all new drug treatments[14] (see Appendix 1 and Appendix 2 to this chapter).

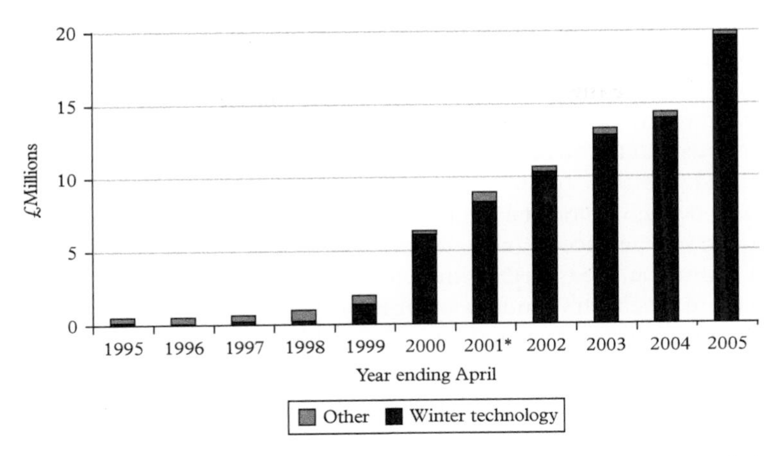

Figure 10.1 *An indication of the magnitude of income from royalties that emerged from work done at the LMB.*

(Source: LMB Quinquennial Report, *Cambridge, 2005,* Vol 1, *p. 28, Figure* 8.1. *Courtesy MRC-LMB)*

10.6 Outline of a New Structural Molecular Biology Technique: The Era of Electron Cryo-Microscopy

Overwhelmingly, ever since W. L. Bragg's pioneering work over a century ago, and the major breakthroughs achieved by Perutz, Kendrew, and Hodgkin in the 1950s and almost all their successors ever since, the structures of 'living molecules'—or, as W. H. Bragg (see Chapter 3) designated them '*Crystals of the Living World*'—have been determined by X-ray diffraction. As mentioned in earlier chapters (see, in particular, Section 5.1), the arrival in the post-1960 era of several other techniques, especially those that utilize synchrotron radiation, has nowadays made it almost routine to solve complicated biological structures by X-ray diffraction, using the principles derived from those pioneered by Perutz, Kendrew, and Hodgkin. Even structures as complicated as those of the ribosome—the large macromolecular complex that translates the genetic code to make proteins—can be solved, as was done by Ramakrishnan and others,[15–17] using the high intensities of X-rays (and their variable wavelengths) available in synchrotrons.

As may be gleaned from *YouTube* and other popular (Internet) sources, every living cell contains millions of ribosomes, which are complex structures consisting of both a large and a small sub-unit, and also incorporates dozens of proteins and a few strands of RNA. Ribosomes are protein factories, in which the genetic instructions encoded in DNA are transcribed into *m*RNA and translated from there into proteins (using *t*RNA and ribosomes). So the proteins that are both the building blocks of living matter and also the agents that enable life to function and flourish are created in the ribosome.[6,15]

X-ray studies, of the kind introduced much earlier by Perutz, Kendrew, and Hodgkin made it possible, inter alia, to study the binding of antibodies of the ribosome and to facilitate the discovery of new medicines.

Of late, however, X-ray crystallography—which was the technique that led to the very existence of structural molecular biology—along with nuclear magnetic resonance and mass spectrometry as powerful supplementary structural techniques, is gradually being superseded by the exciting and readily available technique of electron cryo-microscopy (also termed cryo-electron microscopy). It was the principal technique used by the three individuals who were awarded the 2017 Nobel Prize in Chemistry. So important is electron cryo-microscopy as a structural tool in molecular biology that we outline below its quintessential features and comment briefly on its few weaknesses. As an indicator of its growing importance, Table 10.1 summarizes some salient points.

The total number of electron cryo-microscopy images uploaded to the Electron Microscopy Data Bank (EM DB) has increased from a value of eight in 2002 to 1106 in 2017 (copyright Prof R. Henderson, MRC-LMB).

10.6.1 Nature and examples of electron cryo-microscopy

In Chapter 9, a brief description was given of the electron microscope as an instrument for the determination of structure. There are three attractive advantages associated with electron cryo-microscopy:

1. First, no crystals of the material to be studied are needed. It suffices to have dispersions of the molecules, or aggregates thereof, at very low concentrations (micro- or nanomolar) distributed on an electron microscope grid within an environment of vitreous water i.e. ice film. The material to be studied is, effectively, in solution (in vitreous ice).
2. Second, the technique allows the macromolecules to be structurally characterized under conditions very similar to those that apply in their natural,

Table 10.1 *Numbers of sets of co-ordinates deposited by the Laboratory of Molecular Biology (MRC, Cambridge) in the Protein Data Bank (PDB)*

Year	X-rays	Electron microscopy	Nuclear magnetic resonance
2006–07	66	0	14
2008–09	96	1	19
2010–11	108	0	12
2012–13	104	19	7
2014–15	125	77	11
2016–17	102	81	5

(After R. Henderson, private communication)

biological conditions. (For X-ray diffraction studies, the conformation of the macromolecules is governed by the forces that affect the way in which the molecule crystallizes, and this can be different from the conformation of the molecule under physiological conditions.)

3. Third, it attains near-atomic resolution in the determination of the structures of larger, complex assemblies that are of vital interest to all molecular and cell biologists.

This revolution in electron cryo-microscopy has occurred because of several techno-scientific advances. These include the important development of what has become known as *single-particle* cryo-electron microscopy (EM). The article by Cheng, Glaeser, and Nogales[18] entitled *How Cryo-EM Became So Hot* summarizes the crucial contributions in the arrival of this powerful technique of the following scientists: Saxton and Frank,[19] Taylor and Glaeser,[20] Henderson and Unwin,[21] Dubochet *et al.*[22], Faruqi *et al.*,[23] and Bai *et al.*[24]

It is instructive to emphasize that there are three kinds of electron cryo-microscopy.[25] First, the study of two-dimensional monolayer crystals, such as bacteriorhodopsin, first solved by Unwin and Henderson[21] in 1975. Second, the 'single-particle' method for the study of macromolecules pioneered by Frank and others,[19] based on combining electron microscope images of many of the same molecules, embedded in a vitreous ice layer about 60 nm thick. Each molecule will be lying in a different orientation so that these many random projections may be merged to provide one three-dimensional reconstruction. Third, there is the method of tomography (see Figure 9.7), in which many views of projection of the same large molecule (or virus) are merged into a three-dimensional construction. The cryo-electron tomographic method also enables whole cells to be imaged, a significant advance since it unifies molecular and cell biology.

Because of the additional radiation damage produced when many images are recorded from the same structure using repeated exposures, the resolution of the tomographic method is currently limited to about 2 nm at best, while the two-dimensional crystal method offers the highest resolution. The lower limit on the size of molecules which can be studied by the 'single-particle' method is steadily decreasing.

Dubochet *et al.*'s role was to show how readily any biological specimen could be distributed as essentially single particles in a vitreous ice environment by using liquid ethane for sample preparation. This made the practice of recording diffraction patterns and images straightforward. Frank *et al.*[19,26] followed on from the pioneering work by Crowther and colleagues[27] on viruses, in retrieving the structure from an analysis of the intensities of diffraction from hundreds of thousands of small assemblies and, by adroit choice of algorithms, constructing the detailed structure of the specimen. A user-friendly illustration[28] of the method is shown in Figure 10.2.

With electron cryo-microscopy, it is now possible to solve structures of molecules that previously had proved inaccessible to X-ray diffraction—for example,

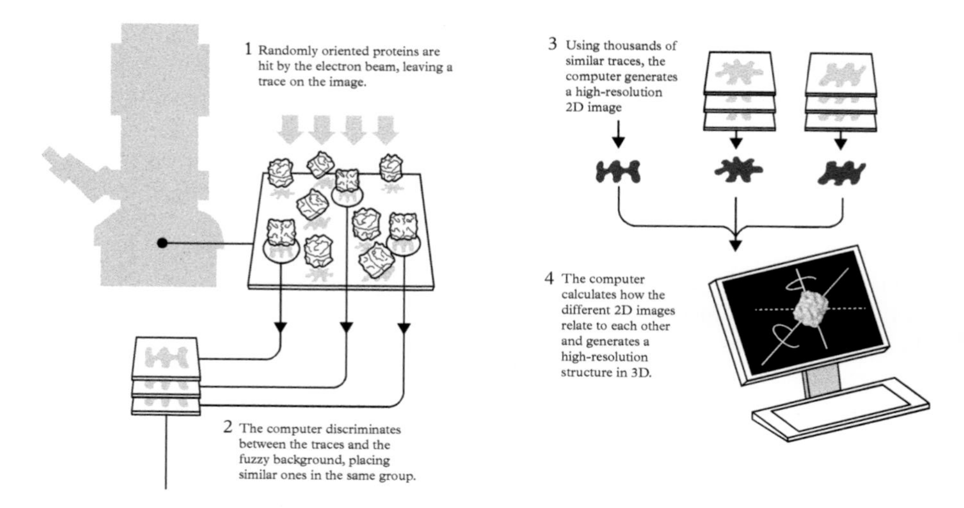

Figure 10.2 *A simplified description of the essence of Frank's image analysis for the determination of three-dimensional structures using the single-particle approach. See text.*[28]
(Courtesy Royal Society of Chemistry)

because of the difficulty in producing suitable crystals. Membrane-bound proteins, for example, are very difficult to study by X-ray diffraction. Additionally, the new cryo-microscopy can solve structures that are too large to study by nuclear magnetic resonance.

It is also possible[25] to follow the likely changes in the shape of molecules with time using single-particle methods. The many molecules in the field of view will actually vary both by orientation and by the small changes in shape or conformation which take place under the conditions of thermal equilibrium present immediately before they were frozen in the ice film. Frank's algorithm, and its development, achieves the remarkable feat of sorting the molecular images both by differences in orientation (from two-dimensional projections) and by small changes in shape. After sorting and merging, the three-dimensional images may be placed side-by-side according to similarity, thereby giving an indication of the continuous changes in shape which may be related to function over a small range of energies.

Henderson,[29] who was largely instrumental in developing the ultra-sensitive direct electron detectors used in electron cryo-microscopy, has explained the quantum leap that has recently occurred in structural biology is due to three factors: improved electron microscopes, better electron detectors, and superior software. Already it is possible to obtain resolutions of approximately 1.6 Å, and recent work in Japan and elsewhere is geared to attaining 1.0 Å resolutions in future structural studies of biological molecules.

In the past four years, this electron microscope technique has generated numerous new structural insights into a wide variety of systems of central importance to molecular and cell biologists and to medical scientists in general, as well as to pharmacologists.[29] It has shed light, for example, on the factors that are involved in diseases such as Alzheimer's disease [30] and the related neuro-disorder known as Pick's disease, which is a syndrome featuring shrinkage of the frontal and temporal anterior lobes of the brain.[31] In both Alzheimer's and Parkinson's disease, the presence of so-called tau proteins[32,33] that have become defective and no longer stabilize microtubules in the central nervous system is thought to be significant. Some six or so tau proteins are known. All six seem to be implicated in Alzheimer's disease; in Pick's disease, however, only three of them are implicated.

There is yet another neurodegenerative disease which is now being explored by electron cryo-microscopy, and it has been the subject of study at the LMB, Cambridge, in association with structural biologists elsewhere; it is known as chronic traumatic encephalopathy (CTE), which is associated with repetitive head impacts or exposure to blast waves. It was first described as punch-drunk syndrome and *dementia pugilistica* in retired boxers as early as 1928 and explicitly identified in 1973 as the aftermath of boxing. No disease-modifying therapies currently exist, and diagnosis requires an autopsy.[34] Figure 10.3 reveals the quality of the high-resolution information that electron cryo-microscopy was able to retrieve about the nature of CTE.[34]

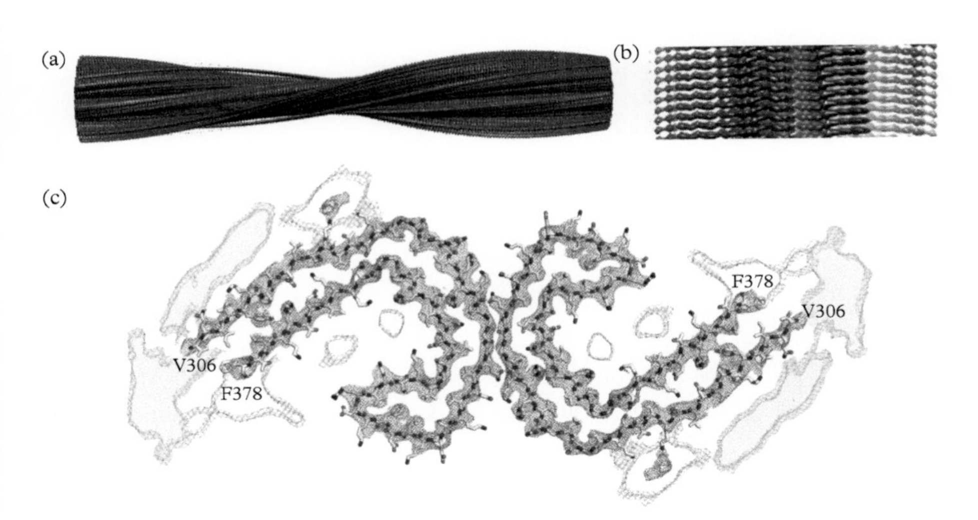

Figure 10.3 *In Alzheimer's disease, degenerating neurons contain characteristic tangles of abnormal paired helical filaments made of tau proteins. (a) General view of part of a filament. (b) Detailed side view showing transverse strands, with a 4.7 Å axial spacing defining the so-called cross-β fold. (c) Section of the map computed from electron cryo-micrographs showing the part of the tau molecules from residue number 306 to residue number 378 that forms the ordered core of the filament.*

(Courtesy Prof R. A. Crowther)

This work is a significant advance in that the increased resolution of the electron cryo-microscopy technique is beginning to disclose functionality to the structures that is disease-specific.

Many other molecular biological problems, such as the nature of the proteinaceous surface layer of prokaryotic cells, have also been greatly clarified[35] by

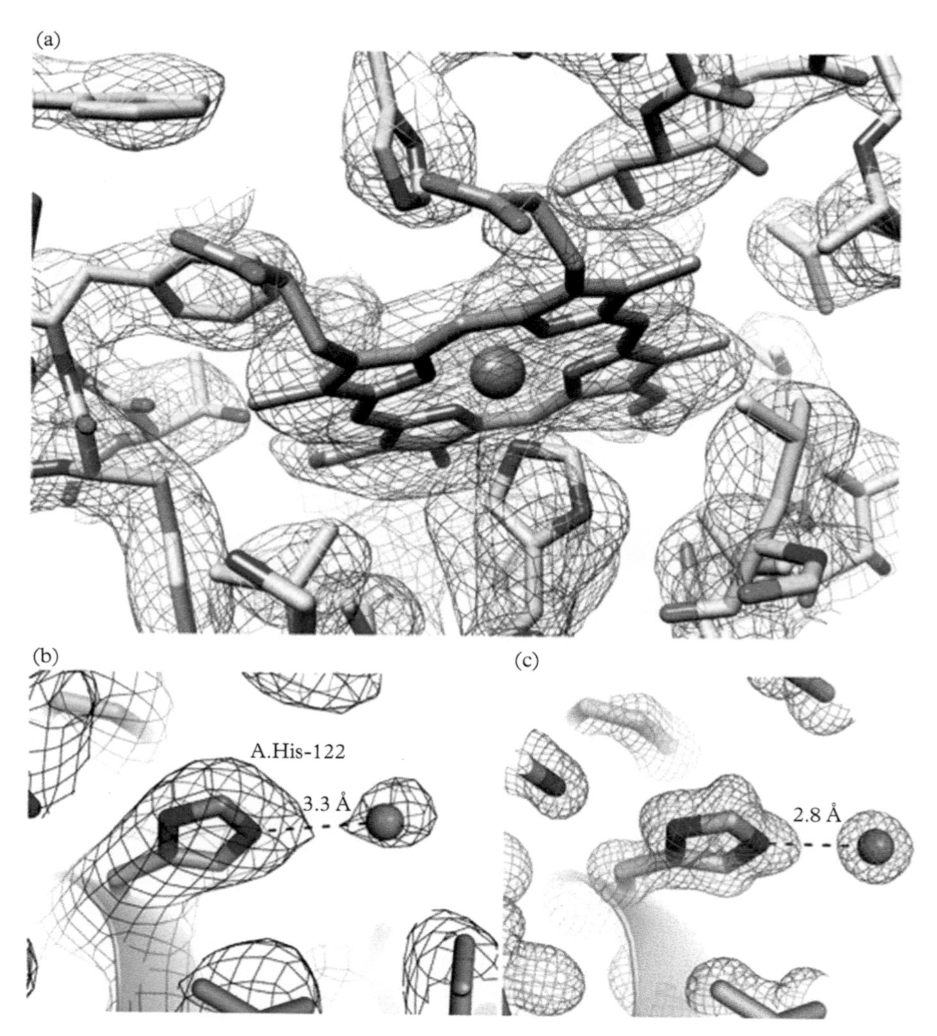

Figure 10.4 *Human haemoglobin at 3.2 Å resolution. (a) Visibility of prosthetic haem (iron atom and proximal histidine—see Chapter 5). (b) Reconstructed three-dimensional density map showing a water molecule in the same location as shown in the X-ray diffraction-derived structure, depicted in the Protein Data Bank (after Khoshouei* et al.[36]*).*

(Courtesy Elsevier)

recent application of electron cryo-microscopy by Jan Loewe *et al.*[36] (A prime example of the new technique is contained in Luisi's work, described in Chapter 9.) In addition, Koshouei *et al.*[37] have re-visited the structure of haemoglobin with electron cryo-microscopy (Figure 10.4).

It is instructive to compare Figure 10.4 with the X-ray diffraction-derived structure published by Kendrew (see Figure 5.6 of this monograph[38]). It has taken some sixty years for the electron microscope technique to rival what could be done hitherto via X-ray diffraction. Now, however, as outlined above, and as demonstrated by the work of Luisi and co-workers, shown in Chapter 9, electron cryo-microscopy is set to supplant X-ray-based methods of structure retrieval—not only because of higher resolution, but also in the ease of operation of the technique.

10.6.2 A cautionary note

While there is currently great enthusiasm for electron cryo-microscopy and almost frenetic use of it in many laboratories of molecular biology throughout the world, with veteran X-ray crystallographers turning to it in lively numbers, there is also call for caution. This has been highlighted in a recent article in *Nature*.[39] At present, the electron cryo-microscopy field lacks the kind of standardized tools for producing robust structural models that X-ray crystallographers developed as their field matured. Leading exponents of the technique, including Henderson (Cambridge), Stark (Göttingen), Lander (Scripps Research Institute), and others from industrial laboratories such as Novartis, have expressed concern about a certain degree of sloppiness that is sometimes associated with high-resolution structures published using the new technique. The manager of the EM DB Dr Patwardhan [at the European Bioinformatics Institute (EBI) in Cambridge] has called for better methods of validating electron cryo-microscopy maps and their associated models.[39,40]

10.7 A Final Thought

It seems appropriate to conclude the final chapter of this monograph by showing the images of three giants of macromolecular structural studies who were instrumental in establishing, more than half a century ago, the excellence of the LMB (see Figure 10.5, panel (a)), alongside three outstanding contemporary successors (see Figure 10.5, panel (b)). Perutz, Kendrew, and Sanger were photographed at Peterhouse on the occasion of Kendrew's eightieth birthday in 1997. They set up and nourished the LMB in such a way that it still attracts universal admiration, as does the work, these days, of Ramakrishnan, Henderson, and Winter, and all their colleagues.

Figure 10.5 *(a) Max Perutz, John Kendrew, and Fred Sanger at the Fellows Garden, Peterhouse, 1997. (b) Three recent Nobel Prize Laureates from the LMB: Venki Ramakrishnan (2009), Richard Henderson (2017), and Greg Winter (2018).*

(a) Courtesy JET Photographic and Master and Fellows of Peterhouse; (b) Courtesy MRC-LMB

APPENDIX 1

Antibodies, Developed at the LMB, and Modern Medicines

HUMIRA®, an acronym for *HU*man *M*onoclonal *A*ntibody *I*n *R*heumatoid *A*rthritis, is the world's best-selling drug. It helps to treat half a million patients across the world. The technologies underpinning this drug, and several other extensively used medicines, including the drugs Lemtrada® for treatment of multiple sclerosis and Herceptin® and Opdivo® for treatment of breast and lung cancers, respectively, emerged from the work of two scientists at the LMB—Greg Winter and the late Michael Neuberger. Their work, in turn, followed on from that of their predecessors César Milstein and Georges Köhler, who were awarded the Nobel Prize in 1984 for the discovery of monoclonal antibodies. (Monoclonal antibodies are antibodies that are made by identical immune cells that are clones of a unique parent cell. Given almost any substance, it is possible to produce monoclonal antibodies that specifically bind to that substance. They can then serve to detect that substance.)

Antibodies are proteins that are used by the immune system to fight foreign invaders in the body such as bacteria, viruses, and toxins. When Milstein, an Argentinian, began his work at the LMB, he was aware that the body produces billions of antibodies. He also knew that antibodies play a crucial role in helping the body fight disease and that each antibody recognizes a specific harmful agent (antigen). Not only can this property be used to protect against the agents of disease, but it also can provide diagnostic tests, for example for cancer cells or for hormones produced in pregnancy or for viruses such as HIV.

Antibodies are produced by a type of white blood cell called B lymphocytes. As shown in the figure below, in their structure, they have a constant region, which is the same for each antibody molecule, and a variable region which is unique to the antibodies produced by a single B lymphocyte.

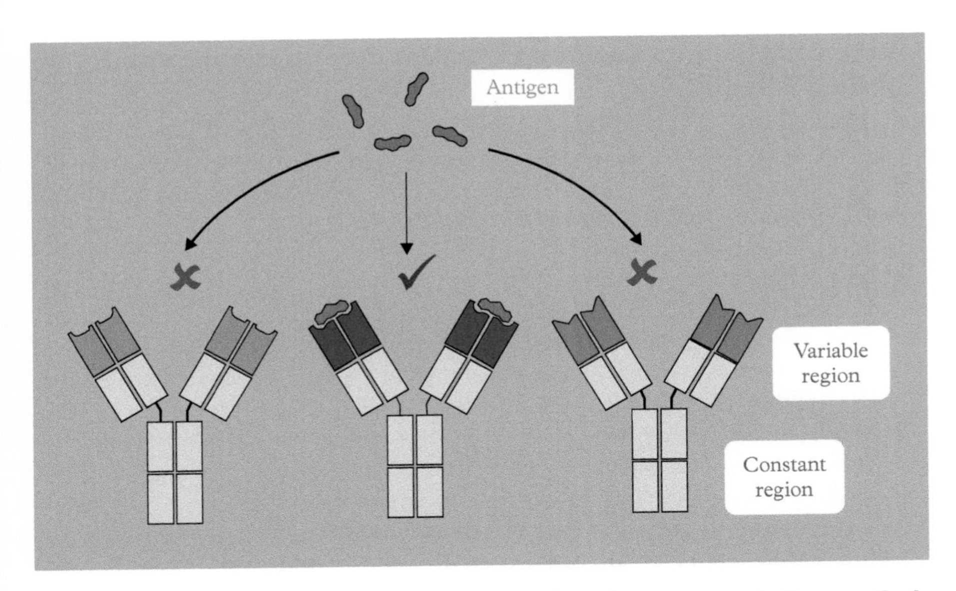

Figure 10.6 *B cells producing an antibody that binds the antigen most strongly (for example, the antibody with the blue variable region in the picture) are selected to expand and produce more of the antibody to help counter the infection.*

(Courtesy MRC-LMB)

Antibodies bind to antigens, fragments or proteins from infectious agents such as viruses and bacteria. Such B cell produces an antibody that can bind one type of antigen. There are billions of B cells, each producing a unique antibody.

B cells producing an antibody that binds the antigen most strongly (for example, the antibody with the blue variable region in the figure) are selected to expand and produce more of the antibody to help counter the infection. Using genetic engineering to modify the proteinaceous antibodies, including a technique called site-directed mutagenesis that earned an alumnus of the LMB—Michael Smith from Canada—the Nobel Prize in 1993, Greg Winter and his colleagues developed monoclonal antibodies suitable for treating human diseases.

To understand how Humira® works against rheumatoid arthritis, we first recall that a protein called tumour necrosis factor (TNF)-α is a molecule that is normally produced by immune cells when fighting an infection. It causes inflammation which helps to clear the infection and ultimately heal the affected tissue. In arthritis, too much TNF-α is produced and this leads to unwanted and persistent inflammation, resulting in tissue damage. Humira® is able to bind TNF-α, thus preventing it from reaching tissues, and therefore stops or limits inflammation.

APPENDIX 2

How the LMB Came to Commercialize their Discoveries and Earn Royalties for the MRC

In mid October 2018, I interviewed the two key scientists who were largely instrumental in enabling the MRC laboratories to gain royalties from the commercialization of their discoveries: Sir Dai Rees and Sir Greg Winter. The following remarks summarize the information that they provided.

- In the mid 1980s, when he was Director of the National Institute of Medical Research (NIMR) in Mill Hill, UK, Sir Dai Rees set up a Collaboration Centre, which encouraged private companies to pursue research at the NIMR, with a view to exploiting research advances for the general good, for the health of the British people, and with the aim of earning royalties for the MRC and addressing the concerns of the personnel who are making exploitable advances. One such advance was the humanization of antibodies using the constant domain region (CDR) of the antibody (see figure in Appendix 1) for grafting (according to the Winter patent). At least three commercially successful antibodies emerged from this work: Actemra®, Tysabri®, and Keytruda®. The royalties for Keytruda® were capitalized recently and form the £500 million funds that LifeArc is now investing in the UK to generate, fund, and champion innovations in antimicrobials, neuroscience, personalized oncology, and respiratory medicine.

- As Chief Executive of the MRC (from 1987 to 1996), Sir Dai (with the support of the LMB Director Sir Aaron Klug) managed to reverse the proposed assignment of the Winter CDR-grafting patent to Celltech in the late 1980s and to institute the LMB-proposed non-exclusive licensing policy for Winter. This policy led to the licensing of many companies and a large antibody pipeline, ultimately humanized antibodies products, including Herceptin®, Lucentis®, Avastin®, Synagis,® and Soliris® (as well as those developed through the Collaborative Centre). The licensing income from humanized antibodies formed the major part of the MRC's commercial income.
- As Chief Executive of the MRC, it was Sir Dai who permitted Sir Greg to form Cambridge Antibody Technology (CAT) in 1989, the first MRC start-up company, setting terms for licensing of the MRC intellectual property that allowed CAT to develop its business model, while protecting MRC interests. CAT created Humira® (sales now $18 billion per annum), of which the royalty share due to the MRC was around two per cent.

Sir Dai emphasized that the admirable ethic adopted by Sir Walter Fletcher in the early days of the MRC, in which he emphasized how improvement in public health—especially the conquering of tuberculosis—was one of the principal aims of the MRC, could be adapted by him (Sir Dai), after seeking permission from the HM Treasury, so as to develop new drugs and to acquire sums of money stemming from their sales.

REFERENCES

1. J. C. Kendrew, 'Bernal (John) Desmond', *Dictionary of National Biography*, **1971**, *86*, p. 53.
2. Quoted in D. M. C. Hodgkin's *Biographical Memoir of Fellows of the Royal Society, J. D. Bernal*, **1980**, 26, 17.
3. K. Bailey, W. T. Astbury, and K. M. Rudall, *Nature*, **1943**, *157*, 716.
4. M. F. Perutz, see Chapter 8, Section 8.2.2.
5. H. E. Huxley (ed.), '*Memories and Consequences. Visiting Scientists at the MRC Laboratory of Molecular Biology*', MRC/LMB, Cambridge, **2013**.
6. V. Ramakrishnan, '*Gene Machine: The Race to Decipher the Secrets of the Ribosome*', Oneworld, London, **2018**.
7. I. Tinoco, see ref [5], pp. 79–84.
8. Max Perutz often said that he was rather shocked to find that both myoglobin and haemoglobin presented themselves as examples of visceral monstrosities!
9. Adenosine triphosphate (ATP) is a complex organic chemical that provides energy to drive many processes in living cells, as in nerve impulses, muscle contraction, and chemical synthesis. It is often called the fuel of life, or the 'molecular unit of currency'. In the process of being consumed in metabolic events, it is converted to adenosine diphosphate (ADP) or adenosine monophosphate (AMP). It is estimated that the human body recycles its own bodyweight equivalent in ATP each day. As of 2015, Sir

John E. Walker is Emeritus Director and Professor at the MRC Mitochondrial Biology Unit at Cambridge.

10. S. de Chadarevian, '*Designs for Life: Molecular Biology after World War II*', Cambridge University Press, Cambridge, **2002**.
11. J. T. Finch, '*A Nobel on Every Floor*', The Medical Research Council, Laboratory of Molecular Biology, Cambridge, p. 238.
12. S. de Chadarevian, *Isis*, **2011**, *102*, 601.
13. See Ref [11], Figure 4, p. 630.
14. It is a fact that drugs based on monoclonal antibodies discovered, studied, and modified at the LMB are now world leaders (see Appendix 2 to this chapter for a summary of the story of commercialization of Winter *et al.*'s. work at the LMB). The royalties from these drugs enabled the new building of the LMB to be erected, and a substantial fraction of the cost of the Francis Crick Institute in London also came from MRC/LMB royalties. One of the reasons why AstraZeneca moved its headquarters from Gothenburg to Cambridge derives from its close links with the LMB.
15. I. Olavsson, A. Liljas, and S. Lidin, '*From a Grain of Salt to the Ribosome*', World Scientific Publishers, Singapore, **2015**, pp. 98–102.
16. N. Ban, P. B. Moore, T. A. Steitz, *et al.*, *Science*, **2000**, *289*, 905.
17. B. T. Wimberly, D. F. Broderson, V. Ramakrishnan, *et al.*, *Nature*, **2000**, *400*, 833.
18. Y. Cheng, R. M. Glaeser, and E. Nogales, *Cell*, **2017**, *171*, 1229.
19. W. O. Saxton and J. Frank, *Ultramicroscopy*, **1977**, *2*, 219.
20. K. A. Taylor and R. M. Glaser, *Science*, **1974**, *186*, 1036.
21. R. Henderson and P. N. T. Unwin, *Nature*, **1975**, *257*, 26.
22. J. Dubochet, M. Adrian, J. J. Chang, *et al.*, *Q. Rev. Biophys.*, **1988**, *21*, 129.
23. R. Faruqi, R. Henderson, M. Pryddecah, P. Allpost, and A. Evans, *Nucl. Inst. Methods Phys. Res.*, **2005**, *A546*, 170.
24. X.-C. Bai, I. S. Fernadez, G. McMullan, and S. H. Scheres, *eLife*, **2013**, 2.
25. J. C. H. Spence, private communication to John Meurig Thomas, December **2018**.
26. J. Frank, *Ultramicroscopy*, **1975**, *1*, 150.
27. R. A. Crowther, *Phil. Trans. R. Soc. B*, **2008**, *363*, 2441 and references therein.
28. P. Broadwith, *Chemistry World*, 4 October **2017**.
29. R. Henderson, *Arch. Biochem. Biophys*, **2015**, *13*, 4989. See also Henderson's Nobel Lecture, *Angew. Chem. Int. Ed.*, **2018**, *57*, 10804.
30. M. Congreve and F. Marshall, *Br. J. Pharmacol.*, **2010**, *159*, 986.
31. A. W. Fitzpatrick, B. Falcon, S. He, *et al.*, *Nature*, **2017**, *547*, 185.
32. B. Falcon, W. Zhang, R. A. Crowther, *et al.*, *Nature*, **2018**, *561*, 137.
33. Tau proteins are abundant in neurons of the central nervous system (CNS) and are less common elsewhere. Pathologies and dementias of the CNS (e.g. Alzheimer's disease) are associated with tau proteins that have become defective and no longer stabilize microtubules in the CNS properly.
34. B. Falcon, J. Ziranov, W. Zhang, *et al.*, *Nature*, **2019**, *568*, 420.
35. T. A. M. Bharat, D. K. Ciziene, G. G. Hardy, *et al.*, *Nat. Microbiol.*, **2017**, *2*, 17059.
36. Jan Loewe is the current Director of the MRC-LMB, Cambridge.
37. M. Khoshorei, R. Danev, J. M. Plitzko, and W. Baumeister, *J. Mol. Biol.*, **2017**, *429*, 2611.
38. J. C. Kendrew, *Science*, **1963**, *139*, 1259 (see, in particular, Figure 10).
39. M. Baker, *Nature*, **2018**, *561*, 565.
40. A. P. Joseph, I. Lagerstedt, A. Patwardhan, M. Topf, and M. Winn, *J. Struct. Biol.*, **2017**, *199*, 12.

Glossary

For the benefit of readers with no prior technical knowledge of structural biology, a few introductory chemical facts, concepts, and phenomena are presented first.

The *valence* (or *valency*) of an element is its combining power with other atoms to form compounds or ions (i.e. charged species). Oxygen is divalent—it combines with two atoms of hydrogen (which is monovalent) to form water (H_2O), that is often symbolized $H/^{O}\backslash H$. Each chemical bond, denoted by –, consists of two shared electrons. A double bond, designated =, has four shared electrons. Since carbon C is quadrivalent, a carbonyl group is represented as:

$$>C{=}O$$

Sulfur is generally divalent, as in H_2S (hydrogen sulphide), and an ether and thio-ether are represented as:

$$CH_3{-}O{-}CH_3\,(\text{methylalcohol or methanol})$$
$$CH_3{-}S{-}CH_3\,(\text{methyl thio-ether})$$
$$CH_3{-}S{-}H\,(\text{thio methanol})$$

Hexane and octane are examples of saturated hydrocarbons (also called alkanes):

$$CH_3{-}CH_2{-}CH_2{-}CH_2{-}CH_2{-}CH_3$$
$$n\text{-}hexane\,(C_6H_{12})$$

Also represented:

```
    H   H H   H H   H
        \/     \/    \/
 H     .C.    .C.   .C.
   \  /   \  /   \ /   \
    C      C      C      H
   / \    / \    / \
  H    H H    H H    H
```

An unsaturated hydrocarbon is *n*-hexene, which has a double bond:

```
 H        3      4      5      6
         CH2 – CH2 – CH2 – CH3
 \1    2/γ     δ      ε      ζ
  C = C
 /α    β\
 H      H
```

A CH_3 group, termed methyl, is an example of a so-called functional group. Other examples are –OH (hydroxyl) or amino $-NH_2$ (since nitrogen is trivalent).

A carboxylic group is COOH, and a fatty acid is one that has a long hydrocarbon chain, like stearic acid, which is $C_{17}H_{35}COOH$ or $CH_3(CH_2)_{16}COOH$. Whereas stearic acid is termed a saturated fatty acid, oleic acid is an unsaturated one, since it contains a C=C (double bond). This is often designated thus:

```
                      O
                      ‖
CH3(CH2)15CH2 ⁄ \ OH
```

Compounds that have the same chemical formula, but a different arrangement of atoms are called *isomers*. These have different properties, as in *n*-hexane and isohexane, which is usually called 2-methylpentane, since there is a methyl group attached to the second carbon atom:

CH_3

H_3C CH_3

When two carbons are joined by a double bond, and there are functional groups (also called substituents) attached to the carbons, the following isomeric possibilities (I and II) exist (taking dichloro ethylene as the example):

Cl Cl Cl H

H H H Cl

I II

These two forms are also called stereoisomers because the atoms in each are arranged differently in space. The term enantiomer, also known as an optical isomer, is one of two stereoisomers that are mirror images of each other that are non-superimposable, as shown in the figure below of lactic acid, the simple formula of which is $CH_3 \cdot CHOH \cdot COOH$.

O O

HO HO

OH OH

H CH_3 H_3C H

L lactic acid D lactic acid

Benzene (C_6H_6) is a ring of six carbon atoms to which are attached hydrogen atoms.

H

H H

H H

H

The circle in the middle signifies its aromatic character, in which there are alternating double and single bonds, i.e.

H

C

H C C H

C C

H C H

H

A phenol is a benzene ring, in which one of the hydrogens is replaced by a hydroxyl (OH) group.

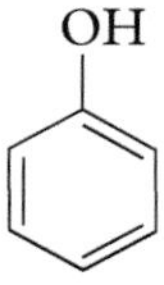

OH

Hexachlorobenzene C_6Cl_6 is:

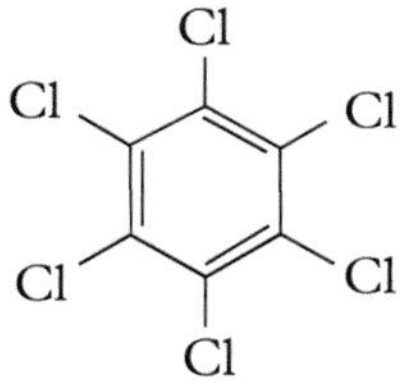

Cl
Cl Cl
Cl Cl
Cl

When a hexagon has no circle, or alternating single and double bonds, inside it, it stands for cyclohexane (C_6H_{12}), an alkane that is more fully represented as:

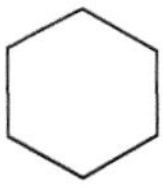

Likewise, cyclopentane is:

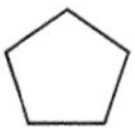

In steroids, such as cholesterol (which is described in detail in Section 3.5.2), the numbering system of the atoms seems intricate. It is as shown in Figure 3.14 of Chapter 3.

Å The angstrom unit. This is 10^{-8} cm or 0.1 nm, and is convenient for describing interatomic distances and the lengths of chemical bonds.
Acetyl This is a functional group derived from acetic acid and is denoted CH_3COO^-.
Acid A substance that releases protons. It is a proton donor.
Actin Actin is a family of globular multi-functional proteins that form microfilaments. It is found in almost all eucaryotic cells.
Activation energy The energy that must be supplied to a system to enable a chemical reaction to proceed.
Active site The region of an enzyme (or any catalyst) where a reactant molecule is converted to another (see Figure 6.9 of Chapter 6).
Adenine A member of the purine class of organic compounds and a nucleobase, and it is one of five nucleobases, the others being: guanine, cytosine, thymine, and uracil. Its chemical structure is:

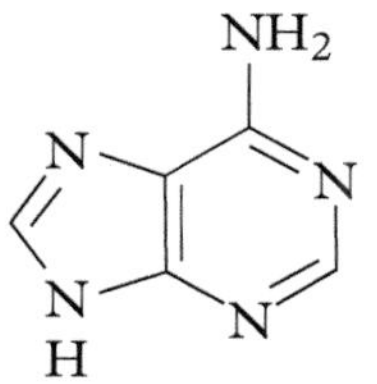

NH_2
N N
N N
H

Aldehyde A compound that contains the functional group.

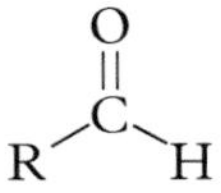

Alkyl group (C_nH_{2n+1}), e.g. C_3H_7

Allosteric proteins Proteins, the biological properties of which are changed when a small molecule is bound to its active site.

Alpha- and beta-carbon atoms The α-carbon (C_α) in organic molecules is the first carbon atom that attaches to a functional group such as carbonyl. The β-carbon is the second carbon atom. A hydrogen atom attached to an α-carbon atom is called an α-hydrogen. Likewise there are β- and γ-carbon and hydrogen atoms.

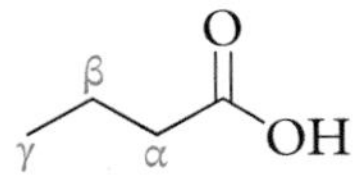

Alpha-helix A common motif in the structure of proteins is a right-hand helix conformation in which every backbone N–H group donates a hydrogen bond to the backbone C=O group of the amino acid located three or four residues earlier along the protein sequence of condensed amino acids (i.e. the residues)—see Figures 4.7, 4.8, and 4.9 in that order. It is customary to use a helical ribbon to represent an α-helix, as in the cartoon section of Figure 4.8.

Amide A grouping formed when an acid combines with an amine.

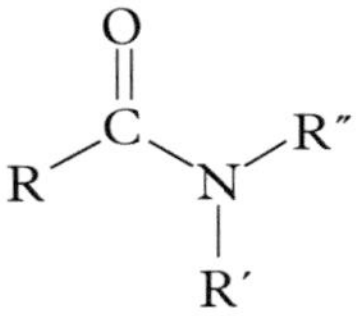

An example is the *peptide bond* that is generated when two amino acids combine with derivation of water, as depicted in Figure 3.6 of Chapter 3.

Amino acids These are the building blocks of all proteins and consist of terminal NH_2 (amino) and COOH (carboxylic) groups. All amino acids, of which there are over 450 known ones, have the same fundamental structure but differ in their side groups R, as shown:

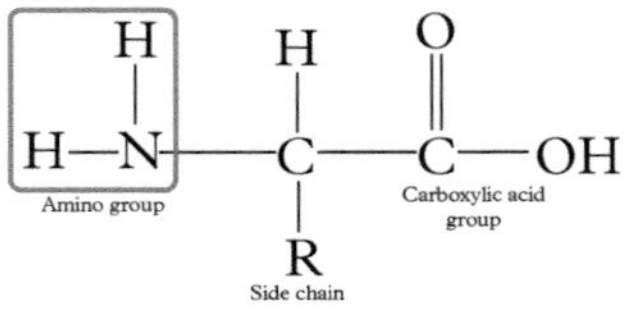

Antibody This is a large Y-shaped protein produced mainly by plasma cells that is used by the immune system to neutralize pathogens (see Appendix 1 to Chapter 10). The antibody recognizes a unique molecule of the pathogen, called an antigen.

Aromatic amino acids These amino acids contain a derivative of a phenyl (C_6H_5 as an R group). Typical examples among the amino acids and in nature are phenylalanine, tyrosine, and tryptophan (see Figure 3.8).

Base-pairing rules The requirement that adenine (A) always forms a base pair with thymine (T), and guanine (G) with cytosine (C).

Blood plasma This is the liquid part of blood that carries blood cells and proteins throughout a living body.

Calciferol This is vitamin D, which is responsible for increasing the intestinal absorption of calcium, magnesium, and phosphate.

Carboxyl group Is a characteristically acidic one, as a result of the dissociation of the H from the OH group:

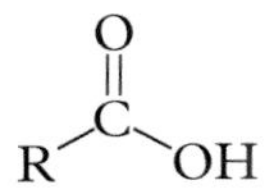

Catalyst A substance that speeds up the attainment of equilibrium in a chemical reaction.

Cell The fundamental unit of life—it is the smallest entity capable of independent reproduction. There are some 27×10^{12} cells in a human body.

Central dogma This is concerned with the relations between DNA, RNA, and proteins. DNA is the template for both its own duplication and the synthesis of RNA, and, in turn, the RNA is the template for protein synthesis.

Chiral A chiral object (like a molecule or ion) is non-superimposable on its mirror image. Mirror images of a chiral molecule are called enantiomers or optical isomers. Individual enantiomers are frequently designated as either right-handed or left-handed. Most biologically significant molecules, and many pharmaceuticals are chiral. The feature that is often the cause of chirality in molecules is the presence of an asymmetric carbon atom where four different functional groups are attached to it.

Chromatin This is a complex of DNA and protein that is found in eucaryotic cells. Its principal function is to package very long DNA molecules into a denser, more compact shape. It plays a key role in cell division.

Chromosome A thread-like structure of nucleic acids and proteins that occurs in the nucleus of most living cells, carrying genetic information in the form of genes.

Chymotrypsin A digestive enzyme which breaks down proteins in the small intestine.

Clone A group of cells all descended from a single common ancestor.

Coenzyme Is the small molecule associated with a protein to form an enzyme.

Covalent bond A strong chemical bond formed by the sharing of electrons.

Cytosine A member of the pyrimidine class of organic compounds and a nucleobase, and it is one of five nucleobases, the others being: guanine, adenine, thymine, and uracil. Its chemical structure is:

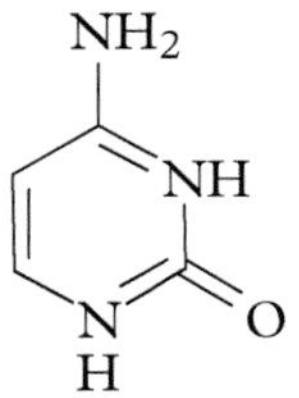

Dalton The unit of weight that is equal to that of a single hydrogen atom. It is the unified atomic mass unit. Some molecules involved in biological properties have molecular weights (also called molar weights) of over a million dalton (>10^6 M Da).

Disulfide bond Two sulfur atoms that are joined by a covalent bond in various amino acids of a protein (see Figures 3.9 and 6.8).

Deoxyribose The five-carbon monosaccharide component of DNA. It differs from ribose in having H at the 2-carbon position, rather than OH.

Electrophile A reagent attracted to electrons. They are usually positively charged entities, having vacant orbitals.

Electrophoresis Is a technique to separate charged molecules (like proteins and DNA). In gel or paper electrophoresis, charged species move through the gel (paper) when an electric current is passed through it.

Enzymes These are molecules of proteins capable of catalysing a chemical reaction. Some 75,000 different enzymes are known. (See Chapter 6 on lysozyme.)

Eucaryote Organisms with cells that have nuclear membranes—membrane-bound organelles (tiny cellular structures that perform specific function)—and characteristic biochemistry.

Ferrous and ferric ions These are the ionized forms of elemental iron Fe. Two electrons have been removed in the ferrous state, symbolized Fe^{2+} (the state of iron in haemoglobin). Ferric ion is Fe^{3+}.

Free energy Energy that has the ability to do work (usually of a chemical kind).

Gene A section along a chromosome that codes for a functional product.

Genotype The genetic constitution of an organism.

Glycogen A polymer formed by so-called glycosidic linkages between adjacent glucose residues (see Figures 6.11 and 6.12).

Glycolipid A lipid is a biomolecule that is soluble in non-polar solvents, and a glycolipid is a lipid that is attached to a carbohydrate through a covalent bond.

Guanine A member of the purine class of organic compounds and a nucleobase, and it is one of five nucleobases, the others being: adenine, cytosine, thymine, and uracil. Its chemical structure is:

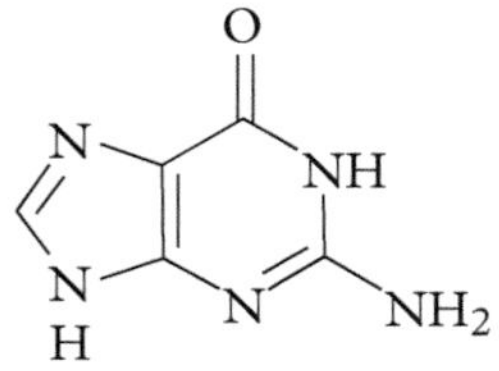

O
N
NH
N
N
NH_2
H

Haemoglobin A protein carrier found in red blood cells (see Chapter 5).

Histone A histone is a highly alkaline protein found in eucaryotic cells. They are the primary protein components of chromatin. They package and order DNA into structural units called nucleosomes (see Figure 9.9).

Hormones A chemical substance synthesized in one organ of the body that stimulates functional activity in cells of other organs or tissues.

Hydrogen bond A weak attractive force between one electronegative atom and a hydrogen atom that is covalently linked to a second electronegative atom.

Hydrolysis The fracture of a molecule into two or more entities by the addition of water, for example, the conversion of an ester (like ethylacetate) to ethanol (ethyl alcohol) and acetic acid.

Hydrophilic Molecules or other entities that readily associate with water.

Hydrophobic Molecules or certain functional groups that are only feebly soluble in water.

Hydrophobic bonding The association of non-polar groups with each other in aqueous solution, owing to the tendency for molecules of water to exclude non-polar molecules.

Icosahedron A regular polyhedron composed of twenty equilateral triangular faces (see Figure 9.5).

In vitro Refers to experiments done in a cell-free system.

L-amino acid L- and D-amino acids can occur in L- (laevo) and D- (dextro) forms, but only L-forms are used in living cells. Every amino acid (except glycine—see Figure 3.9) can occur in two isomeric forms because of the possibility of forming two different enantiomers (stereoisomers around the central carbon atom).

Lipid See *Glycolipid.*

Lysis The rupture of a cell by the destruction of its cell membrane.

Lysozyme An enzyme that degrades the polysaccharides present in the cell wall of certain bacteria.

Messenger RNA (*m*RNA) This is RNA that functions as a template for the synthesis of proteins.

Micron (μ) A unit of length equal to 10^{-3} cm or 10^5 Å.

Mitochondrion (*pl. mitochondria*) This is a double-membrane-bound organelle found in most eucaryotic organisms. Mitochondria generate *messenger RNA* (*m*RNA), RNA that functions as a template for protein synthesis. Most of the cell's supply of adenosine triphosphate (ATP) is used as a source of chemical energy.

Mitosis Process whereby chromosomes duplicate and segregate, accompanied by cell division.

Motility The ability of an organism to move independently using metabolic energy.

Mutagens Any chemical or physical agents that raise the frequency of mutation greatly above the background level.

Mutarotation Is a change in optical rotation because of a change in the equilibrium between two stereoisomers.

Mutation An inheritable change in chromosome.

Myosins Myosins are a superfamily of so-called motor proteins, best known for their roles in muscle contraction.

Nitrogenous base A nitrogen-containing molecule with basic properties; well-known examples are purines and pyrimidines.

Nucleases Enzymes which cleave the phosphodiester bonds in nucleic acid chains.

Nucleic acids These are the biopolymers essential to all known forms of life. It is the term that embraces DNA and RNA. They are composed of nucleotides, which are monomers made of three components: a five-carbon sugar, a phosphate group, and a nitrogenous base. When the sugar is ribose, the polymer is RNA; when it is deoxyribose, it is DNA.

Nucleophile A chemical species that donates an electron pair to an electrophile to form a chemical bond. Ammonia (NH_3) is a classic example.

Nucleoside A nitrogenous base and a five-carbon sugar. A nucleoside plus a phosphate group yields a nucleotide.

Nucleosome This is the structural unit of a eucaryotic chromosome, consisting of a length of DNA coiled around a core of histones.

Nucleotides These are the building blocks of nucleic acids. They consist of a base (one of the following four: adenine, thymine, guanine, and cytosine) plus a molecule of sugar and one of phosphoric acid.

Organelle A membrane-bound structure found in eucaryotic cells, containing enzymes for specific functions.

Pepsin Pepsin is an enzyme that breaks down proteins into smaller fragments (peptides). It is one of the main digestive enzymes and is produced in the stomach.

Peptide bond (*peptide link*) A peptide bond is an amide type of covalent chemical bond linking two consecutive α-amino acids from C_1 of one α-amino acid and N_2 of another along a polypeptide (protein) chain (see Figure 3.6). The amide link and the peptide link are synonymous.

Phage This is the word used for bacteriophage, which is a virus that infects and replicates within bacteria. Phages are the most common entities in the biosphere. It is believed that there are more phages on Earth than all other living organisms combined.

Phosphodiester A molecule that contains the linkage:

$$R - O - \underset{\displaystyle O^-}{\overset{\displaystyle O}{\overset{||}{\underset{|}{P}}}} - O - R'$$

where R and R′ are carbon-containing groups (e.g. nucleosides). P and O are the elements phosphorus and oxygen.

Phospholipids Are a primary component of cell membranes; they are lipids that contain charged hydrophilic phosphate terminal groups.

Phosphorylation Reaction in which a phosphate group is coupled to another molecule.

Photosynthesis Process by which plants, algae, and some bacteria harness sunlight to synthesize organic molecules from CO_2 and water.

Pitch Number of base pairs per turn of the double helix.

Plasmid Small, circular, extra-chromosomal DNA molecule that replicates independently of the genome.

Polypeptide A polymer of amino acids linked together by peptide bonds.

Procaryote Simple unicellular organism, such as a bacterium, with no nuclear membrane.

Purine Purine is a heterocyclic aromatic organic compound that consists of a pyrimidine ring fused to an imidazole ring, as shown in the entry for *adenine*. Purine itself is a heterocyclic aromatic organic compound consisting of a pyrimidine ring fused to an imidazole ring:

Formula $C_5H_4N_4$

Pyrimidines These are molecules of cytosine, thymine, and uracil. Pyrimidine differs from pyridine in having an extra nitrogen in the hexagonal structure. Compare with the entry for *thymine*.

Ribonuclease An enzyme that can cleave the phosphodiester bonds of RNA.

Ribonucleotide A molecule that consists of a purine or pyrimidine base bonded to ribose, which, in turn, is esterified with a phosphate group.

Ribose Is a carbohydrate (formula $C_5H_{10}O_5$). It is a so-called pentose monosaccharide (a simple sugar). Component of RNA.

Ribosomes Small cellular particles (approximately 200 Å in diameter) consisting of *t*RNA and protein. They are the site of protein synthesis.

RNA Ribonucleic acid. A polymeric molecule that is essential for all forms of life (like DNA). It regulates coding and decoding and is an essential feature of genes. Like DNA, RNA is assembled as a chain of nucleotides. Unlike DNA, which is a double helix, it is often found in nature as a single strand folded onto itself.

Secondary protein structure The extended or helical structure of a polypeptide chain (e.g. a plaited sheet or α-helix—see Figure 4.9).

Stereoisomers Molecules that have the same structural formula, but different spatial arrangements of dissimilar groups bonded to a common atom.

Steroids Molecules that are derivatives of a tetracyclic structure composed of a cyclopentane ring fused to a substituted phenanthrene nucleus (see Figures 3.13 and 3.14).

Substrate A molecule, the chemical conversion of which is catalysed by an enzyme.

Sugar Small carbohydrate with a monomer unit (general formula $(CH_2O)_n$). Glucose and fructose are monosaccharides; sucrose is a disaccharide.

Tau proteins (*t proteins*) These are proteins that stabilize microtubules. They are abundant in neurons of the central nervous system (see Figure 10.3).

Tertiary structure of a protein The three-dimensional folding of a polypeptide chain that characterizes a protein in its native state.

Thymine A member of the pyrimidine class of organic compounds and a nucleobase, and it is one of five nucleobases, the others being: guanine, adenine, cytosine, and uracil. Its chemical structure is:

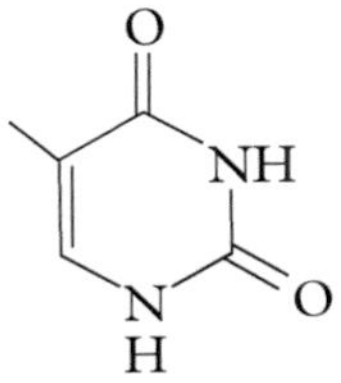

Transfer RNA (*t*RNA) This is an adaptor molecule composed of RNA containing typically some seventy-five to ninety nucleotides in length. It serves as the link between the *m*RNA and the amino acid sequence of proteins. Any of at least twenty structurally similar species of RNA, all of which have a molar mass of approximately 25,000 Da. Each species of *t*RNA can combine covalently with a specific amino acid and to hydrogen bond with at least one *m*RNA.

Trypsin Is an enzyme that is formed in the small intestine. It is found in the digestive system of many vertebrates. A proteolytic enzyme of molecular weight of approximately 23,800 Da. It is secreted by the pancreas, and it cleaves peptide chains (as shown in Figure 3.10) where the basic amino acids arginine and lysine appear.

Uracil A member of the purine class of organic compounds and a nucleobase, and it is one of five nucleobases, the others being: guanine, adenine, cytosine, and thymine. Its chemical structure is:

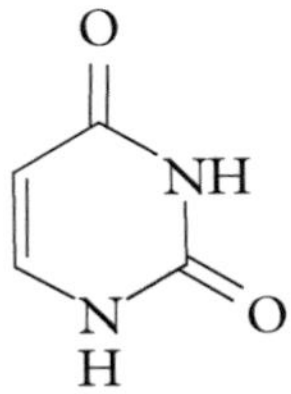

van der Waals force A weak attractive force acting over only very short distances, resulting from attraction of induced dipoles.

Glossary: A Brief Guide to British Institutions and Universities

British universities In the majority of British universities, the method of teaching is similar to that used in the universities of most countries of the world. The lecture is the main means of communication, supplemented by a modest amount of problem classes and tutorials. In Oxford and Cambridge, however, the method of teaching is very different. There are thirty-one colleges in the University of Cambridge and thirty-eight in Oxford. All students (approximately 20,000 in Cambridge and 23,000 in Oxford in 2017) belong to a particular College where they receive a substantial amount of extra tuition over and above what they receive in lectures and laboratories that are organized by the central university authorities. In each College, undergraduates receive detailed extra supervision, by Fellows or teaching officers, often on a 1:1 basis. Problem classes and essays are used routinely. Some of the Fellows in a College may not be members of the University. Both John Kendrew and Aaron Klug, who regularly supervised undergraduates at Peterhouse, were in this category. Some researchers and staff members of the University do not have college posts (as Fellows). J. D. Bernal and Max Perutz were in this category. Several researchers in the Laboratory of Molecular Biology (LMB) were recruited to do supervision in the various Colleges, e.g. John Finch, FRS (see Figure 9.1 in Chapter 9) was in this category.

LMB The Laboratory of Molecular Biology. This is in Cambridge and is operated by the UK Medical Research Council (MRC). Established in 1962.

Master of a College The Master of a College is, in effect, the Head of the establishment. He or she is expected to ensure that all aspects of the College—the tuition of under- and postgraduates, the library, kitchens, the chapel, and physical properties, as well as all the associated administrative staff—run smoothly. The Master chairs the Governing Body of the College, which consists of all permanent Fellows and usually the Bursar and Director of Development. The Master's duties involve ensuring, with the Fellows, that appropriate staff are in position to cover all academic activities. The duties also extend to ensuring that the various student activities, like the various societies that they run (in science, history, music, and politics), function properly.

Each Oxbridge College has halls of residence, dining halls, chapels, and libraries, as well as various sporting facilities. College life greatly facilitates cross-fertilization of subjects and joint research work among its members.

RI The Royal Institution of Great Britain, founded in 1799, largely through the efforts of the American-born Sir Benjamin Thompson (also known as Count Rumford of the Holy Roman Empire), aided by Henry Cavendish and the then President of the Royal Society, Sir Joseph Banks. It is situated on Albemarle Street, Mayfair, London, and its foundational principles were to diffuse knowledge and facilitating the general introduction of useful mechanical inventions and improvements and enhancing the application of science to the common purposes of life. It is an organization open to all members of the general public interested in science. In the early 1800s, two outstanding members of its staff were Humphry Davy and Thomas Young.

Davy discovered sodium, potassium, barium, magnesium, boron, calcium, chlorine, and strontium there. He was also the principal founder of agricultural chemistry and many other disciplines. Thomas Young carried out his famous double-slit experiment there, which resuscitated the wave theory of light. Michael Faraday lived and worked for nearly fifty years there, and he discovered electromagnetic induction, as well as his laws of electrolysis and much else. Later major discoveries were made by James Dewar (the thermos flask), by Lord Rayleigh who discovered argon, and by W. H. Bragg (1923–1942) and W. L. Bragg (1953–1966) who, at different times, made it one of the world's premier centres of research in X-ray crystallography. Kathleen Yardley (later Dame Kathleen Lonsdale) worked there in the Davy-Faraday Research Laboratory (DFRL), established in 1898, for twenty years. She became the first woman to be elected a Fellow of the Royal Society in 1946. Experiments were also carried out there by Rosalind Franklin and Dorothy Hodgkin (both as visiting scientists), and Max Perutz and John Kendrew were Honorary Readers at the DFRL for thirteen years in W. L. Bragg's time as Director of the RI and the DFRL.

The series of annual Christmas Lectures for young people, started by Faraday in 1826, as well as regular Friday Evening Discourses given by men and women distinguished in fields ranging from science to all aspects of culture and learning, continue to be presented there (see Chapters 3 and 6 for further information). The BBC has broadcast the Christmas Lectures for the past sixty years.

The Cavendish Laboratory This is synonymous with the Department of Physics at the University of Cambridge. It was founded in 1872, and the Duke of Devonshire was instrumental in its establishment. The first Cavendish Professor was James Clerk Maxwell. His successors were Lord Rayleigh, J. J. Thomson, Lord Rutherford, Sir Lawrence Bragg, Sir Nevill Mott, Sir Brian Pippard, Sir Samuel Edwards, and the present occupant Sir Richard Friend.

The aerial view of the centre of Cambridge shows the location of the Old Cavendish. It is in its precincts and immediate surroundings where the electron, proton, and neutron were discovered, where the structure of DNA was solved, and where the pioneering work of Perutz and Kendrew was carried out. Nearby, Bernal and Crowfoot (Hodgkin) carried out their pioneering work on '*living*' molecules.

The locations of five colleges are marked (Emmanuel; Corpus Christi; King's; Pembroke; and Peterhouse). Parker's Piece is a public park.

The Royal Society This is the national academy of the UK, founded in 1660 as The Royal Society of London by King Charles II. Fellows of the Royal Society are elected from all countries of the British Commonwealth in all fields of science, medicine, engineering, and latterly computing. Foreign Members from other countries are also elected, famous examples being Benjamin Franklin, Dmitri Mendeleev, Albert Einstein, Alessandro Volta, Carl Fredrick Gauss, and Linus Pauling.

The Bakerian Lecture of the Royal Society is its premier one in the physical sciences. The one given by W. H. Bragg in 1915 greatly influenced the subsequent work of W. L. Bragg, Max Perutz, John Kendrew, and Dorothy Hodgkin, all of whom exploited Fourier analysis to determine electron densities, to which W. H. Bragg drew attention.

Aerial Photograph of Central Cambridge

Readers who wish to pursue the human side of structural and other aspects of molecular biology may wish to read '*The Eighth Day of Creation*' by Horace Freeland Judson (Simon and Schuster, New York, NY, 1979), and also the two books by Georgina Ferry on Dorothy Hodgkin and Max Perutz.

Those who wish to pursue a quantitate account of X-ray diffraction are recommended to read A. M. Glazer's '*Crystallography: A Very Short Introduction*' (Oxford University Press, Oxford, 2016).

Standard university textbooks on '*Biochemistry*' by Berg, Tymoczko, Gatto, and Stryer, and on '*Molecular Biology*' by Bruce A. Alberts *et al.* and J. D. Watson *et al.*, are intended for undergraduates and graduates.

Index

B

I

J

K

L

M

N

Q

R